In Memory of My Brother

PHILIP BROCKMAN
Rocket Scientist
1937–2017

"I began a true rocket scientist.
At the end I conducted research toward making aircraft safer for us all.
I am a man of wind and light."

This Idea Is Brilliant

Also by John Brockman

As Author
By the Late John Brockman
37
Afterwords
The Third Culture: Beyond the Scientific Revolution
Digerati

As Editor
About Bateson
Speculations
Doing Science
Ways of Knowing
Creativity
The Greatest Innovations of the Past 2,000 Years
The Next Fifty Years
The New Humanists
Curious Minds
What We Believe but Cannot Prove
My Einstein
Intelligent Thought
What Is Your Dangerous Idea?
What Are You Optimistic About?
What Have You Changed Your Mind About?
This Will Change Everything
Is the Internet Changing the Way You Think?
Culture
The Mind
This Will Make You Smarter
This Explains Everything
Thinking
What Should We Be Worried About?
The Universe
This Idea Must Die
What to Think About Machines That Think
Life
Know This

As Coeditor
How Things Are (with Katinka Matson)

This Idea Is Brilliant

Lost, Overlooked, and Underappreciated
Scientific Concepts Everyone Should Know

EDITED BY JOHN BROCKMAN

HARPER PERENNIAL

NEW YORK • LONDON • TORONTO • SYDNEY • NEW DELHI • AUCKLAND

HARPER ● PERENNIAL

HarperCollins books may be purchased for educational, business, or sales promotional use. For information, please email the Special Markets Department at SPsales@harpercollins.com.

FIRST EDITION

Library of Congress Cataloging-in-Publication Data has been applied for.

ISBN 978-0-06-269821-6 (pbk.)

18 19 20 21 22   LSC   10 9 8 7 6 5 4 3

CONTENTS

ACKNOWLEDGMENTS

My thanks to Peter Hubbard of HarperCollins and my agent Max Brockman, for their continued encouragement. A special thanks, once again, to Sara Lippincott for her thoughtful attention to the manuscript.

PREFACE: *SCIENTIA* AS A MEME

Richard Dawkins' "meme" became a meme, known far beyond the scientific conversation in which it was coined. It is one of a handful of scientific ideas that have entered the general culture, helping to clarify and inspire. Of course, not everyone likes the idea of spreading scientific understanding. The Bishop of Birmingham's wife is reputed to have said, about Darwin's claim that human beings are descended from monkeys: "My dear, let us hope it is not true, but, if it is true, let us hope it will not become generally known."

Of all the scientific terms or concepts that ought to be more widely known, in order to inspire and expand science-minded thinking in the general culture, perhaps none are more important than "science" itself.

Many people, even many scientists, harbor a narrow view of science—as controlled, replicated experiments performed in the laboratory and consisting quintessentially of physics, chemistry, and molecular biology. The essence of science, however, is best conveyed by its Latin etymology: *scientia*, meaning "knowledge."

The scientific method is simply that body of practices best suited for obtaining reliable knowledge. The practices vary among fields: the controlled laboratory experiment is possible in molecular biology, physics, and chemistry, but it is either impossible, immoral, or illegal in many other fields customarily considered scientific, including all of the historical sciences: astronomy, epidemiology, evolutionary biology, most of the Earth sciences, and paleontology. If the scientific method can be defined as those practices best suited for obtaining knowledge in a particular field, then science itself is simply the body of knowledge obtained by those practices.

Science (that is, reliable methods for obtaining knowledge) is an essential part of psychology and the social sciences, too—especially economics, geography, history, and political science. Along with the broad observation-based and statistical methods of the historical sciences, detailed techniques of such conventional sciences as chemistry and genetics are proving essential for tackling problems in the social sciences. Science, then, is the reliable acquisition of knowledge about anything, whether it be the vagaries of human nature, the role of great figures in history, or the origins of life itself.

It is in this spirit of *scientia* that *Edge*, on the occasion of its 20th anniversary, is pleased to present the *Edge* Annual Question for 2017:

WHAT SCIENTIFIC TERM OR CONCEPT OUGHT TO BE MORE WIDELY KNOWN?

Happy New Year!

John Brockman

Publisher & Editor, *Edge*

This Idea Is Brilliant

THE LONGEVITY FACTOR

YURI MILNER

Physicist; entrepreneur and venture capitalist; science philanthropist

In 1977, the *Voyager* probes were launched toward the outer solar system, each carrying a golden record containing hundreds of sounds and images, from the cry of a newborn baby to the music of Beethoven. In the emptiness of space, they could last for millions of years. By that time, will they be the sole representatives of human culture in the cosmos? Or primitive relics of a civilization that has since bloomed to the galactic scale?

The Drake equation estimates the number of currently communicative civilizations in the Milky Way by multiplying a series of terms, such as the fraction of stars with planets and the fraction of inhabited planets on which intelligence evolves. The final term in the equation doesn't get much attention. Yet it's crucial, not just for the question of intelligent life but also for the question of how to live intelligently. This is L, the *longevity factor*, and it represents the average life span of a technological civilization.

What determines this average? Surely the intelligence of the civilizations. The list of existential threats to humanity includes climate change, nuclear war, pandemics, asteroid collisions, and perhaps artificial intelligence. And all of these threats can be avoided. Some can be addressed here on Earth; others require activity in space, but with the ultimate aim of protecting the planet.

In 1974, Princeton physicist Gerard K. O'Neill published a paper, "The Colonization of Space," which led to the first conference on the subject, sponsored by Princeton and Stewart Brand's Point Foundation, and to O'Neill's highly influential 1976 book, *The High Frontier.* That has been an inspiration for the current generation of visionaries, who advocate steps such as the transfer of heavy industry into orbit, where it can run on solar energy and direct its heat and waste away from Earth, and the colonization of Mars.

However powerful our local solutions, betting everything on one planet would be imprudent. Stephen Hawking has estimated: "Although the chance of a disaster to planet Earth in a given year may be quite low, it adds up over time, and becomes a near certainty in the next 1,000 or 10,000 years. By that time we should have spread out into space, . . ." In the long term, Mars must be a stepping-stone to more distant destinations, because two adjacent planets could be simultaneously affected by the universe's more violent events, such as a nearby supernova. We need to start thinking at the galactic level. The first target might be the Earth-size planet Proxima b, recently discovered orbiting the nearest star to the sun, 4.2 light-years away. Sooner rather than later, we'll have to master propulsion fast enough to make interstellar journeys practical. Perhaps by that time we'll have developed beyond our organic origins. It has been estimated that von Neumann probes—robots that can land on a planet, mine local materials, and replicate themselves—could colonize the entire galaxy within 10 million years.

But even a galactic civilization might face existential threats. According to our current understanding of the laws of physics, in any region of space there's a chance that a "death bubble" forms and then expands at speeds approaching the speed of light. Because the physics inside the bubble would differ from that of ordinary space, as it expanded it would destroy all matter, including life. The chances of this happening in a given year may seem extremely low—perhaps less than 1 in 10 billion. But as

Hawking reminded us, if you wait long enough the improbable is inevitable.

Yet the renowned physicist Ashoke Sen has recently suggested that even in the face of death bubbles there might be an escape route. The loophole is the accelerating expansion of the universe. In 1998, astronomers discovered that all galaxies not strongly bound together by gravity are moving apart ever faster. This accelerating expansion will eventually carry them beyond one another's cosmic horizon—so far that even light from one can never reach another. Thus they can no longer communicate—but, on the bright side, they can also never be swallowed by the same death bubble.

So by splitting into daughter civilizations and putting as much distance between them as possible, a civilization could "ride" the expansion of the universe to relative safety. Of course, another death bubble will eventually pop up within any cosmic horizon, so the remaining civilizations need to keep replicating and parting ways. Their chances of survival depend on how far they can travel: If they can move at a substantial fraction of the speed of light, they'll raise their chances of survival considerably. But even those that dispersed only far enough not to be bound by another galaxy's gravity—about 5 million light-years—might significantly improve their odds.

Such problems may seem remote. But they'll be real to our descendants, if we can survive long enough to allow those descendants to exist. Our responsibility as a civilization is to keep going for as long as the laws of physics allow.

The longevity factor is a measure of intelligence—the ability to predict potential problems and solve them in advance. The question is, How intelligent are we?

THE ILLUSION OF
EXPLANATORY DEPTH

ADAM WAYTZ
Psychologist; Associate Professor, Kellogg School of Management,
Northwestern University

If you asked 100 people on the street if they understand how a refrigerator works, most would say yes. But ask them to produce a detailed, step-by-step explanation of exactly how, and you'd likely hear silence or stammering. This powerful but inaccurate feeling of knowing is what Leonid Rozenblit and Frank Keil in 2002 termed the *illusion of explanatory depth* (IOED), stating, "Most people feel they understand the world with far greater detail, coherence, and depth than they really do."

Rozenblit and Keil initially demonstrated the IOED through multi-phase studies. In a first phase, they asked participants to rate how well they understood certain artifacts—such as a sewing machine, crossbow, or cell phone. In a second phase, they asked participants to write a detailed explanation of *how* each artifact worked and afterward asked them to re-rate how well they understood each one. Study after study showed that ratings of actual knowledge dropped markedly from phase one to phase two, after participants were faced with their inability to explain how the artifact in question operated. Of course, the IOED extends well beyond artifacts: to how we think about scientific fields, mental illnesses, economic markets, and virtually anything we're capable of (mis)understanding.

At present, the IOED is pervasive, given that we have abundant access to information but consume it in a largely superficial way. A 2014 survey found that roughly six in ten Americans read news headlines and nothing more. Major geopolitical issues—civil wars in the Middle East, the latest climate-change research—are distilled into tweets, viral videos, memes, "explainer" Web sites, soundbites on comedy news shows, and daily e-newsletters that get inadvertently re-routed to the spam folder. We consume knowledge widely, but not deeply.

Understanding the IOED helps us combat political extremism. In 2013, Philip Fernbach and colleagues demonstrated that the IOED underlies people's policy positions on issues like single-payer healthcare, a national flat tax, and a cap-and-trade system for carbon emissions. As in Rozenblit and Keil's studies, Fernbach and colleagues first asked people to rate how well they understood those issues and then to explain how each issue worked and finally to re-rate their understanding. In addition, participants rated the extremity of their attitudes on those issues both before and after offering an explanation. Both self-reported understanding of an issue and attitude extremity dropped significantly after the attempts to explain the issue; people who strongly supported or opposed an issue became more moderate. What's more, reduced extremity also reduced willingness to donate money to a group advocating for the issue. These studies suggest that the IOED is a powerful tool for cooling off heated political disagreements.

The IOED provides us with much-needed humility. In any domain of knowledge, often the most ignorant are the most overconfident in their understanding of that domain. Justin Kruger and David Dunning famously showed that the lowest performers on tests of logical reasoning, grammar, and humor are the most likely to overestimate their test scores. Only through gaining expertise in a topic do people recognize its complexity and calibrate their confidence accordingly. Having to explain a phenomenon forces us to confront this complexity and realize our ignorance. At a time when political polarization, income inequality, and urban-rural separation have fractured us regarding social and economic issues, recognizing our modest understanding of those issues is a first step to bridging the divides.

SYNAPTIC TRANSFER

DAVID ROWAN
Editor, *Wired U.K.*

For all its fiber-enabled, live-video-streaming, 24/7-connected promise, our information network encapsulates a fundamental flaw: It is proving a suboptimal system for keeping the world informed. While it embraces nodes dedicated to propagating a rich seam of information, because the Internet's governing algorithms are optimized to connect us to what they believe we're already looking for, we tend to retreat into familiar and comfortably self-reinforcing silos—idea chambers whose feeds, tweets, and updates inevitably echo our preexisting prejudices and limitations. The wider conversation, a precondition for a healthy intellectual culture, isn't getting through. The signals are being blocked. The algorithmic filter is building ever higher walls. Facts are being invalidated by something called "post-truth." And that's just not healthy for the quality of informed public debate that *Edge* has always celebrated.

Thankfully, a solution is suggested by neural networks of a biological kind. Inside our brains, no neuron ever makes direct contact with another neuron; these billions of disconnected cells pursue their own individual agendas without directly communicating with their neighbors. But the reason we can form memories or sustain cogent debate is that the gaps between these neurons are programmed to build connections between them. These gaps, called synapses, connect individual neurons using chemical or electrical signals and thus unite isolated brain cells into a robust central and peripheral nervous system. The synapses transfer instructions between neurons, link our sense receptors to other parts of our nervous system, and carry messages destined for our muscles and glands. Without these gap-bridging entities, our disconnected brain cells would be pretty irrelevant.

We need to celebrate the synapse for its vital role in making connections—and, indeed, to extend the metaphor to the wider worlds of business, media, and politics. In an ever more atomized culture, it's the connectors of silos, the bridgers of worlds, that accrue the greatest value. So we need to promote the intellectual synapses, the journalistic synapses, the political synapses—the rare individuals who pull down walls, who connect divergent ideas, who dare to link two mutually incompatible fixed ideas in order to promote understanding.

Synaptic transfer in its scientific sense can be excitatory (encouraging the receiving neuron to forward the signal) or inhibitory (blocking the receiving neuron from forwarding the signal). Combined, these approaches ensure a coherent and thriving brain-body ecosystem. But as we promote the metaphorical sense of synaptic transfer, we can afford to be looser in our definition. Today we need synapse builders who break down filter bubbles and constrained worldviews by making connections wherever possible. These are the people who further gainful signaling by making unsolicited introductions between those who might mutually benefit; who convene dinner salons and conferences where the divergent may unexpectedly converge;

who, in the Bay Area habit, "pay it forward" by performing favors that transform a business ecosystem from one of hostile competitiveness to one based on hope, optimism, and mutual respect and understanding.

So let's recast the synapse, a term coined a century ago from the Greek for "join together," and promote the term to celebrate the gap-bridgers. Be the neurotransmitter in your world. Diffuse ideas and human connections. And help move us all beyond constrained thinking.

THE GENETIC BOOK OF THE DEAD

RICHARD DAWKINS

Evolutionary biologist; Emeritus Professor of the Public Understanding of Science, University of Oxford; author, *The God Delusion*; co-author (with Yan Wong), *The Ancestor's Tale: A Pilgrimage to the Dawn of Evolution*

Natural selection equips every living creature with the genes that enabled its ancestors—an unbroken line of them—to survive in their environments. To the extent that present environments resemble those of the ancestors, to that extent is a modern animal well equipped to survive and pass on the same genes. The "adaptations" of an animal, its anatomical details, instincts, and internal biochemistry, are a series of keys that exquisitely fit the locks that constituted its ancestral environments.

Given a key, you can reconstruct the lock it fits. Given an animal, you should be able to reconstruct the environments in which its ancestors survived. A knowledgeable zoologist, handed a previously unknown animal, can reconstruct some of the locks that its keys are equipped to open. Many of these are obvious.

Webbed feet indicate an aquatic way of life. Camouflaged animals literally carry on their backs a picture of the environments in which their ancestors evaded predation.

But most of the keys an animal brandishes are not obvious on the surface. Many are buried in cellular chemistry. All of them are, in a sense which is harder to decipher, also buried in the genome. If only we could read the genome in the appropriate way, it would be a kind of negative imprint of ancient worlds, a description of the ancestral environments of the species: *The Genetic Book of the Dead.*

Naturally the book's contents will be weighted in favor of recent ancestral environments. The book of a camel's genome describes recent millennia in deserts. But in there, too, must be descriptions of Devonian seas from before the mammals' remote ancestors crawled out onto land. The genetic book of a giant tortoise most vividly portrays the Galapagos island habitat of its recent ancestors and before that the South American mainland where its smaller ancestors thrived. But we know that all modern land tortoises descend earlier from marine turtles, so our Galapagos tortoise's genetic book will describe somewhat older marine scenes. And those marine ancestral turtles were themselves descended from much older, Triassic land tortoises. And, like all tetrapods, those Triassic tortoises themselves were descended from fish. So the genetic book of our Galapagos giant is a bewildering palimpsest of water, overlain by land, overlain by water, overlain by land.

How shall we read *The Genetic Book of the Dead*? I don't know, and that's one reason for coining the phrase: to stimulate others to come up with a methodology. I have a dim inkling of a plan. For simplicity of illustration, I'll stick to mammals. Gather together a list of mammals who live in water and make them as taxonomically diverse as possible: whales, dugongs, seals, water shrews, otters, yapoks. Now make a similar list of mammals that live in deserts: camels, desert foxes, jerboas, etc. Another list of taxonomically diverse mammals who live up trees: monkeys, squirrels, koalas, sugar gliders. Another list of mammals that live underground: moles, marsupial moles, golden moles, mole rats. Now borrow from the statistical techniques of the numerical taxonomists but use them in a kind of upside-down way. Take specimens of all those lists of mammals and measure as many features as possible—morphological, biochemical, and genetic. Now feed all the measurements into the computer and ask it (here's where I get really vague and ask mathematicians for help) to find features that all the aquatic animals have in common, features that all the desert animals have in common, and so on. Some of these will be obvious, like webbed feet. Others will be non-obvious, and that's why the exercise is worth doing. The most interesting of the non-obvious features will be in the genes. And they will enable us to read *The Genetic Book of the Dead*.

In addition to telling us about ancestral environments, *The Genetic Book of the Dead* can reveal other aspects of history. Demography, for instance. Coalescence analysis performed on my personal genome by my co-author (and ex-student) Yan Wong has revealed that the population from which I spring suffered a major bottleneck, probably corresponding to an out-of-Africa migration event, some 60,000 years ago. Yan's analysis may be the only occasion when one co-author of a book has made detailed historical inferences by reading the Genetic Book of the other co-author.

EXAPTATION

W. TECUMSEH FITCH

Professor of Cognitive Biology, University of Vienna; author, *The Evolution of Language*

Some memes are fortunate at birth: They represent clear new concepts, are blessed with a memorable name, and have prominent intellectual "parents" who ably shepherd them through the crucial initial process of dissemination, clarification, and acceptance. "Meme" itself is one of those lucky memes. Many other memes, however, are less fortunate in one or more of these respects, and through no fault of their own languish, for decades or even centuries, in the shadows of their highborn competitors.

The core evolutionary concept of "change of function" is one of those unfortunate memes. It was one of Darwin's key intellectual offspring in *The Origin*—"the highly important fact that an organ originally constructed for one purpose . . . may be converted into one for a wholly different purpose." It played a central role in Darwin's thinking about the evolution of novelty, particularly when a new function required an already complex organ (e.g., lungs for breathing or wings for flying).

But unlike its more successful sibling memes "natural selection" and "adaptation," Darwin never bothered to name this idea himself. It was left to later writers to coin the term "pre-adaptation," with its unfortunate implicit connotations of evolutionary foresight and planning. And as "pre-adaptation" the meme languished until 1982, when it was adopted, spruced up, and rebaptized as "exaptation" by Stephen Jay Gould and Elisabeth Vrba. The new word's etymology explicitly disowns any teleological implications and focuses attention on the conceptually key evolutionary moment: the change in function.

To illustrate exaptation, consider the many useful organs that are embryologically derived from the branchial arches, which originated as stiffeners for the water-pumping and filtering pharynx of our invertebrate ancestors and then developed into the gill bars of early fish (and still serve that function alone, in a few surviving jawless fish, like lampreys). Each arch is complex, containing cartilage, muscles, nerves, and blood vessels, and there are typically six pairs of them running serially down the neck.

In the first exaptation, the frontmost gill bars were converted into biting jaws in the first jawed fish, ancestral to all living terrestrial vertebrates, while the pairs of arches behind them kept supporting gills. But when these fishy forebears emerged fully onto land and water-breathing gills became superfluous, there was suddenly a lot of prime physiological real estate up for grabs. And like cheap loft space, subsequent evolution has come up with diverse new functions for these tissues.

Numerous novelties stem today from the branchial arches. In humans, our external ears and middle ear bones (themselves derived exaptively from early tetrapod jaw bones) are branchial-arch derivatives, as is our tongue-supporting hyoid skeleton and our sound-producing larynx. Thus, virtually all the hardware used for speech and singing was derived, in multiple exaptive steps and via multiple different physiological functions, from the gill bars of ancestral fish. Such innovative changes in function, shaped and sculpted to their new use by subsequent natural selection, play a central role in the evolution of many novel traits.

As this example illustrates, and Darwin emphasized, change of function is everywhere in biology and thus deserves to be a core part of our conceptual toolkit for evolutionary thinking.

But unfortunately our poor but deserving meme's bad luck was not to end in 1982, because Gould and Vrba were somewhat overzealous in their championship of the concept, implying that *any* trait that had undergone a change in function deserved the name "exaptation." Given how widespread change of function is, this move would rename many or even most adaptations as exaptations in one imperious terminological stroke. By pitting exaptation against adaptation, our poor meme was disadvantaged again, since no one was likely to give up on the latter term.

For exaptation to be a useful term, it should be interpreted (much as Darwin originally suggested) as one important phase in evolution: the initial stage in which old organs are put to new use, for which they will typically be only barely functional. Subsequent "normal" natural selection of small variants will then gradually shape and perfect exaptations to their new function, at which point they become ordinary adaptations again. We can thus envision an exaptive cycle as being at the heart of many new evolutionary traits: first, adaptation for some function, then exaptation for a new function, and finally further adaptive tuning to this new function. A trait's tenure as an exaptation should thus typically be brief, in evolutionary terms: A few thousand generations should suffice for new mutations to appear and shape it to its new function.

I believe exaptation to be a concept of central importance not only for bodily organs but also for the evolution of mind and brain (e.g., for the evolution of language). Much of what we use our brains for in modern times represents a change in function (e.g., piloting airplanes from basic visually-guided motor control, or mathematical thinking from basic precursor concepts of number and geometry). These new cognitive abilities (and many others, like reading) are clearly exaptations, with no further shaping by natural selection (yet). But debate rages about whether somewhat older but still recent human capacities like linguistic syntax have yet been tailored by natural selection to better fulfill their current role. (Proposed cognitive precursors for linguistic syntax include hierarchical social cognition as seen in primates or hierarchical motor control as seen in many vertebrates.)

But before these issues can be productively debated, the long-suffering meme of exaptation must be clearly defined, fully understood, and more widely appreciated. Contemporary theorists' interpretations should fuse the best components of its chequered past: Darwin's concept and Gould and Vrba's term. Only then can exaptation take its rightful place at the high table of evolutionary thought.

THE VIRIAL THEOREM

SETH LLOYD

Quantum mechanic; Professor of Mechanical Engineering, MIT; Miller Fellow, Santa Fe Institute; author, *Programming the Universe*

The word "meme" denotes a rapidly spreading idea, behavior, or concept. Richard Dawkins originally coined the term to explain the action of natural selection on cultural information. For example, good parenting practices lead to children who survive to pass on those good parenting practices to their own children. Ask someone under twenty-five what a meme is, however, and chances are you'll get a different definition: Typing "meme" into Google Images yields page after page of photos of celebrities, babies, and kittens, overlaid with somewhat humorous text. Memes spread only as rapidly as they can reproduce. Parenting is a long-term and arduous task that takes decades to reproduce itself. A kitten photo reproduces in the few seconds it takes to resend. Consequently, the vast majority of memes are now digital, and the digital meaning of "meme" has crowded out its social and evolutionary meaning.

Even in their digital context, memes are still usually taken to be a social phenomenon, selected and re-posted by human beings. Human beings are increasingly out of the loop in the production of viral information, however. Net bots who propagate fake news need not read it. Internet viruses that infect unprotected computers reproduce on their own, without human intervention. An accelerating wave of sell orders issued by high-frequency stock-trading programs can crash the market in seconds. Any interaction between systems that store and process information will cause that information to spread, and some bits spread faster than other bits. By definition, viral information propagates at an accelerating rate, driving stable systems unstable.

Accelerating flows of information aren't confined to humans, computers, and viruses. In the 19th century, physicists such as Ludwig Boltzmann, James Clerk Maxwell, and Josiah Willard Gibbs recognized that the physical quantity called entropy is in fact just a form of information—the number of bits required to describe the microscopic motion of atoms and molecules. At bottom, all physical systems register and process information. The second law of thermodynamics states that entropy tends to increase: This increase of entropy is nothing more nor less than the natural tendency of bits of information to reproduce and spread. The spread of information is not just a human affair, it's as old as the universe.

In systems governed by the laws of gravitation, such as the universe, information tends to spread at an accelerating rate. This accelerating spread of information stems from a centuries-old observation in classical mechanics called the *virial theorem*. The virial theorem (from the Latin *vis*, or "strength," as opposed to the Latin *virus*, or "slimy poison") implies that when gravitating systems lose energy and information, they heat up. A massive cloud of cool dust in the early universe loses energy and entropy and clumps together to form a hot star. As the star loses energy and entropy, radiating light and heat into the cold surrounding space, the star grows hotter, not colder. In our own star, the sun, orderly flows of energy and information between the sun's core (where nuclear reactions take place) and its outer layers result in stable and relatively constant radiation for billions of years. A supermassive star, by contrast, radiates energy and information faster and faster, becoming hotter and hotter in the process. Over the course of a few hundred thousand years, the star burns through its nuclear fuel, its core collapses to form a black hole (an event called the gravothermal catastrophe), and the outer layers of the star explode as a supernova, catapulting light, energy, and information across the universe.

Accelerating flows of information are a fundamental part of the universe; we can't escape them. For human beings, the gravitational instability implied by the virial theorem is a blessing: We wouldn't exist if the stars hadn't begun to shine. The viral nature of digital information is less blessed. Information that reproduces itself twice in a second wins out over information that reproduces only once in a second. In the digital memes ranked as most popular by Google Images, this competition leads to a race to the bottom. Subtlety, intricacy, and nuance take longer to appreciate and so add crucial seconds to the digital-meme-reproduction process, leading to a dominance of dumb and dumber. Any constraint that puts information at a disadvantage in reproducing causes that information to lose out in the meme race. Truth is such a constraint: Fake news can propagate more rapidly than real news exactly because it is unconstrained by reality—

and so can be constructed with reproduction as its only goal. The faulty genetic information contained in cancerous cells can propagate faster than correct genetic information, because cancer cells need not respond to the regulatory signals the body sends them.

Human society, living organisms, and the planets, stars, and galaxies making up the universe all function by the orderly exchange of information. Social cues, metabolic signals, and bits of information carried by the force of gravity give rise to societies, to organisms, and to the structure of the universe. Chaos, by contrast, is defined by the explosive growth and spread of random information. Memes used to be cultural practices that propagated because they benefited humanity. Accelerating flows of digital information have reduced memes to kitten photos on the Internet. When memes propagate so rapidly that they lose their meaning, watch out!

THE SECOND LAW OF THERMODYNAMICS

STEVEN PINKER
Johnstone Family Professor, Department of Psychology, Harvard University; author, *The Sense of Style*

The second law of thermodynamics states that in an isolated system (one that is not taking in energy), entropy never decreases. (The first law is that energy is conserved; the third, that a temperature of absolute zero is unreachable.) Closed systems inexorably become less structured, less organized, less able to accomplish interesting and useful outcomes, until they slide into an equilibrium of gray, tepid, homogeneous monotony and stay there.

In its original formulation, the second law referred to the process in which usable energy, in the form of a difference in temperature between two bodies, is dissipated, as heat flows from the warmer to the cooler body. Once it was appreciated that heat is not an invisible fluid but the motion of molecules, a more general, statistical version of the second law took shape. Now order could be characterized in terms of the set of all microscopically distinct states of a system: Of all these states, the ones we find useful make up a tiny sliver of the possibilities, while the disorderly or useless states make up the vast majority. It follows that any perturbation of the system, whether a random jiggling of its parts or a whack from the outside, will, by the laws of probability, nudge the system toward disorder or uselessness. If you walk away from a sand castle, it won't be there tomorrow, because as the wind, waves, seagulls, and small children push the grains of sand around, they're more likely to arrange them into one of the vast number of configurations that don't look like a castle than into the tiny few that do.

The second law of thermodynamics is acknowledged in everyday life, in sayings such as "Ashes to ashes," "Things fall apart," "Rust never sleeps," "Shit happens," "You can't unscramble an egg," "What can go wrong will go wrong," and (from the Texas lawmaker Sam Rayburn), "Any jackass can kick down a barn, but it takes a carpenter to build one."

Scientists appreciate that the second law is far more than an explanation for everyday nuisances; it is a foundation of our understanding of the universe and our place in it. In 1915 the physicist Arthur Eddington wrote:

The law that entropy always increases holds, I think, the supreme position among the laws of Nature. If someone points out to you that your pet theory of the universe is in disagreement with Maxwell's equations—then so much the worse for Maxwell's equations. If it is found to be contradicted by observation—well, these experimentalists do bungle things sometimes. But if your theory is found to be against the second law of thermodynamics I can give you no hope; there is nothing for it but to collapse in deepest humiliation.

In his famous 1959 lecture "The Two Cultures and the Scientific Revolution," the scientist and novelist C. P. Snow commented on the disdain for science among educated Britons in his day:

A good many times I have been present at gatherings of people who, by the standards of the traditional culture, are thought highly educated and who have with considerable gusto been expressing their incredulity at the illiteracy of scientists. Once or twice I have been provoked and have asked the company how many of them could describe the Second Law of Thermodynamics. The response was cold: it was also negative. Yet I was asking something which is the scientific equivalent of: Have you read a work of Shakespeare's?

And the evolutionary psychologists John Tooby, Leda Cosmides, and Clark Barrett entitled a recent paper on the foundations of the science of mind "The Second Law of Thermodynamics is the First Law of Psychology."

Why the awe for the second law? The second law defines the ultimate purpose of life, mind, and human striving: to deploy energy and information to fight back the tide of entropy and carve out refuges of beneficial order. An underappreciation of the inherent tendency toward disorder, and a failure to appreciate the precious niches of order we carve out, are a major source of human folly.

To start with, the second law implies that *misfortune may be no one's fault*. The biggest breakthrough of the Scientific Revolution was to nullify the intuition that the universe is saturated with purpose—that everything happens for a reason. In this primitive understanding, when bad things happen—accidents, disease, famine—someone or something must have *wanted* them to happen. This in turn impels people to find a defendant, demon, scapegoat, or witch to punish. Galileo and Newton replaced this cosmic morality play with a clockwork universe in which events are caused by conditions in the present, not goals for the future. The second law deepens that discovery: Not only does the universe not care about our desires but in the natural course of events it will appear to thwart them, because there are so many more ways for things to go wrong than to go right. Houses burn down, ships sink, battles are lost for the want of a horseshoe nail.

Poverty, too, needs no explanation. In a world governed by entropy and evolution, it is the default state of humankind. Matter does not just arrange itself into shelter or clothing, and living things do everything they can not to become our food. What needs to be explained is wealth. Yet most discussions of poverty consist of arguments about whom to blame for it.

More generally, an underappreciation of the second law lures people into seeing every unsolved social problem as a sign that their country is being driven off a cliff. It's in the very nature of the universe that life has problems. But it's better to figure out how to solve them—to apply information and energy to expand our refuge of beneficial order—than to start a conflagration and hope for the best.

EMERGENCE

ANTONY GARRETT LISI
Theoretical physicist

Thought, passion, love . . . this internal world we experience, including all the meaning and purpose in our lives, arises naturally from the interactions of elementary particles. This sounds absurd, but it's so. Scientists have attained a full understanding of all fundamental interactions that can happen at scales ranging from subatomic particles to the size of our solar system. There is magic in our world, but it's not from external forces that act on us or through us. Our fates are not guided by mystical energies or the motions of the planets against the stars. We know better now. We know that the magic of life comes from emergence.

It's the unimaginably large numbers of interactions that make this magic possible. To describe romantic love as the timely mutual squirt of oxytocin trivializes the concerted dance of more molecules than there are stars in the observable universe. The numbers are beyond astronomical. There are approximately 100 trillion atoms in each human cell, and about 100 trillion cells in each human. And the number of possible interactions rises exponentially with the number of atoms. It's the emergent qualities of this vast cosmos of interacting entities that make us us. In principle, it would be possible to use a sufficiently powerful computer to simulate the interactions of this myriad of atoms and reproduce all our perceptions, experiences, and emotions. But to simulate something doesn't mean you understand the thing—it only means you understand a thing's parts and their interactions well enough to simulate it. This is the triumph and tragedy of our most ancient and powerful method of science: analysis, understanding a thing as the sum of its parts and their actions. We have learned and benefited from this method, but we have also learned its limits. When the number of parts becomes huge, such as the number of atoms making up a human, analysis is practically useless for understanding the system—even though the system does emerge from its parts and their interactions. We can more effectively understand an entity using principles deduced from experiments at or near its own level of distance scale—its own stratum.

The emergent strata of the world are roughly recapitulated by the hierarchy of our major scientific subjects. Atomic physics emerges from particle physics and quantum field theory; chemistry emerges from atomic physics; biochemistry from chemistry; biology from biochemistry; neuroscience from biology; cognitive science from neuroscience; psychology from cognitive science; sociology from psychology; economics from sociology; and so on. This hierarchical sequence of strata, from low to high, is not exact or linear—other fields, such as computer science and environmental science, branch in and out, depending on their relevance, and mathematics and the constraints of physics apply throughout. But the general pattern of emergence in a sequence is clear: At each higher level, new behavior and properties appear that aren't obvious from the interactions of the constituent entities in the level below but do arise from them. The chemical properties of collections of molecules, such as acidity, can be described and modeled, inefficiently, using particle physics (two levels below), but it's much more practical to describe chemistry, including acidity, using principles derived within its own contextual level and perhaps one level down, with principles of atomic physics. One would almost never think about acidity in terms of particle physics, because it's too far removed.

And emergence is not just the converse of reduction. With each climb up the ladder of emergence to a higher level in the hierarchy, it is the cumulative side effects of interactions of large numbers of constituents that result in qualitatively new properties that are best understood within the context of the new level.

Every step up the ladder to a new stratum is usually associated with an increase in complexity. And the complexities compound. Thermodynamically, this compounding of complexity—and activity at a higher level—requires a readily available source of energy to drive it and a place to dump the resulting heat. If the energy source disappears, or if the heat cannot be expelled, complexity necessarily decays into entropy. Within a viable environment, at every high level of emergence, complexity and behavior is shaped by evolution through natural selection. For example, human goals, meaning, and purposes exist as emergent aspects in psychology favored by natural selection.

The ladder of emergence precludes the need for any supernatural influence in our world. Natural emergence is all it takes to create all the magic of life from building blocks of simple inanimate matter. Once we think we understand things at a high level in the emergence hierarchy, we often ignore the ladder we used to get there from much lower levels. But we should never forget that the ladder is there—that we and everything in our inner and outer world are emergent structures arising in many strata from a comprehensible physical foundation. And we also should not forget an important question this raises: Is there an ultimate fundamental level of this hierarchy, and are we close to knowing it, or is it emergence all the way down?

NATURAL SELECTION

JONATHAN B. LOSOS

Lehner Professor for the Study of Latin America, Professor of Organismic and Evolutionary Biology, Harvard University; Curator in Herpetology, Museum of Comparative Zoology; author, *Improbable Destinies*

It's easy to think of natural selection as omnipotent. As Darwin said, "[N]atural selection is daily and hourly scrutinizing . . . every variation, even the slightest; rejecting that which is bad, preserving and adding up all that is good." And the end result? Through time, a population becomes better and better adapted. Given enough time, wouldn't we expect natural selection to construct the ideal organism, optimally designed to meet the demands of its environment?

If natural selection worked like an engineer—starting with a blank slate and an unlimited range of materials, designing a blueprint in advance to produce the best possible structure— then the result might indeed be perfection. But that's not a good analogy for how natural selection works. As Nobel laureate François Jacob suggested in 1977, the better metaphor is a tinkerer who "gives his materials unexpected functions to produce a new object. From an old bicycle wheel, he makes a roulette; from a broken chair the cabinet of a radio." In just this way, "[e]volution does not produce novelties from scratch. It works on what already exists, either transforming a system to give it new functions or combining several systems to produce a more elaborate one."

What a tinkerer can produce is a function of the materials at hand, and the consequence is that two species facing the same environmental challenge may adapt in different ways. Consider the penguin and the dolphin. Both are speedy marine organisms descended from ancestors that lived on land. Although they have similar lifestyles, chasing down swift prey, they do so in different ways. Most fast-swimming marine predators propel themselves by powerful strokes of their tail, and the dolphin is no exception. But not the penguin—it literally flies through the water, its aquatic celerity propelled by its wings.

Why haven't penguins evolved a powerful tail for swimming like so many other denizens of the sea? The answer is simple. Birds don't have tails (they do have tail feathers but no underlying bones). Natural selection, the tinkerer, had nothing to work with, no tail to modify for force production. What the penguin's ancestor did have, however, were wings, already well suited for moving through air. It didn't take much tinkering to adapt them for locomotion in a different medium.

Sometimes the tinkerer's options are limited and the outcome far from perfect. Take, for example, the panda's "thumb," made famous by Stephen Jay Gould. As opposable digits go, the modified wrist bone is subpar, limited in flexibility and grasping capabilities. But it gets the job done, helping the panda grasp the bamboo stalks on which it feeds.

Or consider another example. The long and flexible neck of the swan is constructed of twenty-five vertebrae. *Elasmosaurus*, a giant marine reptile from the age of dinosaurs, with a contortionist's neck as long as its body, took this approach to the extreme, with seventy-two vertebrae. Pity, then, the poor giraffe, with only seven long blocky vertebrae in its seven-foot bridge. Wouldn't more bones have made the giraffe more graceful, better able to maneuver through branches to reach its leafy fare? Probably so, but the tinkerer didn't have the materials. For some reason, mammals are almost all constrained to have seven cervical vertebrae, no more, no less. Why this is, nobody knows for sure—some have suggested a link between neck-vertebra number and childhood cancer. Whatever the reason, natural selection didn't have the necessary materials; it couldn't increase the number of vertebrae in the giraffe's neck. So it did the next best thing, increasing the length of the individual vertebra to nearly a foot.

There are several messages to be taken from the realization that natural selection functions more like a tinkerer than an engineer. We shouldn't expect optimality from natural selection; it just gets the job done, taking the easiest, most accessible route. And as a corollary: We humans are not perfection personified, just natural selection's way of turning a quadrupedal ape into a big-brained biped. Had we not come along, some other species, from some other ancestral stock, might eventually have evolved hyperintelligence. But having come from different starting blocks, that species probably wouldn't have looked much like us.

DNA

GEORGE CHURCH

Professor of Genetics, Harvard Medical School; Director, Personal Genome Project; co-author (with Ed Regis), *Regenesis: How Synthetic Biology Will Reinvent Nature and Ourselves*

You might feel that "DNA" is already one of the most widely known scientific terms—with 392 million Google hits and Ngram score rising swiftly since 1946 to surpass terms like "bread," "pen," "bomb," "surgery," and "oxygen." "DNA" even beats more general terms, like "genetics" or "inheritance." This super-geeky acronym for deoxyribonucleic acid (by the way, not an acid in nature but a salt) has inspired vast numbers of clichés, like "corporate DNA," and cultural tropes, like crime-scene DNA. It's vital for all life on Earth, responsible for the presence of oxygen in our atmosphere, present in every tissue of every one of our bodies. Nevertheless, knowing even a tiny bit (which could save your life) about your DNA in your own body has probably lagged behind your literacy in sports, fictional characters, and the doings of celebrities.

The news is that you can now read all of your genes in detail for $499 and nearly your complete DNA genome for $999. The nines even make it sound like consumer pricing. ("Marked down to the low, low price of $2,999,999,999. Hurry! Supplies are limited!") But what do you get? If you're fertile, and even if you haven't yet started making babies or are already done, there's a chance you'll produce live human #7,473,123,456. You and your mate could be healthy carriers of a serious disease causing early childhood pain and death, like Tay-Sachs, Walker-Warburg, Niemann-Pick A, or Nemaline myopathy. It doesn't matter if you have no example in your family history, you're still at risk. The cost of care can be $20 million for some genetic diseases, but the psychological impact on the sick child and the family goes far beyond economics. In addition, diagnosis of genetic diseases with adult onset can suggest drugs or surgeries that add many quality-adjusted life years (QALYs).

Would you take analogous chances with your current family (for example, refusing air bags)? If not, why avoid knowing your DNA? As with other (non-genetic) diagnoses, you don't need to learn anything that isn't currently highly predictive and action-able—or involves action you'd be unwilling to take. You might get your DNA–reading cost reimbursed via your health insurance or healthcare provider, but at $499 is this really an issue? Do we wait for a path to reimbursement before we buy a smartphone, a fancy meal, a new car (with air bags), or remodel the kitchen? Isn't a healthy child or ten QALYs for an adult worth it? And before or after you serve yourself, you might give a present to your friends, family, or employees. Once we read human-genome DNA #7,473,123,456, *then* we'll have a widely known scientific concept worth celebrating.

GENETIC RESCUE

STEWART BRAND
Founder, *The Whole Earth Catalog*; Cofounder, The Well; Cofounder, The Long Now Foundation; author, *Whole Earth Discipline*

Wildlife populations are most threatened when their numbers become reduced to the point where their genetic diversity is lost. Their narrowing gene pool can accelerate into what's called an "extinction vortex." With ever fewer gene variants (alleles), the ability to adapt and evolve declines. As inbreeding increases, deleterious genes accumulate and fitness plummets. The creatures typically have fewer offspring, and many of them are physically or behaviorally impaired, susceptible to disease, increasingly unable to thrive. Most people assume they are doomed, but that no longer has to be what happens.

Genetic rescue restores genetic diversity. Conservation biologists are warming to its use with growing proof of its effectiveness. One study of 156 cases of genetic rescue showed that 93 percent had remarkable success. The most famous case was a dramatic turnaround for the nearly extinct Florida panther. By the mid 1990s, only twenty-six were left, and they were in bad shape. In desperation, conservationists brought in eight female Texas cougars, which are closely related to the Florida cats. Five of the females reproduced. The result of the outcrossing was a rapid increase in litter success—424 panther kittens born in the next twelve years. The previous population decline of 5 percent a year reversed to population growth of 4 percent a year. Signs of inbreeding went away, and signs of increasing fitness grew. Scientists noticed, among other things, that the genetically enriched panthers were becoming harder to capture.

Often genetic diversity can be restored by means as straightforward as connecting isolated populations with wildlife corridors or larger protected areas, but new technological capabilities are broadening the options for genetic rescue. Advanced reproductive technology offers an alternative to transporting whole, genetically distinct parents; artificial insemination has brought genetic refreshment to cheetahs, pandas, elephants, whooping cranes, and black-footed ferrets. With the cost of genome sequencing and analysis coming down, we can now examine each stage of genetic rescue at the gene level instead of having to wait for external traits to show improvement. This has already been done with Rocky Mountain bighorn sheep.

Another strategy being considered is "facilitated adaptation." Different populations of a species face different local challenges. When a particular population can't adapt fast enough to keep up with climate change, for example, it may be desirable to import the alleles from a population that has already adapted. With gene editing becoming so efficient (CRISPR, etc.), the desired genes could be introduced to the gene pool directly. If necessary, the needed genes could even come from a different species entirely. That's exactly what's been done to save the American chestnut from the fungus blight that killed 4 billion trees early in the 20th century and reduced the species to functional extinction. Two fungus-resistant genes were added from wheat, and the trees were made blight-proof. They are now gradually returning to their keystone role in America's great Eastern forest.

One further reservoir of genetic variability has yet to be tapped. In museums throughout the world, there are vast collections of specimens of species that have been reduced to genetically impoverished remnant populations in the wild or in captive breeding programs. Those museum specimens are replete with extinct alleles in their preserved (though fragmented) DNA. Ancient-DNA sequencing and analysis is becoming so precise that the needed alleles can be identified, reproduced, and reintroduced to the gene pool of the current population, restoring its original genetic diversity. The long-dead can help rescue the needful living.

POSITIVE FEEDBACKS IN CLIMATE CHANGE

BRUCE PARKER

Visiting Professor, Center for Maritime Systems, Stevens Institute of Technology; author, *The Power of the Sea*

There is very little appreciation among the general public (and even among many scientists) of the great complexity of the mechanisms involved in climate change. Climate change significantly involves physics, chemistry, biology, geology, and astronomical forcing. The present political debate centers on the effect of the increase in carbon dioxide in the Earth's atmosphere since humankind began clearing the forests of the world and (especially) began burning huge quantities of fossil fuels. But this debate often ignores the complex climate system this increase in CO_2 is expected to change—or not change, depending on one's political viewpoint.

The Earth's climate has been changing over the 4.5 billion years of its existence. For at least the past 2.4 million years, the planet has been going through regular cycles of significant cooling and warming in the Northern Hemisphere; we refer to them as ice-age cycles. In each cycle, there's a long glacial period of slowly growing continental ice sheets covering large parts of the Northern Hemisphere—sheets that are miles thick—accompanied by a sea-level drop of 300+ feet. This is eventually followed by a rapid melting of the ice sheets and an accompanying rise in sea level, beginning a relatively short interglacial period, which we're in right now. For some 1.6 million years, the glacial/interglacial cycle averaged about 41,000 years. Evidence of this is seen in the many excellent paleoclimate data sets collected around the world from ice cores, sediment cores from the ocean bottom, and speleothems in caves (stalactites and stalagmites). These data records have recently become longer and with higher resolution—in some cases, down to decadal resolution. Changes in temperature, volume of the ice sheets, sea level, and other parameters over millions of years can be determined from the changing ratios of various isotope pairs (e.g., 18O/16O, 13C/14C, etc.), based on an understanding of how those element isotopes figure differently in various physical, chemical, biological, and geological processes. Changes in atmospheric CO_2 and methane can be measured in ice cores from the Greenland and Antarctic ice sheets.

In these data, the average glacial/interglacial cycle matches the oscillation of the Earth's obliquity (the angle between the Earth's rotational axis and its orbital axis). The Earth's obliquity oscillates between 22.1° and 24.5° on a 41,000-year cycle, causing a very small change in the spatial distribution of insolation (sunlight hitting and warming the Earth). However, beginning about 0.8 million years ago, the glacial/interglacial cycle changed to approximately 82,000 years, and more recently it changed to approximately 123,000 years. This recent change, to longer glacial/interglacial cycles, is referred to by many scientists as the "100,000-year problem," because they don't understand why this change occurred.

But that's not the only thing scientists don't understand. They still don't know why these periods begin and end. The most significant changes in climate over the Earth's recent history are still a mystery. For a while, scientists were also not sure how such a small variation in insolation could lead to the buildup of major ice sheets on the continents of the Northern Hemisphere. Or how it could lead to their melting. They eventually understood that there were *positive feedback mechanisms* within the Earth's climate system that slowly cause these dramatic changes. Perhaps a better understanding of these positive feedback mechanisms will give the public an appreciation of climate change's complexity.

The most agreed upon and easiest to understand positive feedback mechanism involves the reflection of sunlight (albedo) from large ice sheets that build up in a glacial period. Ice and snow reflect more light back into space (causing less warming) than do soil or vegetation or water. Once ice sheets begin to form at the beginning of a glacial period, more sunlight is reflected away, so there's less warming of the Earth; the Earth thus grows colder and the ice sheets expand, covering more soil, vegetation, and water and thus further increasing reflection into space, leading to further cooling and larger ice sheets, etc. There's also a positive feedback during interglacial periods, but in the warming direction: As ice sheets melt, more land or ocean is exposed, which absorbs more light and further warms the Earth, further reducing the ice sheets, leading to more warming, and so on.

CO_2 in the atmosphere also varies over a glacial/interglacial cycle. Its concentration is lower during glacial periods and higher during interglacial periods. There are various processes that can cause CO_2 to decrease during cold periods and increase during warm periods—processes involving temperature effects on CO_2, its solubility in the ocean, changes in biological productivity in the ocean and on land, changes in salinity, changes in dust reaching the ocean, etc. So the debate has been whether CO_2 causes the warmth that produces an interglacial period or whether something else causes the initial warming which then leads to an increase in CO_2. Either way, there's another important positive feedback here involving CO_2, because the temperature changes and the changes in CO_2 are in the same direction.

There are other possible positive feedback mechanisms and much more detail which cannot be included in a short essay. The point here is that, to some degree, climate scientists understand how, through various positive feedback mechanisms, glacial periods get colder and colder as the ice sheets expand, and interglacial periods get warmer and warmer as the ice sheets melt. But the big question remains unsolved: No one has explained how you *switch* from glacials to interglacials and then back to glacials. No one knows what causes a glacial termination (the sudden warming of the Earth and melting of the ice sheets) or what causes a glacial inception (the not-quite-as-sudden cooling of the Earth and growing of the ice sheets). And if you don't understand that, you don't completely understand climate change, and your climate models are lacking a critical part of the climate-change picture.

Carbon dioxide may have been higher at various points in

Earth's history, but there's evidence that it has never risen to its present levels as quickly as it has during humankind's influence. Is it the quantity of CO_2 in the atmosphere that's important or is it the rapidity of its increase? And what influence does such a rate of increase in CO_2 occurring at the end of an interglacial period have on the ice-age cycle? The key to accurately assessing the degree to which we have affected climate change (and especially what the future consequences will be) is to make the climate models as accurate as possible. Which means including all the important physical, chemical, biological, and geological processes. The only way to test those models is to run them in the past and compare their predictions to the paleoclimate data sets we've meticulously acquired. Right now, these models cannot produce glacial terminations and glacial inceptions that accurately match the paleoclimate data. That's the next big hurdle. It would be helpful if those debating anthropogenic global warming could gain a little more understanding of the complexity of what they're debating.

THE ANTHROPOCENE

JENNIFER JACQUET

Assistant Professor of Environmental Studies, NYU; author, *Is Shame Necessary?*

To understand earthquakes in Oklahoma, the Earth's sixth mass extinction, or the rapid melting of the Greenland ice sheet, we need the Anthropocene—an epoch that acknowledges humans as a global geologic force. The Holocene, a cozier geologic epoch that began 11,700 years ago with climatic warming (giving us conditions that, among other things, led to farming), doesn't cut it anymore. The Holocene is outdated because it cannot explain the recent changes to the planet: the now 400-parts-per-million of carbon dioxide in the atmosphere from burning fossil fuels, the radioactive elements present in the Earth's strata from detonating nuclear weapons, or that one in five species in large ecosystems is considered invasive. Humans caused nearly all of the 907 earthquakes in Oklahoma in 2015; they were the result of the extraction process for oil and gas, part of which involves injecting saltwater, a by-product, into rock layers. The Anthropocene is defined by a combination of large-scale human impacts and, as a concept, gives us a sense of both our power and our responsibility.

In 2016, the Anthropocene Working Group, made up of thirty-five individuals, mostly scientists, voted that the Anthropocene should be formalized. This means a proposal should be taken to the overlords of geologic epochs, the International Commission on Stratigraphy, which will decide on whether to formally adopt the Anthropocene into the official geological timescale.

Any epoch needs a starting point, and the Anthropocene Working Group favors a mid-20th century start date, which corresponds to the advent of nuclear technology and the global reach of industrialization—but it won't be that simple. Two geologists, one of whom is the commission chair, pointed out that units of geologic time are defined not by their start date alone but also by their content. They argue that the Anthropocene is more a prediction about what could appear in the future rather than what's here now, because in geologic terms it's "difficult to distinguish the upper few centimeters of sediment from the underlying Holocene." While hardcore geologists were pushing back against formalizing the Anthropocene, a recent article in

Nature argued that social scientists should be involved in determining the Anthropocene's start date. Their role in delineating the geologic timescale would be unprecedented, but then again so is this new, human-led era.

Whether the geologic experts anoint it as an official epoch, enough of society has already decided that the Anthropocene is here. Humans are a planetary force. Not since cyanobacteria has a single taxonomic group been so in charge. Humans have proved that we're capable of seismic influence, of depleting the ozone layer, of changing the biology of every continent—but not, at least so far, capable of living on another planet. The more interesting questions may not be about whether the Anthropocene exists or when it began but about whether we're prepared to take this kind of control.

THE NOÖSPHERE

DAVID CHRISTIAN
Director, Big History Institute and Distinguished Professor in History, Macquarie University, Sydney; author, *Maps of Time: An Introduction to Big History*

The idea of the *Noösphere*, or "the sphere of mind," emerged early in the 20th century. It flourished for a while, then vanished. It deserves a second chance.

The Noösphere belongs to a family of concepts describing planetary envelopes or domains that have shaped Earth's history: biosphere, hydrosphere, atmosphere, lithosphere, and so on. The idea of a distinct realm of mind evolved over several centuries. The 18th-century French naturalist Buffon wrote of a "realm of man" beginning to transform the Earth's surface. Nineteenth-century environmental thinkers like George Perkins Marsh tried to measure the extent of those transformations, and Alexander von Humboldt declared that our impact on the planet was already "incalculable."

The word "Noösphere" emerged in Paris, in 1924, from conversations between the Russian geologist Vladimir Vernadsky and two French scholars, the paleontologist and priest Teilhard de Chardin and the mathematician Édouard Le Roy. In a lecture he gave in Paris in 1925, Vernadsky described humanity and the collective human mind as a new "geological force," by which he seems to have meant a force comparable in scale to mountain building or the movement of continents.

In a 1945 essay, Vernadsky described the Noösphere as "a new geological phenomenon on our planet. In it for the first time, man becomes a *large-scale geological force*." As one of many signs of this profound change, he noted the sudden appearance on Earth of new minerals and purified metals: "That mineralogical rarity, native iron, is now being produced by the billions of tons. Native aluminum, which never before existed on our planet, is now produced in any quantity." In the same essay, published in the year of his death, and fifteen years before Yuri Gagarin became the first human to enter space, Vernadsky wrote that the Noösphere might even launch humans "into cosmic space."

Unlike Gagarin, the idea of a Noösphere did not take off, perhaps because of the taint of vitalism. Both Teilhard and Le Roy were attracted to Henri Bergson's idea that evolution was driven by an *élan vital,* a "vital force." Vernadsky, however, was not tempted by vitalism in any form. As a geologist working in the Soviet Union, he seems to have been a committed materialist.

Today it's worth returning to the idea of a Noösphere in its non-vitalist, Vernadskyian version. Vernadsky is best known for developing the idea of a "biosphere," a sphere of life that has shaped planet Earth on geological timescales. His best known work on the subject is *The Biosphere*. Today the idea seems inescapable, as we learn of the colossal role of living organisms in creating an oxygen-rich atmosphere, in shaping the chemistry of the oceans, and in the evolution of minerals and rock strata such as limestone.

The sphere of mind evolved within the biosphere. All living organisms use information to tap flows of energy and resources, so in some form we can say that "mind" has always played a role within the biosphere. But even organisms with developed neurological systems and brains foraged for energy and resources individually, in pointillesque fashion. It was their cumulative impact that accounted for the growing importance of the biosphere. Humans were different. They didn't just forage for information; they domesticated it, just as early farmers domesticated the land, rivers, plants, and animals surrounding them. Like farming, domesticating information was a collective project. The unique precision and bandwidth of human language allowed our ancestors to share, accumulate, and mobilize information at the level of the community and eventually the species, and to do so at warp speed. And increasing flows of information unlocked unprecedented flows of energy and resources, until we became the first species in 4 billion years to mobilize energy and resources on geological scales. "Collective learning" made us a planet-changing species.

Today, students of the Anthropocene can date when the Noösphere became the primary driver of change on the surface of planet Earth. It was in the middle of the 20th century. So Vernadsky got it more or less right. The sphere of mind joined the pantheon of planet-shaping spheres a little over fifty years ago. In just a century or two, the Noösphere has taken its place alongside the other great shapers of our planet's history: cosmos, earth, and life.

Freed of the taint of vitalism, the idea of a Noösphere can help us get a better grip on the Anthropocene world of today.

THE GAIA HYPOTHESIS

HANS ULRICH OBRIST

Artistic Director, Serpentine Gallery, London; author, editor, *A Brief History of Curating*; co-author (with Rem Koolhaas), *Project Japan: Metabolism Talks*

According to James Lovelock's Gaia hypothesis, the planet Earth is a self-regulated living being. In this captivating theory, the planet, in all its parts, remains in suitable condition for life thanks to the behavior and action of living organisms.

Lovelock, an independent scientist, environmentalist, inventor, author, and researcher, took an early interest in science fiction. This led him to Olaf Stapledon's idea that the Earth itself may have consciousness. From Erwin Schrödinger's *What Is Life?* he picked up the theory of "order-from-disorder"—based on the second law of thermodynamics, according to which entropy only increases (or stays the same) in a closed system (like our universe) and that thus life avoids decaying to thermodynamical equilibrium by feeding on negative entropy. As a researcher at NASA, he worked on developing instruments for the analysis of extraterrestrial atmospheres. This led to an interest in potential life-forms on Mars. He was struck by the idea that to establish whether or not there is life on Mars all one has to do is measure the composition of the gases in the atmosphere.

When I visited Lovelock last year in his home on Chesil Beach, he told me that it was in September 1965 that he had his epiphany. He was in the Jet Propulsion Laboratory with the astronomer Carl Sagan and the philosopher Diane Hitchcock, who was employed by NASA to look at the logical consistency of the experiments conducted there. Another astronomer entered the office with the results of an analysis of the atmosphere on Venus and Mars. In both cases, it was composed almost entirely of carbon dioxide, while Earth's atmosphere also contains oxygen and methane. Lovelock asked himself why the Earth's atmosphere was so different from its two sister planets. Where do the gases come from?

Reasoning that oxygen comes from plants and methane comes from bacteria—both living things—he suddenly understood that Earth must be regulating its own atmosphere. When Lovelock began talking about his theory with Sagan, the astronomer's first response was "Oh, Jim, it's nonsense to think the Earth can regulate itself. Astronomical objects don't do that." But then Sagan said, "Hold on a minute, there's one thing that's been puzzling us astronomers, and that's the cool-sun problem. At the Earth's birth, the sun was 30 percent cooler than it is now, so why aren't we boiling?"

This brought Lovelock to the realization that, as he told me, "if the animal and plant life regulate the CO_2, they can control the temperature." And that was when Gaia entered the building. While subject to criticism as a New Age idea, the Gaia hypothesis was taken to heart by the hard-nosed, empirically-driven biologist and evolutionary theorist Lynn Margulis. Because Lovelock was trained from the medical side, in bacteriology, he tended to think of bacteria as pathogens. He hadn't previously thought of them as the great infrastructure that keeps the Earth going. "It was Lynn who drove that home," he told me. Margulis understood that contrary to so many interpretations, the Gaia hypothesis was not a vision of Earth as a single organism but as a jungle of interlacing and overlying entities, each of which generates its own environment.

Lovelock has tried to persuade humans that they're unwittingly no more than Gaia's disease. The challenge this time isn't to protect humans against microbes but to protect Gaia against the tiny microbes called humans. "Just as bacteria ran the Earth for two billion years and ran it very well, keeping it stabilized," he said, "we are now running the Earth. We're stumbling a bit, but the future of Earth depends on us as much as it depended on the bacteria."

OCEAN ACIDIFICATION

LAURENCE C. SMITH

Department Chair of Geography; Professor of Earth, Planetary, and Space Sciences, UCLA; author, *The World in 2050*

Ocean acidification, a stealthy side effect of rising anthropogenic CO_2 emissions, is a recently discovered, little recognized global climate-change threat that should be more widely known.

Unlike the warming effect on air temperatures that rising atmospheric CO_2 levels cause—which scientists understood theoretically since the late 1800s and began describing forcefully in the late 1970s—the alarm bell for ocean acidification was rung only in 2003, in a brief scientific paper. It introduced the term "ocean acidification" to describe how some of the rising CO_2

levels are absorbed and dissolved into the ocean's surface waters. This has the benefit of slowing the pace of air-temperature warming (thus far, oceans have absorbed at least a quarter of anthropogenic CO_2 emissions) but the detriment of lowering the pH of the world's oceans.

In chemistry notation, dissolving carbon dioxide in water yields carbonic acid ($CO_2 + H_2O \leftrightarrow H_2CO_3$), which quickly converts into bicarbonate (HCO_3^-) and hydrogen ($H+$) ions. $H+$ concentration defines pH (hence the "H" part of pH). Already, the concentration of $H+$ ions in ocean water has increased nearly 30 percent relative to pre-industrial levels. The pH of the world's oceans has correspondingly dropped about 0.1 pH unit and is expected to fall another 0.1 to 0.3 units by the end of this century. These numbers may sound small, but the pH scale is logarithmic, so one unit of pH represents a tenfold change in hydrogen-ion concentration.

The increased abundance of bicarbonate ions leads to decreased availability of calcite and aragonite minerals in ocean water, depriving marine mollusks, crustaceans, and corals of the primary ingredients from which they build their protective shells and skeletons. Highly calcified mollusks, echinoderms, and reef-building corals are especially sensitive. Low pH ocean water can also corrode shells directly.

Of the familiar organisms most affected by these conditions, oysters and mussels are especially vulnerable, but the detrimental effects of ocean acidification go far beyond shelled seafood. They threaten the viability of coral reefs and smaller organisms, like Foraminifera, that comprise the bottom of the food chain for marine food webs. Nor do the problems stop there: New research shows that small changes in ocean water pH alters the *behavior* of fish, snails, and other mobile creatures. They become stunned and confused. They lose sensory responsiveness to odors and sounds and can thus be more easily gobbled up by predators.

How these multiple cascading effects will play out for our planet's marine ecosystem is unknown. What we do know is that as some species suffer, other, more tolerant species will replace them. Spiny sea urchins may do better (up to a point), for example. Jellyfish are especially well positioned to flourish in low pH. In a world of acidifying oceans, we can envision beaches awash not with pretty shells but with the gelatinous stinging corpses of dead jellyfish.

Scientists are now investigating new ways to mitigate the ocean-acidification problem. Certain species of sea grass, for example, may locally buffer pH. Planting or reintroducing sea grasses could provide some relief in protected estuaries and coves. Selective breeding experiments are underway to develop strains of aquatic plants and animals more tolerant of low pH waters. At the extreme end of scientific tinkering, new genetically engineered marine organisms may be on the horizon.

While some of these ideas hold promise for particular locations and species, none can stabilize pH at the global scale. Genetic engineering raises a whole other set of ethical and ecological issues. Furthermore, some of the more hopeful geo-engineering solutions proposed to combat air-temperature warming work by increasing the Earth's reflectivity (e.g., blasting sulfate aerosols into the stratosphere to reflect incoming sunlight back into space). These strategies have problems of their own, like unknown effect on global rainfall patterns, but they might mitigate CO_2-induced air-temperature warming. However, because they do nothing to reduce atmospheric CO_2 levels, they will have zero effect on ocean acidification. Today the only viable way to slow ocean acidification on a global scale is to reduce human-induced CO_2 emissions.

INTERTEMPORAL CHOICE

DAVID DeSTENO
Professor of Social Psychology, Northeastern University

If I offered to give you $20 today or $100 in a year, which would you choose? It's a pretty straightforward question and one that has a logical answer. Unless you need that $20 to ensure your near-term survival, why not wait for the bigger prize if it's sure to come? After all, when was the last time any reputable financial institution offered an investment vehicle guaranteed to quintuple your money in 365 days? Yet if you pose this question to the average person, you'll be surprised to find that most will opt to take the $20 and run. Why? To understand that, and its implications for decisions in many domains of life, we first have to put the framework of the decision in context.

This type of decision—one where the consequences of choices change over time—is known as an *intertemporal choice*. It's the sort of dilemma well studied by economists and psychologists, who often facepalm at the seemingly irrational decisions people make when it comes to investing for the future; but it's a less familiar one, at least in name, to many outside those fields. The "irrational" part of intertemporal decisions derives from the human tendency to excessively discount the value .of future rewards, making it difficult to part with money that could offer pleasure in the moment in order to let it grow and thereby secure greater satisfaction and prosperity in the future.

Troubling as this situation might be for your 401(k), it's essential to recognize that the origin of intertemporal choices and the domains to which this framework can profitably be applied aren't limited to financial ones. In truth, much of human social life—our morality, our relationships—revolves around challenges posed by intertemporal choice. Do I pay back a favor I owe? If not, I'm certainly ahead in the moment, but over time I'll likely lose any future opportunities for cooperation—not only with the person I left hanging but also with any others who learn of my reputation. Should you cheat on your spouse? Although it might lead to pleasure in the short term, the long-term losses to satisfaction, assuming your marriage was a good one, will probably be devastating. Should you spend long hours to hone a skill that would make you valuable to your team or group, rather than spending a summer's day enjoying the weather? Here again, it's the sacrifice of pleasure in the short term to pave the way for greater success in the long term.

Challenges like these—involving cooperation, honesty, loyalty, perseverance, and the like—were the original intertemporal dilemmas our ancestors faced. If they weren't willing to accept smaller benefits in the short term by being less selfish, they wouldn't have many friends or partners with whom to cooperate and sustain themselves in the long one. To thrive, they needed to demonstrate good character, and that meant they needed self-control to put off immediate gratification.

Today, when we think about self-control, we think about marshmallows. But what is the marshmallow test really? In point of fact, it's just a child-friendly version of a dilemma of intertemporal choice: One sweet treat now, or two later? And as Walter Mischel's work showed, an ability to have patience—to solve an intertemporal choice by being future-oriented—predicts success in realms ranging from investing to academics to health to one's social life. In short, being able to delay gratification is a marker of character. People who can do this will be more loyal, more generous, more diligent, and fairer. It's because dilemmas of intertemporal choice underlie social living that they can so easily be applied to economics, not the other way around. After all, self-control didn't evolve to help us manage economic capital; it came about to help us manage social capital.

Recognition of this fact offers two important benefits. First, it provides a framework with which to unify the study of many types of decisions. For example, the dynamics and thus the psychological mechanisms that underlie cheating and compassion will overlap with those that underlie saving and investing. Sacrificing time or energy to help someone else will build long-term capital, just as saving money for retirement does. What's more, this decision framework is a scalable one: The dilemmas posed by climate change, overfishing, and related problems of sustainability are nothing if not intertemporal at base. Solving them requires a collective willingness to forgo immediate profits (or to pay higher prices) in the short term, in order to reap larger, communal gains in the long term.

The second benefit is the expansion of our toolset for solving associated dilemmas. While economists and self-control researchers traditionally emphasize using reason, willpower, and the like to overcome our inherent impatience for pleasure, realizing the intertemporal nature posed by many moral dilemmas suggests an alternate route: the moral emotions. Gratitude makes us repay our debts. Compassion makes us willing to help others. Guilt prevents us from acting in selfish ways. These moral emotions, which are intrinsic to social living, lead people, directly or indirectly, to value the future. They enhance our character, which, when translated to behavior, means they help us to share, to persevere, to be patient, and to be diligent.

Whether it's to save more, to eat less, to be kind, or to reduce a carbon footprint, the resolutions we make will likely require some forbearance. And promoting that forbearance means that all of us, scientists and non-scientists alike, should continue exploring in a multidisciplinary manner the mind's inclination toward selfish, short-term temptations and its many mechanisms to overcome them.

FUTURE SELF-CONTINUITY

BRIAN KNUTSON

Professor of Psychology and Neuroscience, Stanford University

Five years from now, what will your future self think of your current self? Will you even be the same person?

Over a century before the birth of Christ, Bactrian King Milinda challenged the Buddhist sage Nagasena to define identity. The Buddhist responded by inquiring about the identity of the king's chariot: "Is it the axle? Or the wheels, or the chassis, or reins, or yoke that is in the chariot? Is it all of these combined, or is it apart from them?" The King was forced to concede that his chariot's identity could not be reduced to its pieces. Later, the Greek biographer Plutarch noted that the passage of time further complicates definitions of identity. For example, if a ship is restored piece by piece over time, does it retain its original identity? As with Theseus' reconstructed ship, the paradox of identity applies to our constantly regenerating bodies (and their resident brains). Flipping in time from the past to the future, to what extent can we expect our identity to change over the next five years?

Considering the future self to be an entirely different person could have serious consequences. Philosopher Derek Parfit worried that people who regard the future self as distinct should logically have no more reason to care about that future self than about a stranger. By implication, they should have no reason to save money, maintain their health, or cultivate relationships. But perhaps there's a middle ground between self and stranger. To the extent that we imagine our future self to be *similar* to our present self, this sense of "future self-continuity" might predict a willingness to at least consider the interests of the future self.

Beyond rekindling philosophical debates, future self-continuity is a critical (and timely) scientific concept for a number of reasons. First, future self-continuity can be measured. Remarkably, neuro-imaging research suggests that a medial part of the frontal cortex shows greater activity when we think about ourselves versus strangers. When we think about our future self, the activity falls somewhere in between. The closer that activity is to the current self, the more willing individuals are to wait for future rewards (or to show less "temporal discounting" of future rewards). More conveniently, researchers can also simply ask people to rate how similar or connected they feel to their future selves (e.g., in five years). As with neural measures, people who endorse future self-continuity show less temporal discounting and have more money stashed in their savings accounts.

Second, future self-continuity matters. As noted, individuals with greater future self-continuity are more willing to wait for future rewards—not just in the laboratory but also in the real world. Applied research by Hal Hershfield and others suggests that adolescents with greater future self-continuity show less delinquent behavior and that adults with greater future self-continuity act more ethically in business transactions. Future self-continuity may even operate at the group level, since cultures that value respect for elders tend to save more, and nations with longer histories tend to have cleaner environments.

Third, and most important, future self-continuity can be ma-nipulated. Simple manipulations include writing a letter to one's future self; more sophisticated interventions involve interacting with digital renderings of future selves in virtual reality. These interventions can change behavior. For example, adolescents who write a letter to their future selves make fewer subsequent delinquent choices. Adults who interact with an age-progressed avatar allocate more available cash toward retirement plans. While the active ingredients of these manipulations remain to be isolated, enhancing the similarity and vividness of future-self representations seems to help. Scalable future self-continuity interventions may open up new channels for enhancing health, education, and wealth.

The need for future self-continuity continues to grow. On the resource front, people are living longer, whereas job sta-bility is decreasing. In the face of increasing automation, insti-tutional social safety nets are shrinking, forcing individuals to bear the full burden of saving for the future. And yet in the United States, saving has decreased to the point where nearly half the population would have difficulty finding $400 to cover an emergency expense. On the environmental front, global tem-peratures continue to rise to unprecedented levels—along with attendant droughts, increases in sea level, and damage to vul-nerable ecologies. Human choice has a hand in these problems. Perhaps increased future self-continuity—in individuals as well as policymakers—could generate solutions.

Imagine yourself in five years. Did you do everything you could today to make the world a better place—for both your present and future selves? If not, what can you change?

THE CLIMATE SYSTEM

GIULIO BOCCALETTI

Chief Strategy Officer, The Nature Conservancy

The notion of a "climate system" is the powerful idea that the temperature we feel when we walk outside our door every day of the year, the wind blowing in our face while we take a walk, the clouds we see in the sky, the waves we watch rippling on the ocean's surface as we stroll on the beach are all part of the same coherent, interconnected planetary system governed by a small number of deterministic physical laws.

The first explicit realization that planetary coherence is an attribute of many environmental phenomena probably coincided with the great explorations of the 16th and 17th centuries, when Edmond Halley, of comet fame, first postulated the existence of a general circulation of the atmosphere from equator to poles in response to differential heating of the planet by the sun. The reliability of the easterly winds, which ensured safe sailing westward on the great trade routes of the Atlantic, was a telling clue that something must have been explainable. That this had something to do with the shape of the Earth and its rotation required the genius of Halley, who—even without the necessary mathematics, which were only fully available more than a century later—was the first to realize just how powerful a constraint rotation could be on the fluid dynamics of the planet.

But even without such explanations, a sense that there's such a thing as a coherent "climate," in which, for example, weather appears to exhibit coherence over time and space, has followed us throughout history. After all, the saying "Red sun at night, sailor's delight; red sun at morning, sailors take warning" must have captured some degree of coherent predictability to have survived for centuries! Today we can explain that rhyme as predicting mid-latitude weather, and we can quantitatively describe it with a particular solution to the simplified Navier-Stokes equations called a Rossby wave, after Carl-Gustaf Rossby, the founder of modern meteorology.

We're so used to the coherent workings of the climate system that most people don't even think of it as a scientific construct worth knowing. We take for granted—in fact, we think it trivial—that in the mid-latitudes many of us enjoy summer, then autumn, followed by winter, then spring and then summer again, in a predictable sinusoidal sequence of temperature, wind, and rain or snow. That's just our climate, the sort of thing we read in the first page of a tourist guide to a new country we might visit. Yet we explain such climate by the effect on insolation of the tilt of the planet and its revolution around the sun. That same simple solar cycle also produces two rainy seasons in some parts of the tropics and highly predictable yearly monsoons in others. That the seasonal cycle could be such a complex and varied response to such a simple and predictable forcing is nothing short of astounding.

Even more striking is the fact that our scientific understanding of fluid motion on that scale has enabled computational models of the atmosphere—the brainchild, at least in terms of theory, of Lewis Fry Richardson and, in practice, of John von Neumann and Jule Charney, based on relatively simple numerical versions of the Navier-Stokes, continuity, and radiative-transfer equations—to simulate all those phenomena on a planetary scale so precisely that the untrained eye can hardly tell the difference from the real thing. It should be awe-inspiring to realize that the wind blowing through the tree in our garden is part of a coherent system spanning thousands of miles, from the equator to the pole, responding to astronomical forcing; that we can explain that system quantitatively, with equations derived from physical first principles; and that, within certain limits, we can predict its behavior.

Few know that the coherence of the climate system also has an extraordinary function. For example, coherent movement on a planetary scale by both the ocean and the atmosphere ensures that heat is transported poleward from the equator at a peak rate of almost 6 petawatts (or 6 million billion watts—roughly a thousand times all installed power-production capacity in the world), thus ensuring that when people say they're boiling in Nairobi or freezing in Berlin, they can be taken figuratively rather than literally.

A great deal of research is being done on the astonishing idea that almost imperceptible differences in those same solar-forcing conditions—the so-called Milankovitch cycle (tiny adjustments to the amount of sunlight reaching Earth because of small periodic changes in the shape and structure of its orbit)—could result in ice ages, a climatic response of such force that it means the difference between several hundred meters of ice over Chicago and what we have today. And we know that the climate's deterministic coherence doesn't necessarily imply simplicity: It can generate its own resonant modes—such as those associated with El Niño and La Niña, phenomena that require the coupled interaction of ocean and atmosphere—as well as chaotic dynamics.

It's one of the great accomplishments of modern science that we can understand all this by studying our planet and making thousands of measurements from meteorological stations, satellites, buoys, and ice cores, and that we can explain the climate system using the fundamental laws of physics. And this is all the more important because we spend our lives in that system every day. The planetary state of the climate system is what determines how much water we have available at any given time, what kind of crops we can grow, which parts of the world will flood and which parts will be parched to death. It matters, a great deal, to us all.

THE UNIVERSE OF ALGORITHMS

TERRENCE J. SEJNOWSKI
Computational neuroscientist; Francis Crick Professor, Salk Institute for Biological Studies; co-author (with Patricia S. Churchland), *The Computational Brain*

In the 20th century, we gained a deep understanding of the physical world using equations and the mathematics of continuous variables as the chief source of insights. A continuous variable varies smoothly across space and time. Unlike the simplicity of rockets, which follow Newton's laws of motion, there isn't a simple way to describe a tree. In the 21st century, we're making progress understanding the nature of complexity in computer science and biology based on the mathematics of algorithms, which often have discrete rather than continuous variables. An algorithm is a step-by-step recipe that you follow to achieve a goal, not unlike baking a cake.

Self-similar fractals grow out of simple recursive algorithms that create patterns resembling bushes and trees. The construction of a real tree is also an algorithm, driven by a sequence of decisions that turn genes on and off as cells divide. The construction of brains, perhaps the most demanding construction project in the universe, is also guided by algorithms embedded in the DNA which orchestrate the development of connections between thousands of different types of neurons in hundreds of different parts of the brain.

Learning and memory in brains is governed by algorithms that change the strengths of synapses between neurons according to the history of their activity. Learning algorithms have been used recently to train deep neural-network models to recognize speech, translate between languages, caption photographs, and play the game of Go at championship levels. These are surprising capabilities that emerge from applying the same simple learning algorithms to different types of data.

How common are algorithms that generate complexity? The Game of Life is a cellular automaton that generates objects that seem to have lives of their own. Stephen Wolfram wanted to know the simplest cellular-automaton rule that could lead to complex behaviors, so he set out to search through all of them. The first twenty-nine rules produced patterns that would always revert to boring behaviors: All the nodes would end up with the same value and fall into an infinitely repeating sequence or endless chaotic change. But rule thirty dazzled with continually evolving complex patterns. It was even possible to prove that rule thirty was capable of universal computation—that is, the power of a Turing Machine, which can compute any computable function.

One of the implications of this discovery is that the remarkable complexity we find in nature could have evolved by sampling the simplest space of chemical interactions between molecules. Complex molecules should be expected to emerge from evolution and not be considered a miracle. However, cellular automata may not be a good model for early life, and it remains an open question as to which simple chemical systems can create complex molecules. It may be that only special biochemical systems have this property, and this could help narrow the possible set of interactions from which life could have originated. Francis Crick and Leslie Orgel suggested that RNA might have these properties, which led to the concept of an RNA world before DNA appeared, early in evolution.

How many algorithms are there? Imagine the space of all possible algorithms. Every point in the space is an algorithm that does something. Some are amazingly useful and productive. These useful algorithms were once hand-crafted by mathematicians and computer scientists working as artisans. In contrast, Wolfram found cellular automata that produced highly complex patterns via automated search. Wolfram's law states that you don't have to travel far in the space of algorithms to find one that solves an interesting class of problems. This is similar to sending bots to play StarCraft or other games on the Internet that try all possible strategies. According to Wolfram's law, there should be a way to find algorithms, somewhere in the universe of algorithms, that can win the game.

Wolfram focused on the simplest algorithms in the space of cellular automata, a small subspace in the space of all possible algorithms. We now have confirmation of Wolfram's law in the space of neural networks, which are some of the most complex algorithms ever devised. Each deep-learning network is a point in the space of all possible algorithms and was found by automated search. For a large network and a large set of data, learning from different starting points can generate an infinite number of networks roughly equally good at solving the problem. Each data set generates its own galaxy of algorithms, and data sets are proliferating.

Who knows what the universe of algorithms holds for us? There may be whole galaxies of useful algorithms that humans have not yet discovered but can be found by automated discovery. The 21st century has just begun.

BABYLONIAN LOTTERY

COCO KRUMME

Applied mathematician; Lecturer and Postdoctoral Scholar, UC Berkeley

> "Like all men in Babylon, I have been proconsul; like all,
> I have been a slave."
>
> —Jorge Luis Borges, *Lottery in Babylon*

The lottery in Babylon begins as a simple game of chance. Tickets are sold, winners are drawn, a prize awarded. Over time, the game evolves. Punishments are doled out alongside prizes. Eventually the lottery becomes compulsory. Its cadence increases, until the outcomes of its drawings come to underpin everything. Mundane events and life turns are subject to the lottery's "intensification of chance." Or perhaps, as Borges suggests, it's the lottery's explanatory power that grows, as well as that of its shadowy operator, the Company, until all occurrences are recast in light of its odds.

"Babylonian lottery" is a term borrowed from literature, for which no scientific term exists. It describes the slow encroachment of programmatic chance, or what we like to refer to today as "algorithms."

Today, as in Babylon, we feel the weight of these algorithms. They amplify our product choices and news recommendations; they're embedded in our financial markets. While we may not have direct experience building algorithms—or, for that matter, understand their reach, just as the Babylonians never saw the Company—we believe them to be all-encompassing.

Algorithms as rules for computation are nothing new. What's new is the sudden cognizance of their scope. At least three things can be said of the Babylonian lottery, and of our own:

First, the Babylonian lottery increases in complexity and reach over time. Similarly, our algorithms have evolved from deterministic to probabilistic, broadening in scope and incorporating randomness and noisy social signals. A probabilistic computation feels somehow mightier than a deterministic one; we can know it in expectation but not exactly.

Second, while in the beginning all Babylonians understand how the lottery works, over time fewer and fewer do. Similarly, today's algorithms are increasingly specialized. Few can both understand a computational system from first principles and make a meaningful contribution at its bleeding edge.

Third, the Babylonians for some time brushed under the rug the encroachment of the lottery. Because an "intensification of chance" conflicts with our mythologies of self-made meritocracy, we too ignore the impact of algorithms for as long as possible.

So here we are, with algorithms encroaching, few who understand them, and finally waking up. How do we avoid the fate of the Babylonians? Unfortunately, those of us in the centers of algorithm creation are barking up the wrong tree. That algorithms aren't neutral but in many cases codify bias or chance isn't news to anyone who's worked with them. But as this codification becomes common knowledge, we look for a culprit.

Some point to the "lack of empathy" of algorithms and their creators. The solution, they suggest, is a set of more empathetic algorithms to subjugate the dispassionate ones. To combat digital distraction, they'd throttle email on Sundays and build apps for meditation. Instead of recommender systems that reveal what you most want to hear, they'd inject a set of countervailing views.

The irony is that these manufactured gestures only intensify the hold of a Babylonian lottery.

We can no more undo the complexity of such lotteries than we can step back in time. We've swept our lottery's impact under the rug of our mythologies for a good while. Its complexity means that no one who's in a position to alter it understands its workings in totality. And we've seen the futility of building new algorithms to subvert the old.

So what do we do? How could the Babylonians have short-circuited the lottery?

For Borges' Babylonian narrator, there are only two ways out. The first is physical departure. We encounter the narrator, in shackles, aboard a ship about to set sail. Whether his escape is a sign of the lottery's shortcoming or a testament to its latitude remains ambiguous.

The second, less equivocal way out of Babylon is found, as we see, in storytelling. By recounting the lottery's evolution, the narrator replaces enumeration with description. A story, like a code of ethics, is unlike any algorithm. Algorithms are rules for determining outcomes. Stories are guides to decision-making along the way.

Telling the tale ensures that the next instantiation of the lottery isn't merely a newly parameterized version of the old. A story teaches us to make new mistakes rather than recursively repeating the old. It reminds us that the reach of algorithms is perhaps more limited than we think. By beginning with rather than arriving at meaning, a story can overcome the determinism of chance.

Babylon as a physical place, of course, fell apart. Babylon as a story endures.

CLASS BREAKS

Cryptographer, computer security professional; Fellow, Berkman Center for Internet and Society, Harvard Law School; Chief Technology Officer, Resilient; author, *Data and Goliath*

There's a concept from computer security known as a class break. It's a particular security vulnerability that breaks not just one system but an entire class of systems. Examples might be a vulnerability in a particular operating system which allows an attacker to take remote control of every computer running on that system's software. Or a vulnerability in Internet–enabled digital video recorders and webcams which allows an attacker to recruit those devices into a massive botnet.

It's a particular way that computer systems can fail, exacerbated by the characteristics of computers and software. It takes only one smart person to figure out how to attack the system. Once he does, he can write software that automates his attack. He can do it over the Internet, so he doesn't have to be near his victim. He can automate his attack so it works while he sleeps. And then he can pass the ability on to someone (or to lots of people) without the skill. This changes the nature of security failures and completely upends how we need to defend against them.

An example: Picking a mechanical doorlock requires both skill and time. Each lock is a new job, and success at one lock doesn't guarantee success with another of the same design. Electronic doorlocks like the ones now in hotel rooms have different vulnerabilities. An attacker can find a flaw in the design, allowing him to create a key card that opens every door. If he publishes his attack software, then not just the attacker but anyone can now open every lock. And if those locks are connected to the Internet, attackers could potentially open doorlocks remotely; they could open every doorlock remotely at the same time. That's a class break.

It's how computer systems fail, but it's not how we think about failures. We still think about automobile security in terms of individual car thieves manually stealing cars. We don't think of hackers remotely taking control of cars over the Internet. Or remotely disabling every car over the Internet. We think about voting fraud as unauthorized individuals trying to vote. We don't think about a single person or organization remotely manipulating thousands of Internet-connected voting machines.

In a sense, class breaks are not a new concept in risk management. It's the difference between home burglaries and fires (which happen occasionally to different houses in a neighborhood over the course of the year) and floods and earthquakes (which either happen to everyone in the neighborhood or no one). Insurance companies can handle both types of risk, but they're inherently different. The increasing computerization of everything is moving us from a burglary/fire risk model to a flood/earthquake model, in which a given threat either affects everyone in town or doesn't happen at all.

But there's a key difference between floods/earthquakes and class breaks in computer systems: The former are random natural phenomena, whereas the latter are human-directed. Floods don't change their behavior to maximize their damage based on the types of defenses we build. Attackers do that to computer systems. Attackers examine our systems looking for class breaks. And once one of them finds one, they'll exploit it again and again until the vulnerability is fixed.

As we move into the world of the Internet of Things, where computers permeate our lives at every level, class breaks will become increasingly important. The combination of automation and action at a distance will give attackers more power and leverage than they've ever had before. Security notions like the precautionary principle—where the potential of harm is so great that we err on the side of not deploying a new technology without proofs of security—will become more important in a world where an attacker can open all the doorlocks or hack all the power plants. It's not an inherently less secure world, but it's a differently secure world. It's a world where driverless cars are much safer than people-driven cars, until suddenly they're not. We need to build systems that assume the possibility of class breaks and maintain security despite them.

RECURSION

READ MONTAGUE
Neuroscientist; Director, Human Neuroimaging Laboratory and Computational Psychiatry Unit, Virginia Tech Carilion Research Institute; author, *Why Choose This Book? How We Make Decisions*

Recursion resides at the core of all intelligence. Recursion requires the ability to reference an algorithm or procedure and keep the reference distinct from the contents. And this ability to refer is a centerpiece of the way organisms form models of the world around them and even of themselves. Recursion is a profoundly computational idea and lies at the heart of Gödel's incompleteness theorem and the philosophical consequences that flow from that work. Alan Turing's work on computable numbers requires recursion at its center. It's an open question about how recursion is implemented biologically, but one could speculate that it has been discovered by evolution many times.

REFERENTIAL OPACITY

NICHOLAS HUMPHREY

Emeritus Professor of Psychology, London School of Economics; Visiting Professor of Philosophy, New College of the Humanities; Senior Member, Darwin College, University of Cambridge; author, *Soul Dust*

It's commonly assumed that when people make a free choice in an election, the outcome will be what those on the winning side intended. But there are two factors known to cognitive science—and probably not known to politicians—which may well render this assumption false.

The first is the fact of "referential opacity" as it applies to mental states. A peculiar characteristic of mental states—such as believing, wanting, remembering—is that they don't conform to Leibniz's law. This law states that if two things, A and B, are identical, then in any true statement about A, you can replace A with B and the new statement will also be true. So if it's true that A weighs five kilos, it must be true that B weighs five kilos; if it's true that A lives in Cambridge, it must be true that B lives in Cambridge; and so on. The strange thing is, however, that when it comes to mental states, this substitution no longer works. Suppose, for example, the Duke of Clarence and Jack the Ripper were one and the same person. It could be true that you *believe* the Duke of Clarence was Queen Victoria's grandson but not true that you *believe* Jack the Ripper was her grandson.

It's called referential opacity because the identity of the referents in the two linked mental states is not transparent to the subject. It may seem an abstruse concept, but its implications are profound. For one thing, it sets clear limits to what people *really intend* by their words or actions, and therefore on their responsibility for the outcome. Take the case of Oedipus. While it's true that Oedipus decided to marry Jocasta, it's presumably not true that he decided to marry his mother, even though Jocasta and his mother were identical. So Oedipus was wrong to blame himself. Or take the case of Einstein and the atomic bomb. While it's true that Einstein was happy to discover that $e = mc^2$, it's far from true that he was happy to discover the formula for making a nuclear weapon, even if $e = mc^2$ is that formula. He said, "If I had known, I would have been a clockmaker." But he did not know.

Now, what about voters' intentions in elections? Referential opacity can explain why there's so often a mismatch between what voters want and what they get. Take the German elections in 1933. While it's true that 44 percent of voters wanted Hitler to become chancellor, it's clearly not true that they wanted the man who would ruin Germany to become chancellor, even though Hitler was that very man.

So let's turn to the second factor that can cause confusion about the real intentions of voters. This is the phenomenon of "choice blindness," discovered by Lars Hall and Peter Johansson. In a classic experiment, these researchers asked a male subject to choose which he liked better of two photos of young women. They then handed the photo to the subject and asked him to explain the reasons for his choice. But they had secretly switched photos, so the subject was given the photo he didn't choose. Remarkably, most subjects didn't recognize the switch and proceeded, unperturbed, to give reasons for a choice they hadn't made. Hall and Johansson conclude that people's overriding need to maintain a consistent narrative can trump the memory of what has actually occurred.

Consider, then, what may happen when, in the context of an election, choice blindness combines with referential opacity. Suppose the majority of voters choose to elect Mr. A, with never a thought to electing Mr. B. But after the election, it transpires that, unwittingly and unaccountably, they have in fact got Mr. B in place of Mr. A. Now people's need to remain on plot and to make sense of this outcome leads them to rewrite history and persuade themselves it was Mr. B they wanted all along.

I'm not saying the vaunted "democratic choice," the "will of the people," is a mirage. I'm saying that it would be as well if the wider public knew that scientists say that it should be taken with a pinch of salt.

ADAPTIVE PREFERENCE

STEVE FULLER
Philosopher; Auguste Comte Chair in Social Epistemology, University of Warwick, U.K.; author, *Knowledge: The Philosophical Quest in History*

Adaptive preference is a concept that illuminates what happens both in the laboratory and in life. An adaptive preference results when we bend aspiration toward expectation in light of experience. We come to want what we think is within our grasp. More than a simple "reality check," adaptive-preference formation involves disciplining one's motivational structure with the benefit of hindsight. Much of what passes for wisdom in life is about the formation of adaptive preferences.

When the social psychologist Leon Festinger suggested the

idea in the 1950s, it provided a neat account of how people maintain a sense of autonomy while under attack by events beyond their control. He might have been talking about how the U.S. and the U.S.S.R. held their nerve in Cold War vicissitudes, but in fact he was talking about how a millennial religious cult continued to flourish even after its key doomsday prediction had failed to materialize.

In the 1980s, the social and political philosopher Jon Elster brilliantly generalized the idea of adaptive preference in terms of the complementary phenomena of "sour grapes" and "sweet lemons." We tend to downgrade the value of previously desired outcomes as their realization becomes less likely and upgrade the value of previously undesired outcomes as their realization becomes more likely.

The interesting question is whether adaptive-preference formation is rational. Festinger's original case study seemed to imply that it isn't. After a few hours of doubt and despair, the cult regrouped by interpreting the Deity's failure to end the world as a sign that the cult had done sufficient good to reverse its fate. This emboldened its members to proselytize still more vigorously.

One might think that had the cult responded rationally to the failed prophecy, its members would simply have abandoned the belief that they were in a special relationship with the Deity. Instead, the cult did something rather subtle. They did not make the obvious "irrational" move of denying that the prophecy had failed or postponing doomsday to a later date. Rather, they altered their relationship to the Deity, who previously appeared to claim that there was nothing humans could do to reverse their fate. The terms of this renegotiated relationship then gave the cult members a sense of control over their lives, which served to renew their missionary zeal.

This is an instance of what Elster called "sweet lemons," and it's not as obviously irrational as its counterpart, sour grapes. In fact, a sweet-lemon detector, so to speak, may be a key element of the motivational structure of people who are capable of deep learning from negative experience. Such people acquire a clearer sense of what they've truly valued all along, so that they're reinvigorated by adversity.

The phenomenon of sweet lemons is disorienting to the observer, because it highlights just how much we presume that others share our overarching values. We don't simply respect the autonomy of others; we also expect, somewhat paradoxically, that by virtue of their autonomy they'll become more like us. Thus, the post-prophecy behavior of Festinger's cult is confusing, because they carried on in a version of what they'd done previously. They learned from experience, but what they learned was to become more like themselves.

Adaptive preferences are arguably scalable, perhaps even to the level of entire cultures and species. A striking feature of human history is that widespread disruption and destruction don't necessarily result in people avoiding the precipitating behaviors in the future. For example, within a half-century of mass-produced automobiles, the original objections to their introduction had been realized: The cars were a major source of air and noise pollution. The needed roads ravaged the environment and alienated their drivers from nature.

Yet none of that seemed to matter—or at least not enough to lead people to abandon automobiles. Rather, car production worldwide has continued apace while becoming a bit more environment-friendly to avoid the worst envisaged outcomes. For better or worse, we still appear to buy the value package that Henry Ford and others were selling in the early 20th century: We value the car's freedom and speed, not only over the connectedness to nature offered by the horse as presented in Ford's day but also over the relatively low ecological impact offered by mechanized public transport today.

Had Ford not introduced the mass-produced car in the early 20th century, humanity might not have discovered just how much it valued personal mobility. At least, that's how it looks in the sweet-lemons version of adaptive preference. Whether a general policy of sweet lemons lets us survive in the future is an open question. But if we do become extinct, it's likely to have been a by-product of our trying to be better versions of what experience taught us to believe we are.

ANTAGONISTIC PLEIOTROPY

GREGORY BENFORD
Emeritus Professor of Physics and Astronomy, UC Irvine; science fiction writer; author, *The Berlin Project*

Aging comes from evolution. It isn't a bug or a feature of life; it's an inevitable side effect.

Exactly why evolution favors aging is controversial, but plainly it does; all creatures die. It's not a curse from God or imposed by limited natural resources. Aging arises from favoring short-term benefits, mostly early reproduction, over long-term survival when reproduction has stopped.

Thermodynamics doesn't demand senescence, though early thinkers imagined it did. Similarly, generic damage or "wear and tear" theories can't explain why biologically similar organisms show significantly different life spans. Most organisms maintain themselves efficiently until adulthood and then, after they can't reproduce anymore, succumb to age-related damage. Some die swiftly, like flies, and others, like we humans, can live far beyond reproduction.

Peter Medawar introduced the idea that aging was a matter of communication failure between generations. Older organisms have no way to pass on genes that helped them survive if they've stopped having offspring. Nature is a highly competitive place, and almost all animals in nature die before they attain old age. Those who do can't pass on newly arising long-lived genes, so old age is naturally selected against. Genetically detrimental mutations, those would not be efficiently weeded out by natural selection. Hence they would accumulate and perhaps cause all the decline and damage.

It turns out that the genes that cause aging aren't random mutations. Rather, they form tight-knit families that have been around as long as worms and fruit flies. They survive for good reasons.

In 1957, the evolutionary biologist George C. Williams proposed his own theory, called *antagonistic pleiotropy*. If a gene has two or more effects, one beneficial and another detrimental, the bad one exacts a cost later on. If evolution is a race to have the most offspring the fastest, then enhanced early fertility could be selected even if it came with a price tag that included decline and death later on. Because aging was a side effect of necessary functions, Williams considered any alteration of the aging process to be impossible. Antagonistic pleiotropy is a prevailing theory today, but Williams was wrong: We can offset such effects.

Wear and tear can be countered. Wounds heal, dead cells get replaced, claws regrow. Some species are better at maintenance and repair. Medawar disagreed with Williams that there are fundamental limitations on life span. He pointed out that such organisms as sea turtles live great spans, many for more than a century. Aging arises from failure to repair, which can be addressed without unacceptable side effects. Some species, like us, have better maintenance–and–repair mechanisms. These can be enhanced.

Some scientists pursued this by deliberately aging animals, like UC Irvine's Michael Rose. Rose simply didn't let fruit fly eggs hatch until half of each fly generation had died. This eliminated some genes that promoted early reproduction but had bad effects later. Over 700 generations later, his fruit flies live more than four times longer than the control flies. These Methuselahs are more robust than ordinary flies and reproduce more, not less, as some biologists predicted.

Delaying reproduction gradually extends the average lifetime. One side realization: University graduates mate and have children later in life than others. They are then slowly selecting for longevity in those better educated. Education roughly correlates with intelligence. Eventually, longevity will correlate more and more with intelligence.

I bought those Methuselah flies in 2006 and formed a company, Genescient, to explore their genetics. We discovered hundreds of longevity genes shared by both flies and humans. Up-regulating the functioning of those repair genes has led to positive effects in human trials.

So, although aging is inevitable and emerges from antagonistic pleiotropy, it can be attacked. Recent developments point toward possibly major progress. For example, a decade ago, the Japanese biologist Shinya Yamanaka found four crucial genes that reset the clock of the fertilized egg. No matter how old the parents are, their progeny are free of all marks of age; babies begin anew. This is a crucial feature of all creatures. By using his four genes, Yamanaka changed adult tissue cells into cells much like embryonic stem cells. Applying this reprogramming to adult tissue is tricky, but it beckons as a method of rejuvenating our bodies. Although evolution discards us as messengers to our descendants, once we stop reproducing not all is lost. In the game of life, intelligence bats last.

MALADAPTATION

AUBREY DE GREY
Gerontologist; Chief Science Officer and Cofounder, SENS Foundation; author, *Ending Aging*

Many years ago, Francis Crick promoted (attributing it to his longtime collaborator Leslie Orgel) an aphorism that dominates the thinking of most biologists: "Evolution is cleverer than you are." This is often viewed as a more succinct version of Theodosius Dobzhansky's famous dictum: "Nothing in biology makes sense except in the context of evolution." But these two observations, at least in the terms in which they're usually interpreted, are not as synonymous as they first appear.

Most of the difference between them comes down to the concept of maladaptation. A maladaptive trait is one that persists in a population in spite of inflicting a negative influence on the ability of individuals to pass on their genes. Orgel's rule, extrapolated to its logical conclusion, would seem to imply that this can never occur: Evolution will always find a way to maximize the evolutionary fitness of a population. It may take time to respond to changed circumstances, yes, but it won't stabilize in an imperfect state. And yet there are many examples wherein that seems to have occurred. In human health, arguably the most conspicuous case is that the ability to regenerate wounded tissues is lost in adulthood (sometimes even earlier), even though more primitive vertebrates (and, to a lesser extent, even some other mammals) retain it throughout life.

This defiance of Orgel's rule is not, however, in conflict with Dobzhansky's. That's because of the phenomenon of pleiotropy, or trade-offs. Sometimes the advantage gained by optimizing one aspect of fitness is outweighed by some downside that results from the same genetic machinery. The stable state to which the species thus gravitates is then a happy (but not perfectly happy) medium between the two extremes that would optimize the corresponding aspects.

Why is this so important to keep in mind? Many reasons, but in particular it's because when we get this wrong, we can end up making very bad evaluations of the most promising way to improve our health with new medicines. Today, the overwhelming majority of ill-health in the industrialized world consists of the diseases of late life, and we spend billions of dollars trying to alleviate them—but our hit rate in developing even modestly effective interventions has remained pitifully low for decades. Why? Largely because the diseases of old age, being by definition slowly-progressing chronic conditions, are already being fought by the body to the best of its (evolved) ability throughout life, so that any simplistic attempt to augment those preexisting defenses is likely to do more harm than good. The example I gave above, of declining regenerative ability, is a fine example: The body needs to trade better regeneration against preventing cancer, so we'll gain nothing by an intervention that merely pushes that trade-off away from its evolved optimum.

Why, though, does evolution accept these trade-offs? In reality, it doesn't: It's always looking for ways to get closer to the best of both worlds. But that, too, must be considered in the context of how evolution actually works. Some adaptations, even if theoretically possible, just take too long for evolution to find, so what we see is the best that evolution could manage in the time it had. This looks like stability—as if evolution had decided that a particular trade-off was good enough—but it's really just an approximation.

In summary: Follow Orgel when you're coming up with new ideas, but follow Dobzhansky when you're engaging in the essential rigorous evaluation of those ideas.

EPIGENETICS

LEO M. CHALUPA

Neurobiologist; Vice President for Research, George Washington University

Epigenetics is a term that has been around for more than a century, but its usage in the public domain has increased markedly in recent years. In the last decade or so, there have been dozens of articles in newspapers (notably the *New York Times*) and magazines, such as *The New Yorker*, devoted to this topic. Yet when I queried ten people working in the research office of a major university, only one had a general sense of what the term meant, stating that it deals with "how experience influences genes." Close enough, but the other nine had no idea, despite the fact that all were college graduates, two were lawyers, and four others had graduate degrees. Not satisfied by the results of this unscientific sample, I asked an associate dean of a leading medical school how many first-year medical students knew the meaning of this term. He guessed that the majority would know, but at a subsequent lecture when he did a polling of fifty students, only about a dozen could provide a cogent definition. So there you have it—another example of a lack of knowledge by the educated public of a hot topic among the scientific establishment.

What makes epigenetics important, and why is it so much in vogue these days? Its importance stems from the fact that it provides a means by which biological entities, from plants to humans, can be modified by altering gene activity without changes in the genetic sequence. This means that the age-old "nature versus nurture" controversy has been effectively obviated, because experience (as well as a host of other agents) can alter gene activity, so the "either/or" thinking mode no longer applies. Moreover, there's now some tantalizing but still preliminary evidence that changes in gene activity (induced in this case by an insecticide) can endure for a number of subsequent generations. What happens to you today can affect your great-great-great-grandchildren!

As to why epigenetics is a hot research topic, the answer is that major progress is being made in the underlying mechanisms by which gene activity can be modified by specific events. Currently, more than a dozen means by which gene expression or gene repression occurs have been documented. Moreover, epigenetics processes have been linked with early development and normal and pathological aging, as well as with several disease states including cancer. The hope is that a fuller understanding of epigenetics will enable us to control and reverse the undesirable outcomes while enhancing those we deem beneficial to us and future generations.

THE TRANSCRIPTOME

ANDRÉS ROEMER
Diplomat, economist; Cofounder, Ciudad de las Ideas; co-author (with Clotaire Rapaille), *Move UP: Why Some Cultures Advance While Others Don't*

A few months ago I sequenced my genes. The 700-megabyte text file looks something like this:

```
AGCCCCTCAGGAGTCCGGCCACATGGAAACTCCT
CATTCCGGAGGTCAGTCAGATTTACCCTT
GAGTTCAAACTTCAGGGTCCAGAGGCTGATA
ATCTACTTACCCAAACATAGGGCTCACCTTGG
CGTCGCGTCCGGCGGCAAACTAAGAACACGTC
GTCTAAATGACTTCTTAAAGTAGAATAGCGT
GTTCTCTCCTTCCAGCCTCCGAAAAACTCGGAC
CAAAGATCAGGCTTGTCCGTTCTTCGCTAGTGAT
GAGACTGCGCCTCTGTTCGTACAACCAATTTAGG
```

Each individual A, C, G, and T are organic molecules that form the building blocks of what makes me "me"—my DNA. It's composed of approximately 3.3 billion pairs of nucleotides, organized in some 24,000 genes. The information of every living being is codified in this manner. Our shape, our capacities, abilities, needs, and even predisposition to disease are determined largely by our genes.

But this information is only a small percentage (less than 2 percent) of what can be found in the DNA that each of the cells of my body carry. That's the percentage that encodes proteins, the molecules that carry out all the functions necessary for life. The other 98 percent is known as non-coding DNA, and as of this writing we believe only 10 to 15 percent of that has a biological function and shows complex patterns of expression and regulation, while the rest is largely referred to as "junk DNA." This doesn't necessarily mean that most of our DNA is junk; it means only that we still don't know why it's there or what it does. The human genome still has tricks up its sleeve.

The story of how our DNA is expressed and regulated is the story of the transcriptome, and despite all our technological advances its study is still in the early stages. But it already offers enormous potential for better diagnosing, treating, and curing disease.

In order for DNA to be expressed and produce a specific protein, the code must be copied (transcribed) into RNA. These gene readouts are called transcripts, and the transcriptome is the collection of all the RNA molecules, or transcripts, present in a cell. In contrast with the genome, which is characterized by its stability, the transcriptome is constantly changing and can reflect—in real time, at the molecular level—a person's physiology, depending on many factors including the stage of development and environmental conditions.

The transcriptome can tell us when and where each gene is turned on or off in the cells of tissues and organs of an individual. It functions like a dimmer switch, setting whether a gene is 10-percent active or 70-percent active and therefore enabling a much more intricate fine-tuning of gene expression. By comparing the transcriptome of different types of cells, we can understand what makes a specific cell from a specific organ unique, how that cell looks when healthy and working normally, and how its gene activity may reflect or contribute to certain diseases.

The transcriptome may hold the key to the breakthrough we've been waiting for over the last thirty years in gene therapy. There are today two complementary yet different approaches: the replacement or editing of genes within the genome (such as the widely known CRISPR/Cas9 technique) and the inhibition or enhancement of gene expression.

In the latter approach, RNA-based cancer vaccines that activate an individual's immune system are already in clinical trials, with promising results in diseases such as lung or prostate cancer. The vaccination with RNA molecules is a promising and safe approach to let the patient's body produce its own vaccines. By introducing a specific synthetic RNA, the protein synthesis can be controlled without intervening in the genome and by letting the cell's own protein-building machinery work without altering the physiological state of the cell.

This concept will unlock a path to a prosperous future in terms of aging prevention, brain functioning, and stem-cell health, as well as the eradication of cancer, hepatitis B, HIV, or even high cholesterol. The transcriptome has opened our eyes to the mind-staggering complexity of the cell, and when fully fathomed it will allow us to truly achieve our genetic destiny.

POLYGENIC SCORES

ROBERT PLOMIN
Professor of Behavioral Genetics, King's College London

Polygenic scores are beginning to deliver personal genomics from the front lines of the DNA revolution. They make it possible to predict genetic risk and resilience at the level of the individual rather than the level of the family. This has far-reaching implications for science and society.

Polygenic means "many genes." Classic genetic studies over the past century have consistently supported Ronald Fisher's 1918 theory that the heritability of common disorders and complex traits is caused by many genes of small effect. What wasn't realized until recently was just how many and how small these effects are. Systematic gene-hunting studies began a decade ago using hundreds of thousands of DNA differences throughout the genome, a technique called genome-wide association (GWA). The early goal was to break the 1-percent barrier—that is, to be able to detect DNA associations accounting for less than 1 percent of the variance of common disorders and complex traits. Samples in the tens of thousands were needed to detect such tiny effects, after extensively correcting for multiple testing of hundreds of thousands of DNA differences in a GWA study. A great surprise was that these GWA studies able to detect DNA associations accounting for 1 percent of the variance came up empty-handed.

GWA studies would have to break a 0.1-percent barrier, not just a 1-percent barrier. That requires samples in the hundreds of thousands. As GWA studies break that barrier, they're scooping up many DNA differences that contribute to heritability. But what good are DNA associations that account for less than 0.1 percent of the variance? The answer is "not much," if you're a molecular biologist wanting to study pathways from genes to brain to behavior, because this means there are very many minuscule paths.

Associations accounting for less than 0.1 percent of the variance are also of no use for prediction. This is where polygenic scores come in. When psychologists create a composite score—like an IQ score or a score on a personality test—they aggregate many items. They don't worry about the significance or reliability of each item, because the goal is to create a reliable composite. In the same way, polygenic scores aggregate many DNA differences to create a composite that predicts genetic propensities for individuals.

A new development in the last year is to go beyond aggregating a few genome-wide significant "hits" from GWA studies. The predictive power of polygenic scores can increase dramatically by aggregating associations from GWA studies as long as the resulting polygenic score accounts for more variance in an independent sample. Polygenic scores now often include tens of thousands of associations, underlining the extremely polygenic architecture of common disorders and complex traits.

Polygenic scores derived from GWA studies with sample sizes in the hundreds of thousands can predict substantial amounts of variance. For example, polygenic scores can account for 20 percent of the variance of height and 10 percent of the variance in U.K. national exam scores at the end of compulsory education. This is still a long way from accounting for the entire heritability of 90 percent for height and 60 percent for educational achievement; this gap is called *missing heritability*. Nonetheless, these predictions from an individual's DNA alone are substantial. For the sake of comparison, the polygenic score for educational achievement is a more powerful predictor than the socioeconomic status of students' families or the quality of their schools.

Moreover, predictions from polygenic scores have unique causal status. Usually correlations don't imply causation, but correlations involving polygenic scores imply causation in the sense that these correlations aren't subject to reverse causation, because nothing changes inherited DNA sequence variation. For the same reason, polygenic scores are just as predictive at birth or even prenatally as they are later in life.

Like all important findings, polygenic scores have potential for bad as well as good. They deserve to be high on the list of scientific terms that ought to be more widely known, so that this discussion can begin.

REPLICATOR POWER

SUSAN BLACKMORE

Psychologist; Visiting Professor, University of Plymouth, U.K.; author, *Consciousness: An Introduction*

Where does all the design in the universe come from? Around me now, and indeed almost everywhere I go, I see a mixture of undesigned and designed things: rocks, stars, puddles of rain, tables, books, grass, rabbits, my own hands.

The distinction between designed and undesigned isn't commonly made this way. Typically, people are happy to divide rocks from books, on the grounds that rocks weren't designed for a function, whereas books were. Books have an author, a publisher, a printer, and a cover designer, and that means "real" top-down design. As for grass, rabbits, and hands, they do serve functions, but they evolved through a mindless bottom-up process, and that isn't "real" design.

This other distinction—between real design and evolved "design"—is sometimes explicitly stated, but even when it's not, the fear of attributing design to a mindless process is revealed in the scare quotes that evolutionary biologists sometimes put around the word "design." Eyes are brilliantly "designed" for seeing, and wings are "designed" for flying. But they're only "as if" designed. They were produced not top-down by a mind with a plan but bottom-up by an utterly mindless process. Our human "real" design is different.

If the concept of replicator power were better known, this false distinction might be dropped, because we'd see that all design depends on the same underlying process. A replicator is information that affects its environment so as to make new copies of itself. Its power derives from its role as information undergoing the evolutionary algorithm of copying with variation and selection, the process that endlessly increases the available information. Genes are the most obvious example, with the varying creatures they give rise to being known as their vehicles or interactors. It is these that are acted on by natural selection to determine the success or failure of the replicator.

The value of the term "replicator" lies in its generality. This is what Richard Dawkins emphasized in writing about universal Darwinism—applying Darwin's basic insight to all self-replicating information. Whenever there's a replicator in an appropriate environment, design ensues. This is why Dawkins invented the term *meme*, to show that there's more than one replicator evolving on this planet. He also made the claim, which follows from this way of thinking about evolution, that all life everywhere evolves by the differential survival of replicators. I would add that all intelligence everywhere evolves by the differential survival of replicators.

Human design is essentially no different from biological design; both depend on a replicator being copied, whether what's copied is the order of bases in a molecule of DNA or the order of words in a book. In the molecular case, new sequences arise from copying errors, mutations, and recombination. In the writing of a book, new sequences arise from a person recombining familiar words into new phrases, sentences, and paragraphs. In both examples, many different sequences are created and very few survive to be copied again. In both examples, creatively designed products emerge through replicator power. To see human design this way is to drop the assumption that top-down control, intelligence, and planning are essential to creativity—to see that these capacities and the designs they create are consequences, not causes, of evolution.

Accepting this may be uncomfortable, as it means seeing that everything we think we designed all by ourselves was really designed by a clueless, bottom-up process using us as copying machinery. The unease may be similar to that reputedly expressed by the Bishop of Birmingham's wife—that such knowledge belittles us and diminishes our humanity and power. Yet we have (more or less, and in only some parts of the world) learned to embrace rather than fear the knowledge that our bodies evolved by mindless bottom-up processes. This is another step in the same direction.

The significance of recognizing replicator power is that there are other replicators out there and may soon be more. The mindless processes that turned us from an ordinary ape into a speaking, meme-copying ape allowed us to produce tables, books, cars, planes, and—in a crucial step—copying machines. These include writing, leading to printing presses; potters' wheels and woodworking tools, leading to factories; and now computers, leading to the information explosion.

Artificial intelligences, whether confined to desktop boxes and robots or distributed in cyberspace, have been created by replicator power, just as our own intelligence was created by replicator power. They're evolving far faster than we did and may yet give rise to further, even faster replicators. That power won't stop because we want it to. And its products will certainly not be impressed by our claims to be in control or to have designed the machinery on which they thrive. That intelligence will continue to evolve, and the sooner we accept the idea of replicator power, the more realistic we're likely to be about the future of our life with intelligent machines.

FALLIBILISM

OLIVER SCOTT CURRY
Senior Researcher, Director, Oxford Morals Project, Institute of Cognitive and Evolutionary Anthropology, University of Oxford

Fallibilism is the idea that we can never be 100-percent certain we're right and must therefore always be open to the possibility that we're wrong. This might seem a pessimistic notion, but it isn't. Ironically, this apparent weakness is a strength; admitting one's mistakes is the first step to learning from them and overcoming them, in science and society.

Fallibilism lies at the heart of the scientific enterprise. Even science's most well-established findings—the "laws" of nature—are but hypotheses that have withstood scrutiny and testing thus far. The possibility that they may be wrong, or superseded, is what spurs the generation of new alternative hypotheses and the search for further evidence enabling us to choose between them. This is what scientific progress is made of. Science rightly champions winning ideas—ideas that have been tested and have passed—and dispenses with those that have been tested and have failed or are too vague to be tested at all.

Fallibilism is also the guiding principle of free, open, liberal, secular societies. If the laws of nature can be wrong, then how much more fallible are our social and political arrangements? Even our morals, for example, don't reflect some absolute truth—

god-given or otherwise. They, too, are hypotheses—biological and cultural attempts to solve the problems of cooperation and conflict inherent in human social life. They're tentative, provisional, and capable of improvement; and they can be, and have been, improved upon. The awareness of this possibility—allied to the ambition to seize the opportunity it represents, and the scientific ability to do so—is precisely what has driven the tremendous social, moral, legal, and political progress of the past few centuries.

Fallibilism—the notion that we may not be right—doesn't mean we must be entirely wrong. It's not a license to tear everything up and start again. We should respect tried and tested ideas and institutions and recognize that our attempts to improve on them are equally fallible. Nor does fallibilism lead to "anything goes" relativism—the notion that there's no way to distinguish good ideas from bad. On the contrary, fallibilism tells us that our methods for distinguishing better ideas from worse ideas work, and urges us to use them to quantify our uncertainty and resolve it. Better still, to work together. Your opponent is no doubt mistaken, but in all likelihood so are you. So why not see what you can learn from each other and collaborate to see further?

Anyone wishing to understand the world or change it for the better should embrace this fundamental truth.

INTELLECTUAL HONESTY

SAM HARRIS
Neuroscientist; author, *Waking Up: A Guide to Spirituality Without Religion*

Wherever we look, we find otherwise sane men and women making extraordinary efforts to avoid changing their minds.

Of course, many people are reluctant to be *seen* changing their minds, even though they might be willing to change them in private, seemingly on their own terms—perhaps while reading a book. This fear of losing face is a sign of fundamental confusion. Here it's useful to take the audience's perspective: Tenaciously clinging to your beliefs past the point where their falsity has been clearly demonstrated does not make you look good. We've all witnessed men and women of great reputation embarrass themselves in this way. I know at least one eminent scholar who wouldn't admit to any trouble on his side of a debate stage were he to be suddenly engulfed in flames.

If the facts aren't on your side, or your argument is flawed, any attempt to save face is to lose it twice over. And yet many of us find this lesson hard to learn. To the extent that we can learn it, we acquire a superpower of sorts. In fact, a person who surrenders immediately when shown to be in error will appear not to have lost the argument at all. Rather, he will merely afford others the pleasure of having educated him on certain points.

Intellectual honesty lets us stand outside ourselves and think in ways that others can (and should) find compelling. It rests on the understanding that *wanting* something to be true isn't a reason to believe that it *is* true; rather, it's further cause to worry that we might be out of touch with reality in the first place. In this sense, intellectual honesty makes real knowledge possible.

Our scientific, cultural, and moral progress is almost entirely the product of successful acts of persuasion. Therefore, an inability (or refusal) to reason honestly is a social problem. Indeed, to defy the logical expectations of others—to disregard the very standards of reasonableness that *you* demand of *them*—is a form of hostility. And when the stakes are high, it becomes an invitation to violence.

In fact, we live in a perpetual choice between conversation

and violence. Consequently, few things are more important than a willingness to follow evidence and argument wherever they lead. The ability to change our minds, even on important points—*especially* on important points—is the only basis for hope that the human causes of human misery can be finally overcome.

EPSILON

VICTORIA STODDEN

Associate Professor, School of Information Sciences, University of Illinois at Urbana-Champaign

In statistical modeling, the use of the Greek letter *epsilon* explicitly recognizes that uncertainty is intrinsic to our world. The statistical paradigm has two components: data or measurements, drawn from the world we observe, and the underlying processes generating those data. *Epsilon* appears in mathematical descriptions of those underlying processes and represents the inherent randomness with which the data we observe are generated. Through the collection and modeling of data, we hope to make better guesses at the mathematical form of those processes; a better understanding of the data-generating mechanism will enable us to do a better job modeling and predicting the world around us.

That use of *epsilon* is a recognition of the inability of data-driven research to perfectly predict the future, no matter the computing or data-collection resources. It codifies the idea that uncertainty exists in the world itself. We may understand the structure of this uncertainty better over time, but the statistical paradigm asserts it as fundamental.

So we can never expect perfect predictions, even if we manage to take perfect measurements. This inherent uncertainty means that doubt isn't a negative or a weakness but a mature recognition that our knowledge is imperfect. The statistical paradigm is increasingly being used as we continue to collect and analyze vast amounts of data, and as the output of algorithms and models grows as a source of information. We're seeing the effects across society: evidence-based policy, evidence-based medicine, more sophisticated pricing and market-prediction models, social media customized to our online browsing patterns. . . . The intelligent use of the information derived from statistical models relies on understanding uncertainty, as does policymaking and our cultural understanding of this information source. The 21st century is surely the century of data, and we need to correctly understand their use. The stakes are high.

SYSTEMIC BIAS

RICHARD MULLER

Professor of Physics, UC Berkeley; author, *Now: The Physics of Time*

It is said, by those who believe in the devil, that his greatest achievement was convincing the rest of the world that he didn't exist. There are two biases that play a similar role in our search for objective knowledge and our goal of making better decisions. These are optimism bias and skepticism bias. Their true threat comes from the fact that many people are unaware of them, and yet once you recognize these biases you'll see them everywhere.

Optimism is not only infectious but effective. Enthusiasm is often a requirement for success. During World War II, the U.S. Army Corps of Engineers adopted as a motto an aphorism from a French novelist: "The difficult we do quickly. The impossible takes a little longer."

Consider optimism and skepticism biases in the field of energy. Some say solar power is too expensive or too intermittent. Rather than address these objections directly, we can substitute optimism: Let's let loose American can-do and solve those challenges! (In the world today, there are no longer any problems, only challenges.) Electric cars charge too slowly? Look at the computer revolution and lose your confining pessimism! Moore's law will eventually apply to batteries for energy storage just as it worked for electronics. Remember the Manhattan Project! Remember the Apollo Program! We can do anything if we put our minds to it. Yes we can!

While renewables are often treated with optimism, nuclear power is attacked with skepticism and pessimism. The problems are too tricky, too technical, too unknown. We can trust neither industry nor the U.S. government with accident safety. Nuclear energy is dangerous and intractable. And we'll never solve the nuclear-waste issue. But is nuclear power truly more intractable than solar, or is there a hidden bias affecting which arguments we bring forth?

Optimism bias derives not from objective assessment but from a strong like or a deeply felt hope. The opposite of optimism bias, pessimism bias, derives from dislike or fear. Closely related to pessimism bias, but harder to counter and sounding more thoughtful, is skepticism bias. It's remarkably easy to be skeptical—about anything. Try it; the words will come trippingly to your tongue. The skeptic typically sounds more intelligent, more knowing, more experienced, than does the pessimist. If you aren't skeptical, I have a bridge in Brooklyn I'd like to sell you.

Closely related to optimism and skepticism bias is confirmation bias, which refers to the cherry-picking of confirmatory facts and the discarding of those inconvenient to the desired conclusion. Optimism bias is different; it's not the facts that are important but the attitude, the enthusiasm. If the technology is appealing, then of course we can do it! But if we're suspicious, then bring in all the arguments that were ignored: the cost, the difficulty, the lack of trust in industry and authority. Be skeptical.

Skepticism bias is currently affecting discussions of fracking—in particular, the danger of leaked "fugitive" methane, with its potent greenhouse warming potential. The bias takes the form of skepticism that we can fix the leaky pipes and machines; the task is portrayed as too difficult and would require trust. And trust is a tricky and elusive concept, and it, too, can be applied with bias. If you don't like a solution, state that you don't trust it to be implemented properly.

Optimism bias is strong on the issue of electric cars. We hope they'll work, so we support them. Yet skepticism/pessimism is used against their competitor, 100-mpg conventional autos. Are such vehicles truly as difficult as some argue? We learn in physics that, in principle, horizontal transport need not take any energy. But the same people who are optimistic about batteries are often pessimistic about high-mileage gasoline-powered autos. Is that justified?

Biases can be hidden under conviction, an easy substitute for objective analysis. "I just don't believe that ____ (fill in the blank)." Biases often invoke trust or lack of it. Conviction can be psychologically compelling. Optimism bias drove the original flight to the moon but also the largely useless and, to my mind, failed Space Station. Optimism bias gave rise to the U.S. government's wars on both cancer and poverty.

In science, skepticism bias infects referee reports for proposed experimental work. Berkeley experimental physicist Luis Alvarez's major projects were begun before he had solutions to all the technical issues. He was optimistic, but his optimism was based on his own evaluation of what lay within his capabilities. It could be said that he had earned the right to be optimistic by dint of his past successes. His optimism was not a bias; it was based on his demonstrated ability to address challenging new issues as they arise. Optimism can be realistic. And so can pessimism. And skepticism.

The heart of science is in overcoming bias. The difference between a scientist and layman can be summarized as follows: A layman is easily fooled and is particularly susceptible to self-deception. A scientist is easily fooled and is particularly susceptible to self-deception, and knows it. The scientific method consists almost exclusively in techniques to overcome self-deception. The first step in accomplishing this is to recognize that biases exist. The danger of optimism and skepticism biases (like the danger of the devil, for those who believe in such things) is that so many people are unaware of their existence.

CONFIRMATION BIAS

BRIAN ENO
Artist; composer; recording artist; recording producer: U2, Coldplay, Talking Heads, Paul Simon

The great promise of the Internet was that more information would automatically yield better decisions. The great disappointment is that more information actually yields more possibilities to confirm what you already believed anyway.

NEGATIVITY BIAS

MICHAEL SHERMER

Publisher, *Skeptic* magazine; monthly columnist, *Scientific American*; Presidential Fellow, Chapman University; author, *The Moral Arc*

One of the most understated effects in all cognitive science is the psychology behind why negative events, emotions, and thoughts trump by a wide margin those that are positive. This bias was discovered and documented by the psychologists Paul Rozin and Edward Royzman in 2001, showing that across almost all domains of life we seem almost preternaturally pessimistic:

- Negative stimuli command more attention than positive stimuli. In rats, for example, negative tastes elicit stronger responses than positive tastes. And in taste-aversion experiments, a single exposure to a noxious food or drink can cause lasting avoidance of that item, but there's no corresponding parallel with good-tasting food or drinks.
- Pain feels worse than no pain feels good. That is, as the philosopher Arthur Schopenhauer put it, "we feel pain, but not painlessness." There are erogenous zones, Rozin and Royzman point out, but no corresponding torturogenous zones.
- Picking out an angry face in a crowd is easier and faster to do than picking out a happy face.
- Negative events lead us to seek causes more readily than do positive events. Wars, for example, generate endless analyses in books and articles, whereas peace literature is paltry by comparison.

- We have more words to describe the qualities of physical pain (deep, intense, dull, sharp, aching, burning, cutting, pinching, piercing, tearing, twitching, shooting, stabbing, thrusting, throbbing, penetrating, lingering, radiating, etc.) than we have to describe physical pleasure (intense, delicious, exquisite, breathtaking, sumptuous, sweet, etc.).
- There are more cognitive categories for and descriptive terms of negative emotions than positive. As Leo Tolstoy famously observed in 1875: "Happy families are all alike; every unhappy family is unhappy in its own way."
- There are more ways to fail than there are to succeed. It's difficult to reach perfection and the paths to it are few, but there are many ways to fail to achieve perfection and the paths away from it are many.
- Empathy is more readily triggered by negative stimuli than by positive: People identify and sympathize with others who are suffering or in pain more than they do with others who are in a state happier or better off than they are.
- Evil contaminates good more than good purifies evil. As the old Russian proverb says, "A spoonful of tar can spoil a barrel of honey, but a spoonful of honey does nothing for a barrel of tar." In India, members of the higher castes may be considered contaminated by eating food prepared by members of the lower castes, but those in the lower castes don't receive an equivalent rise in purity status by eating food prepared by their upper-caste counterparts.
- The notorious "one drop of blood" rule of racial classification has its origin in the Code Noir, or "Negro Code," of 1685, meant to guarantee the purity of the white race by screening out the tainted blood, whereas, note Rozin and Royzman, "There exists no historical evidence for the positive equivalent of a 'one-drop' ordinance—that

is, a statute whereby one's membership in a racially privileged class would be assured by one's being in possession of 'one drop' of the racially superior blood."

- In religious traditions, possession by demons happens quickly compared to the exorcism of demons, which typically involves long and complex rituals; by contrast in the positive direction, becoming a saint requires a life devoted to holy acts, which can be erased overnight by a single immoral act. In the secular world, decades of devoted work for public causes can be erased in an instant by an extramarital affair, a financial scandal, or a criminal act.

Why is negativity stronger than positivity? Evolution. In the environment of our evolutionary ancestry, there was an asymmetry of payoffs in which the fitness cost of overreacting to a threat was less than the fitness cost of underreacting, so we err on the side of overreaction to negative events. The world was more dangerous in our evolutionary past, so it paid to be risk-averse and highly sensitive to threats, and if things were good, then taking a gamble to improve them a little was not seen as worth the risk.

POSITIVE ILLUSIONS

HELEN FISHER
Biological anthropologist, Rutgers University; Senior Research Fellow, The Kinsey Institute; author, *Anatomy of Love*

What makes a happy marriage? Psychologists offer myriad suggestions, from active listening to arguing appropriately and avoiding contempt. But my brain-scanning partners and I have stumbled on what happens in the brain when you're in a long-term, happy partnership.

In research published in 2011 and 2012, we put seven American men and ten American women (all in their fifties and sixties) into the brain scanner. Their average duration of marriage was 21.4 years, most had grown children, and all maintained that they were still madly in love with their spouse (not just loving but *in love*). All showed activity in several areas of the dopamine-rich reward system, including the ventral tegmental area and dorsal striatum—brain regions associated with feelings of intense romantic love. All also showed activity in several regions associated with feelings of attachment, as well as those linked with empathy and controlling your own stress and emotions. These data show that you can remain in love with a partner for the long term.

More intriguing, we found reduced activity in a region of the cerebral cortex known as the ventromedial prefrontal cortex, which is associated with our human tendency to focus on the negative rather than the positive (among many other functions linked to social judgment). These brain functions may have evolved millions of years ago—perhaps primarily as an adaptive response to strangers who wandered into one's neighborhood. Natural selection has long favored those who responded negatively to the one malevolent intruder rather than positively to myriad friendly guests.

Reduced activity in this brain region suggests that our happily-in-love long-term partners were overlooking the negative to focus on the positive aspects of their marital relationships—something known to scientists as "positive illusions." Looking at our brain-scanning results from other experiments, including long-term lovers in China, we found similar patterns. We humans are able to convince ourselves that the real is the ideal. The neural roots of tolerance, mercy, and pardon may live deep in the human psyche.

RUSSELL CONJUGATION

ERIC R. WEINSTEIN
Mathematician, economist; Managing Director, Thiel Capital

We are told we're entitled to our own opinions but not to our own facts. This leaves out the observation that the war for our minds and attention is now increasingly being waged over neither facts nor opinions but feelings.

In an era when anyone can publish anything, the quest to control information has largely been lost by institutions, with a race on to weaponize empathy by understanding its basis in linguistics and accordingly tweaking the social-media algorithms that now present our world to us. As the theory goes, it's not that we don't have our own opinions so much as that we have too many contradictory ones, and it's generally our emotional state alone that determines on which ones we'll predicate action or inaction.

Russell conjugation (or "emotive conjugation") is a currently obscure construction from linguistics, psychology, and rhetoric which demonstrates how our rational minds are shielded from understanding the junior role that factual information generally plays relative to empathy in our formation of opinions. I frequently suggest it as perhaps the most important idea almost no one appears familiar with, as it showed me just how easily my opinions could be manipulated without any need to falsify facts.

Historically, the idea isn't new and seems to have been first defined by several examples given by Bertrand Russell in 1948 on the BBC without much follow-up work, until it was rediscovered in the Internet Age and developed into a near-data-driven science by pollster Frank Luntz beginning in the early 1990s.

In order to understand the concept properly, you have to appreciate that most words and phrases are actually defined not by a single dictionary description but rather by two distinct attributes:

1. The factual content of the word or phrase.
2. The emotional content of the construction.

Words can be considered synonyms if they carry the same factual content regardless of the emotional content. But this leads to the peculiar effect that the synonyms for a positive word, like "whistle-blower," cannot be used in its place, as they're almost universally negative ("snitch," "fink," and "tattle-tale" being representative examples). This is our first clue that something's wrong with—or at least incomplete about—our concept of "synonym," requiring an upgrade to distinguish words that may be content synonyms but emotional antonyms.

The basic principle of Russell conjugation is that the human mind is constantly looking ahead, well beyond what's true or false, to ask "What is the social consequence of accepting the facts as they are?" While this line of thinking is obviously self-serving, we're descended from social creatures who couldn't safely form opinions around pure facts so much as around how those facts were presented to them by those they trusted or feared. Thus, as listeners and readers, our minds generally mirror the emotional state of the source, while in our roles as authoritative narrators presenting the facts, we maintain an arsenal of language to subliminally instruct our listeners and readers on how we expect them to color their perceptions. Russell discussed this by putting three such presentations of a common underlying fact in the form in which a verb is typically conjugated:

I am firm.
[Positive empathy]

You are obstinate.
[Neutral to mildly negative empathy]

He/She/It is pigheaded.
[Very negative empathy]

In all three cases, Russell was describing people who didn't readily change their minds. Yet by putting these descriptions so close together and without further factual information to separate the individual cases, he forces us to recognize that most of us feel positively toward the steadfast narrator and negatively toward the pigheaded fool, all without any basis in fact.

Years later, the data-driven pollster Frank Luntz stumbled on much the same concept, unaware of Russell's earlier construction. By holding focus groups with new real-time technology that let participants share emotional responses to changes in authoritative language, Luntz made a stunning discovery—one that pushed Russell's construction out of the realm of linguistics and into applied psychology. What he found was extraordinary: Many if not most people form opinions based solely on whatever Russell conjugation is presented to them and *not* on the underlying facts. That is, the same person will oppose a "death tax" while having supported an "estate tax" seconds earlier, even though those taxes are two descriptions of the exact same thing.

Moreover, such is the power of emotive conjugation that we're generally unaware that we hold such contradictory opinions. Thus "illegal aliens" and "undocumented immigrants" may be the same people, but the former label leads to calls for deportation while the latter prompts many of us to consider amnesty programs and paths to citizenship.

If we agree that Russell conjugation keeps us from realizing that we don't hold consistent opinions on facts, we see a possible new answer to a puzzle dating from the birth of the Web: "If the Internet democratized information, why has its social impact been so much slower than many of us expected?" Assuming that our actions are based not on what we know but on how we feel about what we know, we see that traditional media has all but lost control of gatekeeping our information, but not yet of how it's emotively shaded. In fact, it's relatively simple to write a computer program to crawl factually accurate news stories against a look-up table of Russell conjugates, to see the exact bias of every supposedly objective story.

Thus, the answer to the puzzle of our inaction, it seems, may be that we built an information superhighway for everyone but neglected to build an empathy network alongside it to democratize what we feel. We currently get our information from more sources than ever before, but (at least until recently) we've turned to traditional institutions to guide our empathy. Information, as the saying goes, wants to be free. But we fear that authentic emotions will get us into trouble with our social group, and so we continue to look to others to tell us what is safe to feel.

EMPATHIC CONCERN

DANIEL GOLEMAN

Psychologist; author, *Focus: The Hidden Driver of Excellence*

Empathy has gotten a bad reputation of late, largely undeserved. That negative spin occurs because people fail to understand the nuanced differences between three aspects of empathy.

The first kind, *cognitive empathy*, lets me see the world through your eyes—take your perspective and understand the mental models making up your lens on events. The second kind, *emotional empathy*, means I feel what you feel; this empathy gives me an instant felt sense of another person's emotions.

It's the third kind, *empathic concern,* that leads us to care about another person's welfare, to want to help them if they're in need. Empathic concern forms a basis for compassion. The first two, while essential for intimate connection, can also become tools in the service of pure self-interest. Marketing and political campaigns that manipulate people's fears and hatreds require effective cognitive and emotional empathy—as do con men. And, perhaps, artful seductions. It's empathic concern—caring about the other person's welfare—that puts these two kinds of empathy in the service of a greater good.

Brain research at the University of Chicago by Jean Decety and at the Max Planck Institute for Human Cognitive and Brain Sciences by Tania Singer has established that each of these varieties of empathy engages its own unique neural circuitry. Neocortical circuitry, primarily, undergirds cognitive empathy. Emotional empathy stems from the social networks that facilitate rapport and tune us in to another person's pain—my brain's pain circuitry activates when I see that you're in pain.

The problem with empathic concern: If your suffering makes me suffer, I can feel better by tuning out. That's a common reaction and a major reason so few people go through the whole arc from attunement and emotional empathy to caring and helping a person in need.

Empathic concern draws on the mammalian circuitry for parental caretaking—the love of a parent for a child. Research finds that loving-kindness meditation—where you wish well-being for a circle expanding outward from yourself and your loved ones, people you know, strangers, and finally the entire world—boosts feelings of empathic concern and strengthens connectivity within the brain's caretaking circuits. By contrast, a longitudinal study in Norway found that seven-year-olds who showed little empathic concern for their mothers had an unusually high incidence of being jailed as felons in adulthood.

When we think of empathy as a spur to prosocial acts, it's empathic concern we have in mind. When we think of the cynical uses of empathy, it's the other two, which can be twisted in the service of pure self-interest.

NAÏVE REALISM

MATTHEW D. LIEBERMAN

Professor of Psychology, UCLA; author, *Social: Why Our Brains Are Wired to Connect*

The comedian George Carlin once noted, "Anyone driving slower than you is an idiot and anyone going faster than you is a maniac." The obscure scientific term explaining why we see most other people as unintelligent or crazy is *naïve realism*. Its origins trace back to at least the 1880s, when philosophers used the term to suggest that we should take our perceptions of the world at face value. In its modern incarnation, it has almost the opposite meaning. Stanford social psychologist Lee Ross uses the term to indicate that although most people do take their perceptions of the world at face value, this is a profound error that regularly causes virtually unresolvable conflicts between people.

Imagine three drivers in Carlin's world—Larry, Moe, and Curly. Larry is driving 30 mph, Moe is driving 50 mph, and Curly is driving 70 mph. Larry and Curly agree that Moe's driving was terrible but are likely to come to blows over whether Moe is an idiot or a maniac. Meanwhile, Moe disagrees with both, because it's obvious to him that Larry is an idiot (which Curly agrees with) and Curly is a maniac (which Larry agrees with). As in ordinary life, Larry, Moe, and Curly each fail to appreciate that their own understanding of the others is hopelessly tied to their own driving, rather than reflecting something objective about the other person.

Naïve realism occurs as an unfortunate side effect of an otherwise adaptive aspect of brain function. Our remarkably sophisticated perceptual system performs its countless computations so rapidly that we're unaware of all the special-effects teams working in the background to construct our seamless experience. We "see" so much more than is in front of us, thanks to our brain's automatically combining sensory input with our expectations and motivations. This is why a bicycle partly hidden by a wall is instantly "seen" as a normal bicycle, without a moment's thought that it might be only part of a bicycle. Because these constructive processes happen behind the scenes in our mind, we mistake our perception for reality—a mistake we're often better off for having made.

When it comes to perceiving the physical world, we appear to mostly see things the same way. When confronted with trees, shoes, and gummy bears, our brains construct these things for us in similar enough ways that we can agree on which to climb, which to wear, and which to eat. But when we move to the social domain of understanding people and their interactions, our "seeing" is driven less by external input and more by expectation and motivation. Because our mental construction of the social world is just as invisible to us as our construction of the physical world, our idiosyncratic expectations and motivations are much more problematic in the social realm. In short, we are just as confident in our assessment of Donald Trump's temperament and Hillary Clinton's dishonesty as we are in our assessment of trees, shoes, and gummy bears. In both cases, we're quite certain we're seeing reality for what it is.

And this is the real problem. This isn't a heuristics and biases problem, where our simplistic thinking can be corrected when we see the correct solution. This is about "seeing" reality. If I am seeing reality for what it is and you see it differently, then one of us has a broken reality detector and I know mine isn't broken. If you can't see reality as it is—or, worse yet, see it but refuse to acknowledge it—then you must be crazy, stupid, biased, lazy, or deceitful.

Absent a thorough appreciation for how our brain ensures that we'll end up as naïve realists, we can't help but see complex social events differently from one another, with each of us denigrating the other for failing to see what is so obviously true. Although there are real differences separating groups of people, naïve realism might be the most pernicious undetected source of conflicts and their durability. Israel versus the Palestinians, the American political left and right, the fight over vaccines and autism—in each case, our inability to appreciate our own miraculous construction of reality is preventing us from appreciating the miraculous construction of reality happening all around us.

MOTIVATED REASONING

DAVID PIZARRO
Associate Professor of Psychology, Cornell University

Why, in an age when the world's information is easily accessible at our fingertips, is there still widespread disagreement about basic facts? Why is it so hard to change people's minds about truth, even in the face of overwhelming evidence?

Perhaps some inaccurate beliefs result from an increase in the intentional spreading of false information, a problem exacerbated by the efficiency of the Internet. But false information has been spread pretty much since we've been able to spread information. More important, the same technologies that allow for the efficient spreading of false information also enable us to fact-check more efficiently. For most questions, we can find a reliable, authoritative answer easier than anyone ever could in all of human history. In short, we have more access to truth than ever. So why do false beliefs persist?

Social psychologists offer a compelling answer to this question: The failure of people to alter their beliefs in response to evidence is the result of a deep problem with our psychology. In a nutshell, psychologists have shown that how we process information conflicting with our beliefs is fundamentally different from how we process information consistent with those beliefs—a phenomenon called "motivated reasoning." When we're exposed to information that meshes well with what we already believe (or with what we want to believe), we're quick to accept it as factual. We readily categorize this information as another piece of confirmatory evidence and move along. But when we're exposed to information contradicting a cherished belief, we tend to pay more attention, scrutinize the source of information, and process the information carefully and deeply. Unsurprisingly, this allows us to find flaws in the information, dismiss it, and maintain our (potentially erroneous) beliefs. The psychologist Tom Gilovich captures this process elegantly, describing our minds as guided by two different questions, depending on whether the information is consistent or inconsistent with our beliefs: "Can I believe this?" or "Must I believe this?"

This goes not just for political beliefs but for beliefs about science, health, superstitions, sports, celebrities, and anything else you might be inclined (or disinclined) to believe. There's plenty of evidence that this bias is universal, not just a quirk of highly political individuals on the right or left, a symptom of the very opinionated, or a flaw of narcissistic personalities. In fact, I can easily spot the bias in myself, with minimal reflection. When presented with medical evidence on the health benefits of caffeine, for instance, I congratulate myself on my coffee-drinking habits. When shown a study concluding that caffeine has negative health effects, I scrutinize the methods ("Participants weren't randomly assigned"), the sample size ("Forty college-age males? Please!"), the journal ("Who's even heard of this publication?"), and anything else I can.

A bit more reflection on this bias, however, and I admit I'm distressed. It's possible that because of motivated reasoning I've acquired beliefs that are distorted, biased, or just plain false. I could have retained these beliefs even while maintaining a sincere desire to find out the truth of the matter, exposing myself to the best information I could find on a topic and making a real effort to think critically and rationally about the information I found. Another person, with a different set of preexisting beliefs, may come to the opposite conclusion following these same steps, with the same sincere desire to know the truth. In short, even when we reason about things carefully, we may be deploying this reasoning selectively without realizing it. We may hope that just recognizing our motivated reasoning can help us defeat it. But I know of no evidence indicating that it will.

SPATIAL AGENCY BIAS

SIMONE SCHNALL

Director, Cambridge Embodied Cognition and Emotion Laboratory;
Reader in experimental social psychology; Fellow, Jesus College,
University of Cambridge

Progressive, dynamic, and forward-thinking—these are personal qualities highly sought after in nearly all social circles and cultures. Do you want to be seen in such positive terms whenever people come across your picture? An intriguing line of psychological research suggests how to accomplish just that: When caught on film, you need to pay attention to the direction in which you're facing. People looking toward the right are perceived as more powerful and agentic than those who look left. In other words, how a person is represented in space shapes perceivers' automatic impressions, as though we imagined the depicted person moving from left to right along an imaginary path taking them from the present to future accomplishments.

This principle of *spatial agency bias* also figures in how simple actions are interpreted. For example, a soccer goal is considered more elegant, or an act of aggression more forceful, when the actor moves from left to right, compared to the mirror sequence occurring in the opposite direction. Similarly, in advertising, cars are usually shown facing right, and when they are, participants in research studies judge them to be faster and therefore more desirable.

Spatial position can also be indicative of social status. Historical analyses of hundreds of paintings indicate that when two people appear in the same picture, the more dominant, powerful person is usually facing to the right. For example, relative to men, women are more often displayed showing the left cheek, consistent with gender roles that consider them as less agentic. Traditionally, weak and submissive characters have been assigned to their respective place by where they're situated in space. From the 15th to the 20th century, however, this gender bias in paintings has become less pronounced, paralleling increasingly modern views of women's role in society.

Where does the spatial agency bias come from? Is there some innate reason for preferring objects and persons facing the right— perhaps because 90 percent of people are right-handed? Or is there a learning component involved? Cross-cultural studies indicate that there is variability indeed. For example, for Arab and Hebrew speakers the pattern is reversed: People and objects facing left are seen as more dynamic and agentic. This suggests a provocative possibility—namely, that the spatial agency bias develops as a function of writing direction. As we move across the page, we progress from what has happened to what is not yet, from what is established to what could still be. Years of experience with printed matter determine how we expect actions to unfold. Thought therefore follows language, in a rather literal sense.

So, next time you take that selfie, make sure it reflects you from the right perspective!

COUNTING

PETER NORVIG

Director of Research, Google Inc.; co-author (with Stuart Russell),
Artificial Intelligence: A Modern Approach

John McCarthy, the late cofounder of the field of artificial intelligence, wrote, "He who refuses to do arithmetic is doomed to talk nonsense." It seems incongruous that a professor who worked with esoteric high-level math would be touting simple arithmetic, but he was right; in many cases, all we need to avoid nonsense is the simplest form of arithmetic—counting.

In 2008, the U.S. government approved a $700 *billion* bank bailout package. A search for the phrase "$700 *million* bailout" reveals hundreds of writers who were eager to debate whether this was prudent or rash but who couldn't distinguish the difference between $2 per citizen and $2,000 per citizen. Knowing the difference is crucial to understanding the efficacy of the deal and is just a matter of counting.

Consider the case of a patient who undergoes a routine medical screening and tests positive for a disease affecting 1 percent of the population. The screening test is known to be 90-percent accurate. What's the chance that the patient actually has the disease? When a group of trained physicians was asked, their average answer was 75 percent. They reasoned that it should be somewhat less than the 90-percent accuracy of the test because the disease is rare. But if they'd bothered to count, they would have reasoned like this: On average, out of every 100 people, one will have the disease and ninety-nine won't. The one will have a positive test result, and so will about ten of the ninety-nine (because the test is 10-percent inaccurate). So we have a pool of eleven people who test positive, of which only one actually has the disease, so the chance is about $\frac{1}{11}$ or 9 percent. This means that highly trained physicians are reasoning poorly, scaring patients with an estimate that's much too high, all because they didn't bother to count. The physicians might argue that they were trained to do medicine, not probability theory, but, as Pierre Laplace said in 1812, "Probability theory is nothing but common sense reduced to calculation," and the basis of probability theory is just counting: "The probability of an event is the ratio of the number of cases favorable to it, to the number of all cases possible."

In their report on climate change, the Intergovernmental Panel on Climate Change stated the consensus view that "most of the observed warming over the last 50 years is likely to have been due to the increase in greenhouse gas concentrations." Yet some criticized this consensus, saying that scientists are still uncertain. Who's right? Naomi Oreskes took it upon herself to resolve the question by counting. She searched a scientific database using the keywords "climate change" and scrutinized the 928 abstracts that matched. She found that 25 percent of the articles did not address the consensus (because, for example, they were about regional climate rather than global), but that *none* of the 928 rejected the consensus. This is a powerful form of counting. What's more, I don't need to take Professor Oreskes' word for it: I did my own experiment, sampling 25 abstracts (I didn't have the patience to do 928) and I, too, found that none of them rejected the consensus.

This is a powerful tool. When faced with a complex issue, you can resolve the question not by examining your political predispositions and arguing for whatever agrees with them but by examining the evidence, counting the number of favorable cases and comparing it with the number of those that aren't. You don't need to be a mathematical wizard, just apply what you learned as a toddler: Count!

ON AVERAGE

KAI KRAUSE
Software pioneer, philosopher; author, *A Realtime Literature Explorer*

Humans have fewer than two legs . . .

. . . on average.

It takes a moment to realize the logic of that sentence.

Just a single one-legged pirate lowers the average for all of humankind to just a fraction under two. Simple and true—but also counterintuitive.

A variation moving the average upward instead: "Billionaire walks into a bar."

And everyone is a millionaire . . . on average.

That's all rather basic statistics, but as obvious as it might seem, among the abundance of highly complex concepts and terms in this essay collection, many scientific truths are not easily grasped in everyday life, and the basic tools of understanding are woefully underutilized by the general public. They're badly taught—if indeed they are part of the curriculum at all.

Everyone today is surrounded by practical math. Leaving high school, it should be standard-issue knowledge to understand how credit cards, compounding interest, or mortgage rates work. Or percentage discounts, goods on sale, and, even more basic, a grasp of numbers in general: millions, billions, trillions.

How far the moon is from Earth, the speed of light, the age of the universe.

Following the news, do people really get what all those zeros mean in the national debt? Or the difference between gross national product and gross domestic income?

Do they know the population of the U.S. versus that of, say, Nigeria? (Hint: Nigeria is projected to overtake the U.S. as the third largest nation, reaching 400 million by 2050.)

If you ask around among your family and friends, who could give you a rational description of quantum computing, the Higgs boson, or how and why bitcoin mining works?

Millions of people are out there playing lotteries. They may have heard that the odds are infinitesimal, but often the inability to deal with such large numbers or tiny fractions turns into an intuitive reaction. I have had someone tell me, in all honesty, "Yes it's a small chance, but I figure it's like fifty-fifty. Either I win or I don't."

Hard to argue with that logic.

And yet the cumulative cost of such seemingly small matters is tremendous. Smoking a pack a day can add up to more than leasing a car. Conversely, the historically low interest rates do allow leveraging possibilities. Buying a house, doing your taxes—it all revolves around a certain comfort level with numbers and percentages, which a surprisingly large portion of us are shrugging off or approaching haphazardly.

Not quite getting the probabilities in throwing dice and drawing cards is what built Vegas, but what about the odds in rare diseases or the chances of accidents or crime?

The general innumeracy (nod to Hofstadter and Paulos) has far-reaching effects. There's a constant and real danger in being manipulated, be it by graphs with truncated y axes, or pie charts that go beyond 100 percent, or hearing news of suddenly greater chances to die of some disease, or the effectiveness of medication, or so many of the tiny footnotes in advertising claims. You need a minimum level of awareness to sort through these things. Acquiring common sense needs to include math.

So, which scientific term or concept ought to be more widely known? I would plead to start with a solid foundation for the basics of science and math, and raising the awareness, improving the schooling, bettering the lives of our kids . . . on average.

NUMBER SENSE

KEITH DEVLIN

Mathematician; Executive Director, H-STAR Institute, Stanford University; author, *Finding Fibonacci*

When I graduated with a bachelor's degree in mathematics from one of the most prestigious university mathematics programs in the world (King's College London) in 1968, I had acquired a set of skills that guaranteed full employment wherever I chose to go for the then foreseeable future—a state of affairs that had existed ever since modern mathematics began some 3,000 years earlier. By the turn of the new millennium, however, just over thirty years later, those skills were essentially worthless, having been effectively outsourced to machines that did it faster and more reliably. In a single lifetime, I experienced a dramatic change in the nature of mathematics and the role it plays in society.

The shift began with the introduction of the digital arithmetic calculator in the 1960s, which rendered obsolete the need for humans to master the ancient art of mental arithmetical calculation. Over the succeeding decades, the scope of algorithms that were developed to perform mathematical procedures steadily expanded, culminating in the creation of desktop- and cloud-based mathematical computation systems that can execute pretty well any mathematical procedure, solving—accurately and in a fraction of a second—any mathematical problem formulated with sufficient precision (a bar that allows in all the exam questions I and other math students faced throughout our school and university careers).

So, what then remains in mathematics that people need to master? The answer is the set of skills needed to make effective use of those powerful new (procedural) mathematical tools that we can access from our smartphones. Whereas humans once had to master the computational skills required to carry out various procedures (adding and multiplying numbers, inverting matrices, solving polynomial equations, differentiating analytic functions, solving differential equations, etc.), what's needed today is a sufficiently deep understanding of those procedures and the concepts they're built on, in order to know when and how to use those digitally-implemented tools effectively, productively, and safely.

The most basic of the new skills is *number sense*. The other important one is mathematical thinking, but whereas the latter is important only for those going into STEM careers, number sense is a crucial 21st-century skill for everyone. Definitions of "number sense" generally conform to that of educators Russell Gersten and David Chard: "fluidity and flexibility with numbers, a sense of what numbers mean, and an ability to use mental mathematics to negotiate the world and make comparisons." Likewise, Marilyn Burns, in *About Teaching Mathematics*, describes students with a strong number sense this way: "[They] can think and reason flexibly with numbers, use numbers to solve problems, spot unreasonable answers, understand how numbers can be taken apart and put together in different ways, see connections among operations, figure mentally, and make reasonable estimates." In 1989, the National Council of Teachers of Mathematics identified five components characterizing number sense: number meaning, number relationships, number magnitude, operations involving numbers and referents for number, and referents for numbers and quantities.

Although to outsiders mathematics teaching designed to develop number sense can seem "fuzzy" and "imprecise," it has been well demonstrated that children who don't acquire number sense early in their mathematics education struggle throughout their subsequent school and college years and generally find themselves cut off from any career requiring some mathematical ability.

That outsiders' misperception is understandable. Compared to the rigid, rule-based, right-or-wrong precision of the math taught in my schooldays, number sense (and mathematical thinking) does seem fuzzy and imprecise. But fuzziness and imprecision is exactly why it's such an important aspect of mathematics in an era when the rule-based, precise part is done by machines. The human brain compares miserably to the digital computer when it comes to performing rule-based procedures. But the human mind is capable of something computers can't begin to do and maybe never will: *understanding*. Desktop-computer- and cloud-based mathematics systems provide useful tools to solve the mathematical aspects of real-world problems. But without a human in the driver's seat, those tools are useless. And high on the list of "driving" skills is number sense.

If your children are in the K–12 system, there's one thing you should ensure they've mastered in math class by the time they graduate: number sense. Once they have that, whatever concepts or procedures they run into at the college level and later on can be mastered quickly and easily, as and when required.

FERMI PROBLEMS

SETH SHOSTAK

Senior Astronomer, SETI Institute; author, *Confessions of an Alien Hunter*

It's a familiar peeve: The public doesn't understand science or its workings. Society would be stronger and safer if only the citizenry could judge the reliability of climate-change studies, the benefits of vaccines, or even the significance of the Higgs boson.

This plaint sounds both familiar and worthy, but to lament the impoverished state of science literacy is to flog an expired equine. It's easy to do but neither new nor helpful.

Of course, that's not to say we shouldn't try to ameliorate the situation. The teaching and popularization of science are the Lord's work. But one can easily allow the perfect to become the enemy of the good. Rather than hope for a future in which everyone has a basic understanding of atomic theory or can evaluate the statistical significance of polls, I'm willing to aspire to a more conditional victory. I would appreciate a populace able to make order-of-magnitude estimates.

This is a superbly useful skill and one that can be acquired by young people with no more than a bit of practice. Learning the valences of the elements or the functions of cellular organelles, both topics in high school science, requires memorization. Estimating the approximate magnitude of things does not.

And *mirabile dictu*, no personal electronics are required. Indeed, gadgets are a hindrance. Ask a young person, "How much does the Earth weigh?" and he'll pull out his phone and look it up. The number will be just that—a number—arrived at without effort or the slightest insight. But this is a number that can be approximated in one's head with no more than middle-school geometry, a sense of the approximate weight of a rock or a car, and about a minute's thought. The rough-and-ready answer might be wrong by a factor of 2 or 3, but in most cases that will be adequate for the purpose at hand.

To scientists, such questions are known as Fermi problems, after the famous physicist who encouraged colleagues to make back-of-the-envelope calculations. Apparently, one such problem posed by Enrico Fermi himself—and reputedly used by Google when interviewing potential employees—is "How many piano tuners are there in Chicago?"

Answering this requires making reasonable guesses about such things as the fraction of households having pianos, how long it takes to tune them, etc. But anyone can do that. No background in advanced mathematics is required, just the self-confidence to take on the problem.

If we want the public to make smart choices about such issues as the relative dangers of football versus driving, how long it will take to burn away the entire Amazon basin, or whether it's safer to inoculate your child or not, we're fabulizing if we think that hearing the answers in a news report will suffice. Just as skills are developed by practice, not by reading, so too will an ability (and a readiness) to make an order-of-magnitude estimate produce a deep and long-lasting understanding.

Aside from its utility, this skill rewards its practitioners with personal gratification. It's an everyday demonstration that quantitative knowledge about the world isn't just handed down from on high. Observation, simplifying assumptions ("Let's approximate a chicken by a sphere. . . ."), and the simplest calculation can get us close to the truth. It's not solely the province of experts with tweed jackets and a wall covered with sheepskins.

Schoolteachers have long tried to promote an interest in science by maintaining that curiosity and logical thinking are characteristics of us all ("Everyone's a scientist"). But this felicitous idea is generally followed up with course curricula that are warehouses of facts. Being able to make order-of-magnitude computations with nothing more than one's wits would be far more satisfying, because then you'd know—not because someone else told you but because you worked it out yourself.

We've inadvertently let a device in everyone's back pocket become the back of the book for any question requiring a quantitative answer. It needn't be so, and it shouldn't.

EXPONENTIAL

BRUNO GIUSSANI
European Director, Global Curator, TED

It isn't clear whether it was rice or wheat. We're also not sure of the origin of the story, for there are many versions. But it goes something like this: A king was presented with a beautiful game of chess by its inventor. So pleased was the king that he asked the inventor to name his own reward. The inventor modestly asked for some rice (or wheat). The exact quantity would be calculated through the simplest formula: Put a grain on the first square, two on the second, four on the third, and so on, doubling the number of grains until the sixty-fourth and last square. The king readily agreed, before realizing that he'd been deceived. By mid-chessboard, his castle was barely big enough to contain the grains, and just the first square of the other half would double that.

The story has been used by 13th-century Islamic scholars

on up to scientist/author Carl Sagan and social-media videographers to explain the power of exponential sequences, wherein things begin small, very small, but once they start growing they grow faster and faster. To paraphrase Ernest Hemingway, they grow slowly, then suddenly.

The idea of *exponential* and its ramifications should be better known and understood (and the chessboard fable is a useful metaphor), because we live in an exponential world. This has been the case for a while, but so far we've been in the first half of the board; things are radically accelerating now that we're entering its second half.

The "second half of the chessboard" is a notion put forth by Ray Kurzweil in his 1999 book *The Age of Spiritual Machines*. He suggests that while exponentiality is significant in the first half of the board, when we approach the second half its effects are enormous, things get crazy, and the acceleration eludes most humans' grasp.

In *The Second Machine Age* (2014), Andrew McAfee and Erik Brynjolfsson looked at Kurzweil's suggestion in relation to Moore's law. In 1965, reflecting on the first few years of silicon-transistor development, computer pioneer Gordon Moore made a prediction that computing power would double roughly every eighteen to twenty-four months for a given cost (I'm simplifying here): In other words, it would grow exponentially. His prediction held for decades, with huge technological and business effects, although the pace has been slowing in recent years. It's worth mentioning that Moore's was an insight, not a physical law, and that we're likely moving away from transistors to a world of quantum computing, which relies on qubits instead to perform calculations.

McAfee and Brynjolfsson reckon that if we put the starting point of Moore's law in 1958, when the first silicon transistors were commercialized, and we follow the exponential curve, in digital technology terms we entered the second half of the chessboard sometime around 2006. For context, consider that the first mapping of the human genome was completed in 2003, the operating systems for today's smartphones were launched in 2007, and scientists at Yale created the first solid-state quantum processor in 2009.

Hence we find ourselves somewhere in the first, maybe the second, square of the second half of the chessboard. This helps make sense of the dramatically fast advances happening in science and technology, from smartphones and language translation and the blockchain to big analytics and self-driving vehicles and artificial intelligence, from robotics to sensors, from solar cells to biotech to genomics and neuroscience, and more.

While each of these fields is growing exponentially on its own, the combinatorial effect—the accelerating influence each has on others—is prodigious. Add to that the capacity of AI systems for self-improvement and we're talking about almost incomprehensible rates of change.

To stay with the original metaphor, this is what entering the second half of the chessboard means: Until now, we were accumulating rice grains at an increasingly fast pace, but we were still within the confines of the king's castle. The next squares will inundate the city, and then the land, and then the world. And there are thirty-two squares to go, so this won't be a brief period of transformation. It will be a long, deep, unprecedented upheaval. These developments may lead us to an age of abundance and a tech-driven renaissance, as many claim and/or hope, or uncontrollably down a dark hole, as others fear.

Yet we still live, for the most part, in a world that doesn't understand and isn't made for exponentiality. Almost every structure and method we've developed to run our societies—

governments, democracy, education and healthcare systems, legal and regulatory frameworks, the press, companies, security and safety arrangements, even science management itself—is designed to function in a predictable, linear world, and sudden spikes or downturns are seen as crises. Thus it's unsurprising that the exponential pace of change is causing, almost daily, all sorts of disquiet and stresses—political, social, and psychological.

How do we learn to think exponentially without losing depth, careful consideration, nuance? How does a society function in a second-half-of-the-chessboard reality? What do governance and democracy mean in an exponential world? How do we rethink everything, from education to legal frameworks to notions of ethics and morals?

It starts with a better understanding of exponents and of the metaphor of the second half of the chessboard. And with applying second-half thinking to pretty much everything.

IMPEDANCE MATCHING

W. DANIEL HILLIS

Physicist, computer scientist; Co-founder, Applied Invention and Applied Minds; author, *The Pattern on the Stone*

"Impedance" is a measure of how a system resists the inflow of energy. Often a system can be changed to accept energy more efficiently by adding an element called an impedance matcher. Most people have never heard of impedance matchers, but once you know about them, you begin to see them everywhere: in the shape of a trumpet, the antireflective coating on a lens, the foam spikes on the inside of a recording booth.

A familiar example of an impedance matcher is the transmission of an automobile, which couples the energy from the engine into the wheels by converting the relatively fast rotation of the engine into the slower, stronger rotation required to propel the car. The transformer on an electric pole solves an analogous problem, converting the high-voltage electricity on the transmission lines to the high-current electricity required to power your home. Lots of machines, from jet engines to radios, depend on impedance matchers to move energy from one part of the system to another.

Some of the most interesting impedance matching occurs when energy comes in the form of a wave. You've probably noticed in a swimming pool that waves from a splash reflect off the sides of the wall. Because there's an impedance mismatch between the water and the wall, the wave energy can't couple into the wall, so it reflects back. If you watch a vertical seawall, you'll notice that the reflected wave adds to the height of the original wave, creating a splash almost twice the height of the wave. This doesn't happen when a wave rolls up a gently sloped beach, because the slope acts as an impedance matcher, allowing the wave's energy to flow into the sand. The foam spikes on the inside of an anechoic recording booth serve much the same function. The sound waves are much larger than the spikes, so they make the foam gradually denser as it gets closer to the wall. This couples the sound energy into the absorbing wall without reflection, so there's no echo.

The flare at the end of a trumpet works on the same principle, in reverse. Without the flare, most of the sound energy would just be reflected back into the trumpet instead of coupling to the outside air. A speaking cone, or megaphone, is another example. The cone acts as an impedance matcher, connecting the sound energy of the voice into the air. This kind of impedance matcher is so effective that it allows the tiny motion of an old Victrola record needle to fill a dance hall with music.

Impedance matching works with light waves as well as sound. The antireflective coating on a camera lens is similar in function to the foam on the sound booth. A mirror, on the other hand, works like the seawall, creating a reflection with an impedance mismatch. By putting only a few molecules of aluminum on the glass, we can tune the optical properties of the surface to create an impedance mismatch, a "half-silvered" mirror reflecting only a little bit of the light.

On a larger scale, we can think of atmospheric carbon dioxide as an undesired impedance matcher coupling the infrared light waves of the sun into the planet. Someday we may decide to cool our Earth by adding tiny particles of dust to the stratosphere, tuning the optical surface to reflect away a tiny part of the infrared waves from the sun. The impedance mismatch between the atmosphere and sunlight would create a kind of half-silvered mirror to keep us cooler by reflecting away the unwanted energy flowing into our planet.

HOMEOSTASIS

MARTIN LERCHER

Professor of Computational Cell Biology, Heinrich-Heine University, Düsseldorf; co-author (with Itai Yanai), *The Society of Genes*

Life as we know it requires some degree of stability both internally and externally. Consider a bacterium like *E. coli*, which needs to maintain its internal copper concentrations within a narrow range. Too much copper would kill the cell, while too little would impede important metabolic functions that rely on copper atoms as catalytic centers of enzymes. Keeping copper concentrations within the required range—copper homeostasis—is achieved through a negative feedback loop: The bacterium has internal sensors that react to suboptimal copper levels by changing the production rate of proteins that pump copper out of the cell. This feedback system has its limits, though, and most bacteria succumb to too much copper in their environment—storing water in copper containers is an age-old strategy to keep it fresh.

Human cells require not just elaborate systems to achieve internal homeostasis of many types of molecules but also a precisely tuned environment. Our cells can count on a working temperature of close to 37°C, measured by thermometers in the brain and maintained through behavior (e.g., wardrobe adjustments) as well as through sweating and the regulation of blood flow to the limbs. Our cells can also count on a constant supply of nutrients through the bloodstream, including glucose, measured in the pancreas and regulated through insulin secretion, and oxygen, measured in the major blood vessels and the kidneys and maintained through adjustments of the activity of breathing muscles and the production of red blood cells. Again, homeostasis is achieved by negative feedback loops: In response to measured deviations from a desired level, our body initiates responses that move us back toward the target value.

That homeostasis plays a major role in human health was recognized by the ancient Greek philosophers. Hippocrates believed that health represented a harmonious balance of the "elements" making up the human body, whereas disease was a state of systematic imbalance. For many important diseases, this is an accurate description. In type 1 diabetes, for example, the pancreas cells responsible for the blood-glucose-level measurements are destroyed and the homeostasis system breaks down. Chronic diseases, however, are often initially compensated by homeostatic systems; e.g., anemia caused by an accelerated breakdown of red blood cells can be compensated by an increased production of those cells, as long as the body has enough raw materials for this.

Complex systems can hardly be stable without at least some level of homeostasis. The Earth's biosphere is a prime example. Surface temperatures and atmospheric CO_2 levels are both affected by biological activities. Higher atmospheric CO_2 leads to increased plant growth, causing an increased consumption of CO_2 and thus maintaining homeostasis. Higher temperatures cause increased phytoplankton growth in the oceans, which produces airborne gases and organic matter that seeds cloud droplets; more and denser clouds, in turn, lead to an increased reflection of sunlight back into space and thus contribute to temperature homeostasis. These systems also have their limits, though; like an *E. coli* bacterium with too much copper, our planet's homeostasis systems, on their own, may be unable to overcome the current onslaught of human activities on global temperatures and CO_2 levels.

ASHBY'S LAW OF REQUISITE VARIETY

JOHN NAUGHTON

Senior Research Fellow, Centre for Research in the Arts, Social Sciences, and Humanities; Director, Wolfson College Press Fellowship Programme, University of Cambridge; author, *From Gutenberg to Zuckerberg*

W. Ross Ashby was a British cybernetician working in the 1950s who became interested in the phenomenon of homeostasis—

the way in which complex systems operating in changing environments succeed in maintaining critical variables (for example, internal body temperature in biological systems) within tightly defined limits. Ashby came up with the concept of *variety* as a measurement of the number of possible states of a system. His "law" of requisite variety states that for a system to be stable, the number of states its control mechanism can attain (its variety) must be greater than or equal to the number of possible states that the system can attain.

Ashby's law was framed in the context of his interest in self-regulating biological systems, but it was rapidly seen as relevant to other kinds of systems. The British cybernetician and operational research practitioner Stafford Beer, for example, used it as the basis for his concept of a *viable system* in organizational design. In colloquial terms, Ashby's law has come to be understood as a simple proposition: If a system is to deal successfully with the diversity of challenges its environment produces, then it needs to have a repertoire of responses (at least) as nuanced as the problems thrown up by the environment. So a viable system is one that can handle the variability of its environment. Or, as Ashby put it, only variety can absorb variety.

Until comparatively recently, organizations coped with environmental challenges mainly by measures to reduce the variety they had to cope with. Mass production, for example, reduced the variety of its environment by limiting the range of choice available to consumers. Product standardization was essentially an extrapolation of Henry Ford's slogan that customers could have the Model T in any color as long as it was black. But the rise of the Internet has made variety reduction increasingly difficult. By any metric you choose—numbers of users and publishers, density of interactions between agents, pace of change (to name just three)—our contemporary information ecosystem is orders of magnitude more complex than it was forty years ago. And its variety, in Ashby's terms, has grown in proportion to its complexity. Given that variety reduction seems unfeasible in this new situation, many of our organizations and social systems—ones that evolved to cope with much lower levels of variety—may no longer be viable. For them, the path back to viability means finding ways of increasing their variety. And the big question is whether and how they can do so.

VARIETY

LEE SMOLIN

Theoretical physicist, Perimeter Institute, Waterloo, Ontario; author, *Time Reborn*

Leibniz was famously satirized by Voltaire, in *Candide*, as saying that ours is the best of all possible worlds. What Leibniz actually wrote, in 1714, in his *Monadology*, was a good deal more interesting. He did argue that God chose the one real world from an infinitude of possible worlds by requiring it to have "as much perfection as possible." But what is often missed is how he characterized degrees of perfection. Leibniz defined a world with "as much perfection as possible" to be one having "as much variety as possible, but with the greatest order possible."

I believe that Leibniz's insight of a world that optimizes variety, subject to "the greatest order possible," is a powerful concept that could be helpful for current work in biology, computer science, neuroscience, physics, and numerous other domains including social and political theory and urban planning.

To explain why, I have to define variety. I believe we ought to see variety as a measure of complexity that applies to systems of relationships. These are systems of individual units, each of which has a unique set of interactions or relationships with the other units in the system. Leibniz saw the universe as just such a system of relationships. In a Leibnizian world, an object's properties aren't intrinsic to it; rather, they reflect the relationships or interactions the object has with other objects.

Systems of relationships are often visualized as graphs or networks. Each element is represented by a node, and two nodes are related when they're connected by a line. We know of a great many systems, natural and artificial, that can be represented by such a network; these include ecosystems, economies, the Internet, social networks, etc.

What does it mean for such a relational world to have a high variety? I would argue that variety is a measure of how unique is each element's role in the network. In a Liebnizian world, each element has a *view* of the rest which summarizes how it's related to the other elements. An element's perspective tells us what the system looks like from its point of view. Variety is a measure of how distinct these different views are.

Leibniz expressed this almost poetically. "This interconnection (or accommodation) of all created things to each other, brings it about that each simple substance has relations that express all the others, and consequently, that each simple substance is a perpetual, living mirror of the universe."

He then reaches for a striking metaphor, of a city. "Just as the same city viewed from different directions, appears entirely different and, as it were, multiplied perspectively, in just the same way it happens that, because of the infinite multitude of simple substances, there are, as it were, just as many different universes, which are, nevertheless, only perspectives on a single one."

One can almost hear Jane Jacobs in this, when she praises a good city as one with many eyes on the street.

A system of relations can be said to have its maximal variety when the different views are maximally distinct from each other. A city has low variety if the views from many of the houses are similar. A city has high variety if it's easy to tell, just by looking out the window, which street you're on.

An ecosystem is a system of relations, such as who eats whom. A niche is a situation characterized by what you eat and who eats you. An economy is a system of relations including who buys what from whom. The variety of an ecosystem is a measure of the extent to which each species has a unique niche. The variety of an economy measures the uniqueness of each firm's role in the market.

It's commonly asserted that ecosystems and economies evolve to higher degrees of complexity. But to develop these ideas, we need a precise notion of complexity. Negative entropy doesn't suffice to measure complexity, because the network of chemical reactions in our bodies and a regular lattice both have low entropy. We want a measure of complexity that recognizes that chemical-reaction networks are far more complex than either random graphs or regular lattices.

I would suggest that variety is a helpful notion of complexity because it distinguishes the truly complex from the regular. In a high-variety network, it's easy to know where you are by looking around at your neighborhood. In other words, the less information you need about the neighborhoods to distinguish each node from the rest, the higher the variety. This captures a notion of complexity distinct from and perhaps more useful than negative entropy.

Having defined variety, we can go back and try to imagine what Leibniz meant by maximizing variety but "with the greatest order possible." Order can mean subject to law. Can there then be a law of maximal variety?

Such a law might be emergent in a complex system such as an economy or ecology. This might arise as follows: When the variety is maximal, the network is most efficient, because there's a maximum amount of cooperation and a minimum of redundancy. Entities compete not to dominate a single niche but to invent new ways to cooperate by inventing new niches, which have a novel interrelation to the rest.

ALLOSTASIS

TOR NØRRETRANDERS
Writer, lecturer, science journalist; co-author (with Olafur Eliasson), *Light! On Light in Life and the Life in Light*

How to respond to change: by trying to keep your inner state constant, or by adjusting your inner state according to external change?

Staying constant inside is the classic idea of homeostasis, pioneered by the physiologist Claude Bernard in 1865; the term was coined by physiologist Walter Cannon in 1926. Homeostasis describes an essential feature of all living things—that they define an inside and keep it stable in an unstable environment. Body temperature is a classic example. Homeostasis isn't very dynamic or Darwinian; the business of living creatures is not to optimize their interior state but to survive whether or not the internal state is stable.

Therefore, the concept of *allostasis* was created in the 1980s by the neuroscientist Peter Sterling and co-workers. The word "allostasis" means a changing state, whereas homeostasis means staying in (about) the same state. The idea of allostasis is that the organism will change its inner milieu to meet challenges from the outside. Blood pressure isn't constant; it will be higher if the organism has to be very active and lower if it doesn't. Constancy isn't the ideal; the ideal is to have the relevant inner state for the particular outer state.

Stress reaction is an example of allostasis: When there's a tiger in the room, all resources available must be mobilized. Blood pressure and many other parameters go up quickly. All depots are emptied. The emergency stress reaction is a plus for survival, but only when there's a stressor to meet. If the reaction is permanent, it's dangerous.

Allostasis also brings out another important physiological feature: looking ahead in time. While homeostasis is about conserving a state (thus looking back in time), allostasis looks forward: What will be the most relevant inner state in the next moment? The role of the brain is essential in allostasis, because it predicts the environment and allows for adjustment, so that blood pressure or blood-glucose level can become relevant to what's up.

Although born in physiology, the idea of allostasis in the coming years may become important as an umbrella for trends currently fermenting in our understanding of the mind. States of mind exemplify the role of relevance: It's not always relevant to be in a good mood. When the organism is challenged, negative emotions are highly relevant. But if they're there all the time, negative emotions become a problem. When there's no challenge, it's more relevant to have positive emotions that will broaden your perspective and build new relationships, as described by psychologist Barbara Fredrickson.

Reward prediction has, over the past decades, become a key

notion in understanding perception and behavior in both robots and biological creatures. Navigation is based on predictions and prediction errors rather than on a full mapping of the entire environment. The world is described inside-out by throwing predictions at it and seeing how they work out. "Controlled hallucinations" has become a common phrase describing this process of imagining or predicting a spectrum of perceptional scenarios—made subject to selection by experience. Much like the scientific process of making hypotheses and testing them. Prospection, originally described by the social psychologist Daniel Gilbert in 2005, allows us to imagine several possible futures and observe the internal emotional reaction to them. Anticipating allostasis.

Allostasis is an important concept for science because it roots the future-oriented aspects of the mind in bodily physiology. It is an important concept in everyday life, because it points to the importance of embracing change.

THE BRAINSTEM

EDUARDO SALCEDO-ALBARÁN
Philosopher; Director, Scientific Vortex Inc.

Today, neurobiology provides increasing evidence of how our reason, although powerful, isn't as constant as most social scientists assume. Now we know that inner-brain areas related to homeostatic and basic physiological functions are more permanent and relevant for surviving and evolving than external areas of the cortex related to cognitive faculties.

Complex cognitive skills do enrich our mental life but aren't critical for regulating physiological functions basic to our existence. We can live with a poorer mental life—for instance, following damage to our external cortex—but we can't live without the basic regulation of our heartbeat or respiratory rhythm following damage to our brainstem.

In fact, as the neuroscientist Antonio Damasio has explained, the brainstem potentially houses the origin of consciousness, the complex mental representation of the "self" we continuously experience in first person. Even slight damage in sub-areas of the brainstem lead to comatose and vegetative states and the permanent lack of consciousness. Since the brainstem's activity is related to basic physiological and homeostatic functions, its activity is constant, allowing the continuum construct of the "self." Although our external appearance changes over our lifetime, our internal organs and biological functions remain mostly unchanged. If our consciousness is therefore grounded on those constant biological functions, it too will remain mostly constant.

While regulating basic physiological functions, the brainstem sends conscious signals to our "self" through automatic emotions and specifically through the conscious feelings of those emotions. Emotions are therefore the initial process linking our physiological needs—such as eating or breathing—and the conscious self; the last part of the process consists of the feelings of those emotions, in the form of enriched mental experiences of joy, beauty, or sorrow, for instance.

Emotions as a reflection of our physiological needs and homeostatic functions—and feelings, as conscious experiences of those emotions—are permanent and preponderant in our life because they link our physiology with our conscious self through the brainstem.

Acknowledging functions and dynamics of the brainstem and its sub-areas is useful for understanding human nature, the one driven mostly by spontaneous, automatic, and capricious emotions and feelings. Daniel Kahneman and other theorists— mainly a small fraction of psychologists and neuro-economists— have called attention to biases and capricious and irrational behaviors, but in most social sciences such knowledge is still uncommon.

Acknowledging the origin and influence of emotions in human behavior doesn't mean abdicating to irrationality but merely using our powerful rationality to understand the critical role in our behavior played by dynamics in the brainstem.

THE PRINCIPLE OF LEAST ACTION

JANNA LEVIN

Professor of Physics and Astronomy, Barnard College, Columbia University; author, *Black Hole Blues and Other Songs from Outer Space*

Complexity makes life interesting. A universe of just hydrogen is bland, but the helpful production of carbon in stellar cores allows for all kinds of chemical connections. A universe of just two dimensions is pretty limited, but we live in at least three and enjoy the greater range of motion and possible spatial permutations. When I'm sitting on a bench in my friend's garden in California, there's a lot to look at. The visual information filling my field of view is complicated. The dry winter leaves trace vortices in the air's motion. Plants respire, and we breathe, and the neural connections fire, and it's all complex and interesting. The physicist's job is to see through the overwhelming intricacy and find the rallying, organizing principle.

Everything in this garden, from the insects under the rocks

to the blue dome overhead to the distant stars washed out by the sunlight, can be traced to a remarkably lean origin in a Big Bang. Not to overstate the case. There's much we don't understand about the first trillionth of a trillionth of a trillionth of a trillionth of a second after the inception of our universe. But we can detail the initial three minutes with decent confidence and impressive precision.

Our ability to comprehend the early universe that took 13.8 billion years to make my friend's yard is the direct consequence of the well-known successes of unification. Beginning with Maxwell's stunning fusion of electricity and magnetism into one electromagnetic force, physicists have reduced the list of fundamental laws to two. All the matter forces—weak, electromagnetic, and strong—can be unified in principle (though there are some hitches). Gravity stands apart and defiant, so that we haven't yet realized the greatest ambition of theoretical physics: the Theory of Everything, the one physical law that unifies all forces, that pushes and prods the universe to our current complexity. But that's not the point.

The point is that a fundamental law is expressible as one mathematical sentence. We move from that single sentence to the glorious Rube Goldberg machine of our cosmos by exploiting my favorite principle, that of least action. To find the curves in spacetime, which are due to matter and energy, you must find the shortest path in the space of possibilities. To find the orbit of a comet around a black hole, you must find the shortest path in the curved black-hole spacetime.

More simply, the principle of least action can be stated as a principle of least resistance. If you drop a ball in midair, it falls along the shortest path to the ground, the path of least resistance under the force of gravity. If the ball does anything but fall along the shortest path, if it spirals around in widening loops and goes back up in the air, you would know that other forces were at work—a hidden string, or gusts of air. And those additional forces would drive the ball along the path of least resistance in their mathematical description. The principle of least action is an old one. It allows physicists to share the most profound concepts in human history in a single line. Take that one mathematical sentence and calculate the shortest paths allowed in the space of possibilities and you'll find the story of the origin of the universe and the evolution of our cosmological ecosystem.

"THE BIG BANG"

JOHN C. MATHER

Senior Astrophysicist, Observational Cosmology Laboratory, NASA's Goddard Space Flight Center; Senior Project Scientist, the James Webb Space Telescope; Nobel laureate

The name "Big Bang" has been misleading scientists, philosophers, and the general public since Sir Fred Hoyle coined the term on the radio in 1949. It conjures up the image of a giant firecracker, an ordinary explosion happening at a place and a time, a collection of material suddenly beginning to expand into the surrounding empty space. But this is so exactly opposite to what astronomers have observed that it's shocking that we still use the name—and not the least bit surprising that some people object to it. Einstein didn't like it at first but became convinced. Hoyle never liked it at all. People might like it better if they knew what it meant.

What astronomers actually have observed is that distant galaxies all appear to be receding from us at a speed roughly proportional to their distance. We've known this since 1929, when Edwin Hubble drew his famous plot. From this we conclude a few simple things. First, we can get the approximate age of the universe by dividing the distance by the speed; the current value is around 14 billion years. The second and more striking conclusion is that there's no center of this expansion, even though we seem to be at the center. We can imagine what an astronomer living in a distant galaxy would see; she, too, would conclude that the universe appears to be receding from her location. The upshot is that there's no sign of a center of the universe. So much for the "giant firecracker." A third conclusion is that there's no sign of an edge of the universe, no place where we run out of either matter or space. This is what the ancient Greeks recognized as infinite, unbounded, without limits. This is also the exact opposite of a giant firecracker, for which there's a moving boundary between the space filled with debris and the space outside it. The actual universe appears to be infinite now, and if so it has probably always been infinite. It's often said that the whole universe we can now observe was once compressed into a volume the size of a golf ball, but we should imagine that the golf ball is only a tiny piece of a universe that was infinite even then. The unending infinite universe is expanding into itself.

There's another way in which the giant-firecracker idea misleads us, because even scientists often talk about the "universe springing into existence." Well, it didn't, as far as we can tell. The opposite is true. There's no first moment of time, just as there's no smallest positive number. In physics, we have equations and laws of nature describing how one situation changes into another, but we have no equations that show how true nothingness turns into somethingness. So, since the universe did not spring into existence, it has always existed, though perhaps not in its current form. That's true even though the apparent age of the universe isn't infinite but only very large. And even though there's no first moment of time, we can still measure the age.

There's plenty of mystery left. What are those equations of the early universe which might describe what came before the atoms we see today? We're pretty confident that we can imagine times in the early universe when temperatures and pressures were so high that atoms would have been ripped apart into the particles we've manufactured at our Large Hadron Collider. We have a Standard Model of cosmology and we have a Standard Model of particles. But the mysteries include: Why is there an asymmetry between matter and antimatter, such that the whole observable universe is made of matter? What is dark matter? What is dark energy? What came before the expansion and made it happen, if anything did? We've got the idea of cosmic inflation, which might be right. What are space and time themselves? Einstein's general theory of relativity tells us how they're curved, but scientists suspect that this isn't the whole story, because of quantum mechanics and especially quantum entanglement.

Stay tuned! There are more Nobel prizes to be earned.

MULTIVERSE

MARTIN REES

Former President of the Royal Society; Emeritus Professor of Cosmology and Astrophysics, University of Cambridge; Fellow, Trinity College; author, *What We Still Don't Know*

An astonishing concept has entered mainstream cosmological thought: It deserves to be more widely known. Physical reality could be hugely more extensive than the patch of space and time traditionally called "the universe." Our cosmic environment could be richly textured but on scales so vast that our astronomical purview is restricted to a tiny fraction. We're not aware of the "grand design"—any more than plankton whose "universe" was a spoonful of water would be aware of the world's topography and biosphere. We may inhabit a multiverse.

However powerful our telescopes are, our vision is bounded by a horizon—a shell around us, delineating the distance light can have traveled since the Big Bang. But this shell has no more physical significance than the circle delineating your horizon if you're in the middle of the ocean. There are billions of galaxies within our horizon, but we expect far more galaxies lying beyond. We can't tell just how many. If space stretched far enough, then all combinatorial possibilities would be repeated. Far beyond the horizon, we could all have avatars. And it may be some consolation that when we make a bad decision, there's another one of us, far beyond our horizon, who has made a better one.

So the aftermath of "our" Big Bang could encompass a stupendous volume. But that's not all. "Our" Big Bang could be just one island of spacetime in an unbounded cosmic archipelago. A challenge for 21st-century physics is to answer two questions. First, are there many "Big Bangs" rather than just one? Second—and this is even more interesting—if there are many, are they all governed by the same physics? Or is there a huge number of different vacuum states—each the arena for different microphysics and therefore offering differing propensities for spawning life?

If the answer to this latter question is "Yes," there will still be overarching laws governing the multiverse—maybe a version of string theory. But what we've traditionally called the laws of nature will be just local bylaws. Even though it makes some physicists foam at the mouth, we then can't avoid the A-word—"anthropic." Many domains could be stillborn or sterile; the laws prevailing in them might not allow any kind of complexity. We therefore wouldn't expect to find ourselves in a typical universe. Ours would belong to the unusual subset where there was a "lucky draw" of cosmic numbers conducive to the emergence of complexity and consciousness. Its seemingly designed or fine-tuned features wouldn't be surprising.

Some claim that unobservable entities aren't part of science. But it's hard to defend that hard-line view. For instance, unless we're in some special central position and our universe has an "edge" just beyond the present horizon, there will be some galaxies lying beyond our horizon—and if the cosmic acceleration continues, they will remain beyond forever. Not even the most conservative astronomer would deny that these never-to-be observable galaxies (which, as I've mentioned, could hugely outnumber those we can see) are part of physical reality. They're part of the aftermath of our own Big Bang. But why should they be accorded higher epistemological status than unobservable objects that are the aftermath of other Big Bangs? So it's surely a genuine scientific question to ask whether there's one Big Bang or many.

Fifty years ago, we weren't sure whether there was a Big Bang at all. (Fred Hoyle and other "Steady Statesmen" still contested the idea.) Now we can confidently describe cosmic history back to the ultra-dense first nanosecond. So it's not overoptimistic to hope that in fifty more years we may have a "unified" physical theory, corroborated by experiment and observation in the everyday world, that tells us what happened in the first trillionth of a trillionth of a trillionth of a second, when inflation is postulated to have occurred. If that theory predicts multiple Big

Bangs, we should take that prediction seriously, even though it can't be directly verified (just as we take seriously general relativity's predictions for the unobservable insides of black holes because the theory has survived many tests in domains we can observe).

Some physicists don't like the multiverse. They'd be disappointed if some of the key numbers they're trying to explain turn out to be mere environmental contingencies governing our local spacetime patch—no more "fundamental" than the parameters of Earth's orbit around the sun. But that disappointment would surely be outweighed by the revelation that physical reality was grander and richer than hitherto envisioned. In any case, our preferences are irrelevant to the way physical reality actually is—so we should surely be open-minded.

Indeed, there's an intellectual and aesthetic upside. If we're in a multiverse, it would imply a fourth and grandest Copernican revolution. We've had the Copernican revolution itself, then the realization that there are billions of planetary systems in our galaxy, then that there are billions of galaxies in our observable universe. But now that's not all. The entire panorama that astronomers can observe could be a tiny part of the aftermath of "our" Big Bang, which is itself just one Bang among a perhaps infinite ensemble.

We may, by the end of this century, be able to say with confidence whether or not we live in a multiverse and how much variety its constituent universes display. The answer to this question will determine how we should interpret the biofriendly universe in which we live.

GRAVITATIONAL RADIATION

GINO SEGRÈ

Professor Emeritus of Physics, University of Pennsylvania; co-author (with Bettina Hoerlin), *The Pope of Physics: Enrico Fermi and the Birth of the Atomic Age*

Humans, as well as most other mammals, detect from birth onward electromagnetic radiation in the form of visible light. But it wasn't until the 1887 experiments by Heinrich Hertz that scientists realized that what they were seeing was nothing but a frequency band of the electromagnetic radiation generated by accelerating electric charges. These experiments confirmed the prediction that James Clerk Maxwell made less than three decades earlier.

The realization opened the door to studying electromagnetic radiation of all frequencies, from radio waves to X-rays and beyond. It also stimulated scientists to ask whether an entirely different form of radiation might ever be observed. Accelerating masses rather than electric charges would be its source, and it would be called gravitational radiation.

Maxwell's equations showed what form electromagnetic radiation would take. Albert Einstein's 1915 general theory of relativity offered a prediction for gravitational radiation. But since the gravitational force is so much weaker than the electromagnetic force, there was doubt that a direct observation of this radiation would ever be possible.

Yet the seemingly impossible has taken place! In an experiment of almost unimaginable difficulty, the LIGO (Laser Interferometer Gravitational-wave Observatory) reported on September 14, 2015, at 9:50:45 GMT, detection of a signal that corresponded to two black holes. They were 1.3 billion light-years away, one with 36 solar masses and the other with 29, spiraling into each other to form a more massive black hole. Theory predicted that the equivalent of 3 solar masses had been emitted as gravitational waves in the last second of the two holes' merger.

The general theory of relativity tells us that gravity is a cur-vature in space caused by the presence of masses and energy. The signal for gravitational radiation (or, alternatively, gravitational waves) is therefore a deformation in the space of the detector the waves are passing through. The LIGO detectors, one in Liv-ingston, Louisiana, and the other in Hanford, Washington, are L-shaped structures containing a vacuum tube in which a laser beam travels back and forth from one arm to the other. In dis-torting space, the gravitational waves from the spiraling black holes altered the relative length of the detectors' two arms by approximately a billionth of a billionth of a meter, an incredibly small distance but sufficient to change the interference pattern of the laser light recombining at the detectors' nexus.

The need for two detectors was obvious. Despite every effort to eliminate background, such a minuscule effect would be hard to take seriously unless observed simultaneously in two widely separated locations. This was the case: The detection at Living-ston and Hanford, which are a little over 3,000 kilometers apart, was separated by only 7 milliseconds.

Four hundred years of observational astronomy, from Gali-leo's telescope to the Hubble Space Telescope, has enriched our view of the universe immeasurably. The study of gravitational radiation offers the potential for the next step. The universe be-comes transparent to electromagnetic radiation only when the universe has cooled sufficiently for atoms to form, which happened some 380,000 years after the Big Bang. Gravitational radiation suffers no such limitations. Someday we may even use it as a tool to observe the inflationary expansion the universe is presumed to have undergone immediately after the Big Bang.

We're at the dawn of a new era in astronomy. If all goes well and the importance of studying gravitational radiation is appreciated, twenty years from now we'll anxiously be awaiting reports from LISA (Laser Interferometer Space Antenna), whose interferometers are separated by 5 million kilometers.

THE NON-RETURNABLE UNIVERSE

ANDREI LINDE
Harald Trap Friis Professor in Physics, Stanford University; inventor of eternal chaotic inflation; recipient, 2014 Kavli Prize in Astrophysics

Thanks to online shopping, buying things now is much easier than before. If you don't like something, you can always return it for a full refund. A physicist might say that one can turn the arrow of time back for unused purchases. A cosmologist may comment that we use a similar time-reversal method in our research.

Indeed, to understand the origin of the universe, we may study its present evolution and then solve the Einstein equations back in time. What we find is similar to what happens when we play a movie back. At present, galaxies move away from one another. Playing the movie back shows them moving closer together. Going further back in time, we see that at some point the density of matter becomes infinitely large. This is the cosmological singularity. Solving the same equations forward in time, starting from the singularity, we find that all matter appears from the singularity in a huge explosion called the Big Bang. When the original cosmic fire cools down, matter condenses into galaxies, and they fly away from one another.

The possibility of going back and forth in time is a powerful tool that helps us visualize the evolution of the universe. The resulting picture is so convincing that many of us still use it in our lectures. It tells us, essentially, that our universe is a gift sent to us from the cosmological online retailer 14 billion years ago. We can track its delivery from its origin to the present time, running our calculations all the way back to the Big Bang.

Of course, we know that we cannot reverse the arrow of time, because of the second law of thermodynamics. But the standard lore of the Big Bang theory was that the expansion of the universe was nearly adiabatic and therefore approximately reversible. In particular, it was assumed that the total number of elementary particles in the universe didn't change much during cosmological evolution.

Some parts of this picture, however, were problematic. One may wonder, for example, who paid the bill for sending us the more than 10^{90} elementary particles populating our universe, and who made the universe uniform and suitable for life?

In the beginning of the 1980s, it was found that one can solve these problems if, soon after the Big Bang, there was a short stage of exponentially fast expansion of the universe called inflation. This idea went through numerous modifications, and now we don't even think the universe was born in fire. Instead, it could have been created from a tiny vacuumlike speck of matter with special properties, weighing less than a gram and containing no elementary particles at all. Normal matter emerged only later, when the universe became exponentially large and its original vacuumlike state decayed.

In the last thirty-five years, many predictions of this theory have been verified by cosmological observations; nevertheless, it's somewhat difficult to get used to it, in part because it doesn't match the conventional picture of the time-reversible universe.

Indeed, let's try following the cosmological evolution back in time in this scenario. At first, galaxies move toward one another and the universe becomes very dense, just as in the standard Big Bang theory. But then something really weird happens. Suddenly, at the moment corresponding to the end of inflation, played back in time, all 10^{90} elementary particles populating our part of the universe *completely disappear!* . . . And then—not a flash, not a sound, nothing, except the exponentially shrinking and disappearing empty universe.

Thus, in this scenario, the movie playing back in time has a nonsensical ending. Making 10^{90} elementary particles instantly vanish is impossible!

But there's nothing wrong about it if one moves in the proper time direction. According to inflationary theory, *all* elementary particles in the universe were created by conversion of vacuum energy to matter during the decay of the original vacuumlike state. The theory of this process is well developed, but the process is irreversible: Once particles are created, they cannot be un-created.

This is a part of the mechanism that makes the new theory work: We don't need anybody to send us a huge container with more than 10^{90} elementary particles; a tiny package with 1 gram of matter is more than enough. On its way to us, this package started growing, unwrapping itself, producing billions of galaxies containing hundreds of billions of stars. We cannot, even in our imagination, follow this package back to its origin and watch all matter dissolving into nothingness. Our universe is unreturnable, so our only choice is to accept this gift and use it to the best of our abilities.

THE BIG BOUNCE

PAUL J. STEINHARDT

Albert Einstein Professor in Science; Director, Princeton Center for Theoretical Science, Princeton University; co-author (with Neil Turok), *Endless Universe*

2016 was the Year of the "Big Bounce."

Everyone has heard of the Big Bang—the idea that about 14 billion years ago the universe emerged from nothingness through some sudden quantum event into an expanding space-time filled with hot matter and radiation. Many know that the Big Bang alone cannot explain the remarkably uniform distribution of matter and energy observed today or the absence of curves and warps that one might expect after a sudden quantum event. To account for these observations, an enhancement has been added—a brief epoch of superluminal expansion known as inflation, which immediately followed the Bang. Adding inflation to the Big Bang was supposed to explain how the turbulent and twisted conditions following a Bang could have been stretched out, leaving behind a smooth universe except for a pattern of tiny variations in temperature and energy.

But neither the Big Bang nor inflation are proven ideas, and there are good reasons to consider an alternative hypothesis in which the Big Bang is replaced by a Big Bounce. In a universe created by a Big Bang followed by inflation, there immediately arise a number of obvious questions:

- What caused the Big Bang?
- If the Big Bang is a quantum-dominated event in which space and time have no certain definition, how does the universe ever settle down to a classical spacetime described by Einstein's general theory of relativity in time for inflation to begin?
- Even if the universe manages to settle into a classical spacetime, why should it do so in the exponentially fine-tuned way required for inflation to begin?
- Why don't we observe large-amplitude gravitational waves? If some way were found for inflation to begin, one would expect that the same high-energy inflationary processes that produced fluctuations in temperature and density also generated gravitational waves with large enough amplitude to have been detected by now.
- How does inflation end? The current idea is that inflation is eternal and that it transforms the universe into a "multi-mess" consisting of infinitely many patches or universes that can have any conceivable properties with no principle to determine which is more probable.

All these questions, which have been known for decades and which theorists have failed to answer despite best efforts, immediately become moot if the Big Bang is replaced by a Big Bounce. The universe need not ever be dominated by quantum physics, and the large-scale structure of the universe can be explained by a non-inflationary process occurring during the period of contraction leading up to the Bounce. This includes avoiding the multi-mess and producing fluctuations in temperature and density without producing large-amplitude gravitational waves or isocurvature fluctuations that would conflict with observations.

If that's the case, then why haven't astrophysicists and cosmologists jettisoned the Big Bang and embraced the Big Bounce? The answer in part is that many astrophysicists and cosmologists understand inflation as it was first introduced in the 1980s, when it was sold as a cure-all, and do not appreciate how thorny the questions I just listed really are. But perhaps the bigger reason is that before 2016 there was no theory of the Bounce itself, and so no Big Bounce theory to compare to. Attempts to construct examples of Bounces consistent with quantum physics and general relativity typically led to instabilities and mathematical pathologies that made them implausible. Some theorists even believed they could prove that Bounces are impossible.

2016 was the Year of the Big Bounce because, depending on how one counts them, at least four different theories for producing stable, nonpathological Bounces were introduced by different groups around the world. Each uses different sets of reasonable assumptions and principles, and each suggests that a smooth transition from contraction to expansion is possible. This isn't the place to go into details, but let me briefly describe one specific case, discovered by Anna Ijjas and me at the Princeton Center for Theoretical Science in 2016. In this theory of the Big Bounce, quantum physics is always a minor player, even near the Bounce. At each moment, the evolution of the universe is well described by classical equations that are well defined and can be solved on a computer using the same sort of tools of numerical general relativity that were introduced to study mergers of black holes and employed in the recent discovery of gravitational waves by the LIGO collaboration. With this approach, the Big Bounce becomes calculable and prosaic.

And once one knows that a Big Bounce is possible, it's hard to go back to considering the Big Bang again. The notion that time has a beginning was always strange, and, as the preceding list of questions illustrates, it has created more problems in explaining the universe than it has solved.

The time is ripe for the Big Bounce to become the new meme of cosmology.

AFFORDANCES

DANIEL C. DENNETT
Philosopher; Austin B. Fletcher Professor of Philosophy, Codirector, Center for Cognitive Studies, Tufts University; author, *From Bacteria to Bach and Back*

Psychologist James J. Gibson introduced the term *affordance* way back in the seventies. The basic idea is that the perceptual systems of any organism are designed to "pick up" the information relevant to its survival and ignore the rest. The relevant information is about *opportunities* "afforded" by the furnishings of the world: Holes afford hiding in, cups afford drinking out of, trees afford climbing (if you're a child or a monkey or a bear, but not a lion or a rabbit), and so forth. Affordances make a nicely abstract category of behavioral options that can be guided by something other than blind luck—in other words, by information extracted from the world. Affordances are what the environment "offers the animal . . . either for good or ill," according to Gibson, and "the information is in the light." (Gibson, like most psychologists and philosophers of perception, concentrated on vision.) While many researchers and theoreticians in the fledgling interdisciplinary field of cognitive science found Gibson's basic idea of affordances compelling, Gibson and his more radical followers managed to create a cultlike aura around his ideas, repelling many otherwise interested thinkers.

The huge gap in Gibson's perspective was his refusal even to entertain the question of how this "direct pickup" of information was accomplished by the brain. As a Gibsonian slogan put it, "It's not what's in your head; it's what your head is in." But as a revisionary description of the *purpose* of vision and the other senses, and a redirection of theorists' attention from retinal *images* to three-dimensional optical *arrays* (the information is in the light) through which the organism moves, Gibson's idea of affordances has much to recommend it—and we'll do a better job of figuring out how the neural machinery does its jobs when we better understand the jobs assigned to it.

Mainly, Gibson helps us move away from the halfway-only theories of consciousness that see the senses as having completed their mission once they've created "a movie playing inside your head," as David Chalmers has put it. There sometimes seems to be such a movie, but there is no such movie, and if there were, the task of explaining the mind would have to include explaining how this inner movie was perceived by the inner sense organs that then went about the important work of the mind: helping the organism discern and detect the available opportunities— the affordances—and acting appropriately on them. To identify an affordance is to have achieved access to a panoply of expectations that can be exploited, reflected upon (by us and maybe some other animals), used as generators of further reflections, etc. Consciousness is still a magnificent set of puzzles but appears as less of a flatfooted mystery when we think about the fruits of cognition with Gibson's help. The term is growing in frequency across the spectrum of cognitive science, but many users of the term seem to have a diminished appreciation of its potential.

ENACTIVISM

AMANDA GEFTER

Writer on physics and cosmology; consultant, *New Scientist* magazine;
author, *Trespassing on Einstein's Lawn*

What exactly do brains do? The usual answer is that they form mental representations of the world on the far side of the skull. Brains, that is, create internal virtual worlds—their best or most useful simulations of the real external world, one that exists independently of any of them but within which they all reside.

The problem is that fundamental physics denies the existence of this observer-independent world. From quantum physics in the early 20th century to the black-hole firewall debate raging today, physicists have found that we tangle ourselves in paradox and violate laws of physics when we attempt to compile multiple viewpoints into a single spacetime. The state of a physical system, we've learned, can only be defined relative to a given observer. (Here "observer" doesn't mean consciousness but a physical system able to act as a measuring device—yet one that itself must enjoy only a relational existence.) Slices of spacetime accessed by different observers cannot be considered broken shards of a single, shared world but, rather, as self-contained and incommensurable versions of reality, each a universe unto itself.

In other words, there's no third-person view of the world. There's one world per observer, and no more than one at a time.

What happens, then, to the concept of representation? What is it that brains are doing if there's no observer-independent world out there for them to represent?

149

One possibility is that instead of representing the world, brains *enact* one.

The term "enactivism" was introduced by philosopher/neuroscientist Francisco Varela and colleagues in the 1990s, but it has taken on fresh significance in light of an emerging set of ideas at the forefront of cognitive science, including embodied cognition, the Bayesian brain, active inference, and the free-energy principle—ideas emphasizing top-down, generative perception wherein observers actively shape the worlds they perceive. The old passive view of perception is giving way to an active one, just as the Newtonian observer was replaced by the participator of modern physics. Suddenly the words of physicist John Archibald Wheeler apply to cognitive science: "We used to think that the world exists out there, independent of us, we the observer safely hidden behind a one-foot-thick slab of plate glass, not getting involved, only observing. However, we've concluded that that isn't the way the world works. We have to smash the glass, reach in."

According to enactivism, observer and world co-evolve, hoisting each other up by their bootstraps through reciprocal interaction. Perception and action are inextricably and cyclically linked: Our perceptions guide our actions, and our actions determine what we perceive. Their symmetry ensures an even more essential symmetry, that between observer and world. The observer's actions form the world's perceptions, and the world's actions form the observer's perceptions. Labels such as "observer" and "observed," "inside" and "outside," thus become profoundly interchangeable, removing any need to resort to mystical, magical, or hopelessly vague talk about consciousness as something over and above the physical world.

What the enactivist perspective leaves us with, ontologically speaking, is an observer and world that exist relative to each other and not in any absolute way. The enacted world cannot be described from a third-person perspective. It's observer-dependent, rendered in the first person, no more than one at a time. Which, of course, is exactly the ontology prescribed by fundamental physics.

Does it really matter if cognitive science aligns with fundamental physics? For the everyday purposes of research and practice in neuroscience, representation works well enough, just as our belief in a single, shared physical world mostly suffices, whether we're driving to work or launching rockets to the moon. It's only when we push to the outermost edges of either discipline—when we deal with tiny distances or intense gravity in physics, or when we ask about the fundamental nature of consciousness in cognitive science—that the cracks in the third-person perspective begin to show and a more fundamental theory is needed.

Fundamental physics and cognitive science have long been embroiled in a perpetual game of chicken or egg: The brain arises from the physical world and yet everything we know of the physical world we know only through the brain. Each serves as foundation for the other. As long as we restrict our inquiry to one at a time, something critical is left unexplained. If we want to understand reality as a whole, we need to understand both sides of the coin and how they're fused together. Physics finds that the world is observer-dependent but remains silent on the nature of the observer. Cognitive science finds that the observer is an active participant but never questions the nature of the physical world. Enactivism might just be the concept we need to begin to piece the two sides together.

PALEONEUROLOGY

JUAN ENRIQUEZ

Managing Director, Excel Venture Management; co-author (with Steve Gullans), *Evolving Ourselves*

On a brisk May day in 1967, Tilly Edinger crossed a peaceful, leafy Cambridge, Massachusetts, street for the last time. Ironic, given that she had survived challenge after challenge. In the 1920s, after becoming a paleontologist against her father's, and the profession's, wishes, Edinger and a few others began systematically measuring the fossilized heads of various animals and human ancestors. The idea was to understand the evolution of the cranial cavity and attempt to infer changes in brain anatomy, thus birthing paleoneurology. Then she lost everything, fleeing Frankfurt just after *Kristallnacht*. As Hitler wiped out most of her relatives, she painstakingly rebuilt her life in the U.S. Although a lifelong friend and correspondent of Einstein, she led a somewhat reclusive existence, and, as occurred with Rosalind Franklin, she was somewhat underestimated and unappreciated. On May 6th, she left Harvard's Museum of Comparative Zoology and, having lost most of her hearing in her teenage years, she never heard the car coming. Thus ends the first chapter of the field of paleoneurology.

Once the initial, crude skull-measuring methods, such as sand- and water-displacement, were established, and after a few fossils were measured, there wasn't a whole lot of rapid progress. Few scholars bet their careers on the new field, ever fewer graduate students signed up. Progress was fitful. Gradually, measurements improved. Liquid latex, Dentsply, and plasticine gave way to exquisite 3D computer scans. But the field remained largely a data desert. One leading scholar, Ralph Holloway, estimated the entire global collection of hominid measurable skulls at around 160—that is, "one brain endocast for every 235,000+ years of evolutionary time." Lack of data eventually led to a civil war over measurement methods and conclusions, waged between two of the core leaders of paleoneurology, Professors Dean Falk and Holloway.

Nevertheless, the fundamental question paleoneurology seeks to address—"How do brains change over time?"—goes straight to the core of why we're human. Now as various technologies develop we may be able to get a whole lot savvier about how brains have changed. Ancient DNA and full genomes are beginning to fill in some gaps. Some even claim you can use genomes to predict faces. Perhaps soon we could get better at partially predicting brain development just from sequence data. And, alongside new instruments, Big Data emerging from comparative neurology and developmental neurology experiments provide many opportunities to hypothesize answers to some of the most basic lagoons in paleoneurology. Someday we may even revive a Neanderthal and find out why, given their bigger brains and likely at least comparable intelligence, they didn't survive.

There's a second, more fundamental reason that paleoneurology might become a common term. Brains once changed slowly. That's no longer the case. The comparative study of brains over time becomes ever more relevant as we place incredible evolutionary pressures on the most malleable of our organs.

How we live, eat, absorb information, and die have all become radically different: a daytime hunter-gatherer species became a mostly dispersed, settled, agricultural species. And then, in a single century, we became mostly an urban species. Studying rapid changes in brains, animal and human, gives us a benchmark to understand how changes occurred over thousands of years, then hundreds of years, and even over the past few decades.

Paleoneurology should retool itself to focus on changes occurring in far shorter time spans, on the rapid rewiring that can result in explosions of autism, on the effects of drastic changes in diet, size, and weight. We need a historic context for the evolution occurring as our core brain inputs shift from observing nature to reading pages and then digital screens. We have to understand what happens when brains that evolved around contemplation, observation, and boredom interrupted by sudden violence, are now bombarded from every direction as our phones, computers, tablets, TVs, tickers, ads, and masses of humans demand an immediate assessment and response. We're de-facto outsourcing and melding parts of our memories with external devices, like our PDAs.

What remains a somewhat sleepy, slow-moving field should take up the challenge of understanding enormous change in short periods of time. It's possible that a 1950s brain, for better and worse, might look Jurassic when compared, on a wiring and chemical level, with a current brain. The same might be true for animals, like the once shy pigeons that migrated from farms into cities and became aggressive pests.

Given that one in five Americans is now taking a mind-altering drug, the experiment continues and accelerates. Never mind fields like optogenetics, which alters and reconnects our brains, thoughts, memories, fears, using light stimuli projected inside the brain. Eventual implants will radically change brain design, inputs, and outputs. Having a baseline, a good paleo-neurological history of brain design, for ourselves and many other basic species, may teach us a lot about what our brains were and where they came from, but even more important, what they're becoming. I bet this is a challenge Tilly Edinger would have relished.

COMPLEMENTARITY

FRANK WILCZEK
Theoretical physicist, MIT; Nobel laureate; author, *A Beautiful Question: Finding Nature's Deep Design*

Complementarity is the idea that there can be different ways of describing a system, each useful and internally consistent, which are mutually incompatible.

Complementarity first emerged as a surprising feature of quantum theory, but I, following Niels Bohr, believe it contains wisdom that's much more widely applicable.

Here's how it works in quantum theory: The ultimate description of a system is its wave function, but the wave function isn't something we can observe directly. It's like a raw material that must be processed and sculpted in order to build something usable (i.e., observable). There are many things we might choose to build from a lode of iron ore—a sword or a plowshare, for example. Both might be useful, in different circumstances. But either uses up the raw material and precludes building the other. Similarly, we can process the wave function of a particle to predict things about its position, or alternatively to predict things about its velocity, but not both at the same time (Heisenberg's uncertainty principle).

Quantum theory is complicated, but reality as a whole is even more complicated. In dealing with it, people can and do take many different approaches: scientific, legal, moral, artistic, religious, etc. Each of those approaches can be useful or rewarding in different circumstances. But they involve processing the full complexity of reality in radically different ways that are often deeply incompatible (for an example, see below). Complementarity is the wisdom to recognize that fact and to welcome it. Walt Whitman, at the apex of his *Song of Myself*, embraced the spirit of complementarity, crowing, "Do I contradict myself? Very well then, I contradict myself. I am large, I contain multitudes."

An important example of complementarity arises around the concept of legal responsibility. Generally speaking, we don't hold children or insane people responsible for otherwise criminal acts, because they cannot control their behavior. Yet science, on the face of it, suggests that human beings are physical objects whose behavior is fully determined by physical laws. And that's a useful perspective to take if we want to design reading glasses or drugs, for example. But from that perspective, nobody really controls their behavior. The point is that the scientific description of human beings, in terms of the wave function of the quarks, gluons, electrons, and photons making them up, is not a useful way to describe the way they act. The perception that we exercise will and make choices is based on a coarser but more usable description of humans and their motivations, which comes naturally to us and guides our legal and moral intuitions.

Understanding the importance of complementarity stimulates imagination, because it gives us license to think differently. It also suggests engaged tolerance, as we try to appreciate apparently strange perspectives that other people have come up with. We can take them seriously without compromising our own understandings, scientific and otherwise.

THE SCHNITT

MICHAEL GAZZANIGA

Neuroscientist; Director, SAGE Center for the Study of the Mind, UC
Santa Barbara; author, *Tales from Both Sides of the Brain*

For all our scientific efforts, there remains a gap: the *Schnitt*.
The gap between the quantum and the classical, the living and
the nonliving, the mind and the brain. How on earth is science
going to close this gap—what the physicist Werner Heisenberg
called the *Schnitt* and what the theoretical biologist Howard
Pattee calls the "epistemic cut"? Some maintain that the gaps
only reflect a current failure of knowledge. Others think the
gaps will never be closed—that they are, in fact, unclosable.

Pattee, who has been working at the problem for fifty years,
believes he has a handle on it. I think he does, too. The clue to
grasping his idea goes all the way back to understanding the difference between nonliving and living systems. To understand this
difference, biologists need to fully embrace a gift from modern
physics, the idea of complementarity.

Accepting this unexpected gift isn't easy. Einstein himself
wouldn't accept it until Niels Bohr forced him to. The discovery
of the quantum world meant that the classical world of physics
had a new partner that had to be considered when explaining
stuff. Suddenly, the world of reversible time, the notion of dogmatic determinism, and the aspiration to a grand theory of the
universe were on the rocks. Bohr's idea, the principle of complementarity, maintains that quantum objects have complementary (paired) properties, only one of which can be exhibited and

measured—and thus known—at a given point in time. That's a big blow to a scientist, and to physicists it's perhaps the most devastating. As theoretical biologist Robert Rosen pointed out, "Physics strives, at least, to restrict itself to 'objectivities.' It thus presumes a rigid separation between what is objective, and falls directly within its precincts, and what is not. . . . Some believe that whatever is outside is so because of removable and impermanent technical issues of formulation. . . . Others believe the separation is absolute and irrevocable."

This is where Howard Pattee picks up the story. Pattee argues that complementarity demands that life be seen as a layered system in which each layer has (indeed, demands) its own vocabulary. On one side of the *Schnitt*, there's the firing of neurons. On the other, there are symbols, the representations of the physical, which also have a physical reality. Only one side of the *Schnitt* can be evaluated at a time, although both are real and physical and tangible. Here we have Bohr's complementarity on a larger scale—two modes of mutually exclusive description making up a single system from the get-go.

There's no spook in the system introduced here, and Pattee calls on the venerable mechanisms of DNA to make his point. DNA is a primeval example of symbolic information (the DNA code) controlling material function (the action of the enzymes), just as John von Neumann had predicted must exist for evolving, self-reproducing automatons. However, it's also the old chicken-and-egg problem, with fancier terms: Without enzymes to break apart the DNA strands, DNA is simply an inert message that cannot be replicated, transcribed, or translated. Yet without DNA, there would be no enzymes!

Staring us right in the face is a phenomenal idea. A hunk of molecules—molecules that have been shaped by natural selection—makes matter reproducible. These molecules, which can be stored and recalled, are a symbol, a code for information that describes how to build a new unit of life. Yet those same molecules also physically constrain the building process. DNA is both a talker and a doer, an erudite outdoorsman. There are two realities to this thing. Just as light is a wave and a particle at the same time. Information and construction, structure and function, are irreducible properties of the same physical object, existing in different layers with different protocols.

While I find this a tricky idea, it's utterly simple and elegant. Pattee has given us a schema and a way to think about how, using nothing but physics (including the principle of complementarity), life comes out of nonliving stuff. The schema, way up the evolutionary scale, also accounts for how the subjective mind can emerge from objective neurons. Pattee suggests that instead of approaching conscious cognition as either information processing or neural dynamics, there's a third approach available. Consciousness is not reducible to one or the other. Both should be kept on the table. They're complementary properties of the same system. I say, "Hats off to Pattee!" We brain scientists have our work cut out for us.

MATTER

HANS HALVORSON
Stuart Professor, Department of Philosophy, Princeton University

The concept of matter ought to be more widely known.

You might wonder whether I misunderstood the *Edge* Question? Did I think it was asking, "What scientific term or concept is *already* widely known?" For if there's any scientific concept that's widely known, it's the concept of matter—that stuff from which all things are made.

But no, I didn't misread the question. While every intelligent person has heard of matter, few people know the scientific meaning of the word. What we have here is an example of a concept that was first used in ordinary life but has come to be explicated in the development of science.

So, what does science tell us about matter?

As you probably know, there's an age-old debate about whether things are ultimately made out of particles or are excitations or waves in some continuous medium. (In fact, the philosopher Immanuel Kant found this debate so tedious he declared it irresolvable in principle.) But many of us were told that the wave/particle debate was solved by quantum physics, which says that matter has both particle-like and wave-like aspects.

Things got a bit weird when Niels Bohr said, "There is no quantum reality," and when Eugene Wigner said, "There is no reality without an observer." What the hell is going on here? Has matter disappeared from physics? Has physics really told us that mind-independent matter doesn't exist?

Thank goodness for the renegades of physics. In the 1960s, people like John Bell and David Bohm and Hugh Everett said, "We don't buy the story being told by Bohr, Wigner, and their ilk. In fact, we find this talk of 'observer-created reality' to be confusing and misleading." These physicists then went on to argue that there is a quantum reality and that it exists whether or not anyone's there to see it.

But what is this quantum reality like? It's here that we have to stretch our imagination to the breaking point. It's here that we have to let science expand our horizon far beyond what our eyes and ears can teach us.

To a first approximation, what really exists, at the very

bottom, is quantum wave functions. But we must be careful not to confuse an assertion of mathematical existence with an assertion of physical existence. A quantum wave function is a mathematical object—a function that takes numbers as inputs and spits out numbers as outputs. Thus, to speak accurately, we ought to say that a quantum wave function *represents* matter, not that it *is* matter. But how does it accomplish that representation? In other words, what are the things that exist, and what properties are being attributed to them? It's at this point that the situation becomes a bit unclear, a bit scholastic. There are many questions we could ask about what it means to say that wave functions exist. But what's the use, because quantum mechanics isn't really true.

We now know that quantum mechanics is not literally true—at least, not if Einstein's relativity theory is true. In the middle of the 20th century, physicists saw that if you combine relativity with quantum mechanics, then wave functions cannot be localized; the result is that strictly speaking there aren't any localized material objects. What there are, they said, are quantum fields—nebulous quantum entities spreading themselves throughout all of space.

But don't get too excited about quantum fields, because they have their own problems. It was already suspected in the 1960s that quantum fields aren't quantum reality in itself; rather, they're a sort of observer-dependent description of that reality, in the same way that saying a car is moving at 45 miles per hour is an observer-dependent description of reality. In fact, it was proved by the German physicist Hans-Jürgen Borchers that many distinct and incompatible quantum-field descriptions correspond to any one situation. A similar result has recently been demonstrated by the Michigan philosopher David Baker. The upshot is that you've got to take quantum fields with a grain of salt:

They're a human contrivance that gives just one perspective on reality.

In summary, particles, in the traditional sense of the word, do not exist. Nor do quantum wave functions really exist. Nor do fields exist, either in the traditional sense of the word or in the quantum-theoretic sense of the word.

These facts can seem depressing. Apparently matter, in itself, is always hiding behind the veil of our descriptions of it.

But don't despair. Note what's been happening here. The description of matter as particles was helpful but not exactly correct. The description of matter as a wave function is even more accurate but has limitations. Our best current description of matter is in terms of quantum fields, but the quantum fields aren't yet the thing itself.

At each stage, our description of matter has become more nuanced, more widely applicable, and more useful. Will this process come to an end? Will we ever arrive at the *one true* description of the basic constituents of the universe?

Who's to say? But as long as each generation outdoes the previous one, what more could we want?

SUBSTRATE INDEPENDENCE

MAX TEGMARK
Theoretical physicist, cosmologist, MIT; Scientific Director,
Foundational Questions Institute; Cofounder, Future of Life Institute;
author, *Our Mathematical Universe*

What do waves, computations, and conscious experiences have in common that provides crucial clues about the future of intelligence? They all share an intriguing ability to take on a life of their own, independent of their physical substrate.

Waves have properties such as speed, wavelength, and frequency, and we physicists can study the equations they obey without needing to know what substance they're waves in. When you hear something, you're detecting sound waves caused by molecules bouncing around in the mixture of gases we call air, and we can calculate all sorts of interesting things about these waves—how their intensity fades as the square of the distance, how they bend when they pass through open doors, how they reflect off walls and cause echoes, etc.—without knowing what air is made of.

We can ignore all details about oxygen, nitrogen, carbon dioxide, etc., because the only property of the wave's substrate that matters, and enters into the famous wave equation, is a single number that we can measure: the wave speed, which in this case is about 300 meters per second. Indeed, this wave equation, which MIT students are now studying, was first discovered and put to great use long before physicists had established that atoms and molecules even existed!

Alan Turing famously proved that *computations* are substrate-independent as well: There's a vast variety of different computer architectures that are "universal" in the sense that they can all perform the exact same computations. So if you were a conscious, superintelligent character in a future computer game, you'd have no way of knowing whether you ran on a desktop, a tablet, or a phone, because you would be substrate-independent.

Nor could you tell whether the logic gates of the computer were made of transistors, optical circuits, or other hardware, or even what the fundamental laws of physics were. Because of this substrate independence, shrewd engineers have been able to repeatedly replace the technologies inside our computers with significantly better ones without changing the software—making computation twice as cheap roughly every couple of years for over a century, cutting computer cost a whopping million million million times since my grandmothers were born. It's precisely this substrate independence of computation that implies that artificial intelligence is possible: Intelligence doesn't require flesh, blood, or carbon atoms.

These examples illustrate three important points.

First, substrate independence means not that a substrate is unnecessary but that most of its details don't matter. You obviously can't have sound waves in a gas if there's no gas, but any gas whatsoever will suffice. Similarly, you obviously can't have computation without matter, but any matter will do, as long as it can be arranged into logic gates, connected neurons, or some other building block enabling universal computation.

Second, the substrate-independent phenomenon takes on a life of its own, independent of its substrate. A wave can travel across a lake, even though none of its water molecules do—they mostly bob up and down.

Third, it's often only the substrate-independent aspect that we're interested in: A surfer usually cares more about the position and height of a wave than about its detailed molecular composition, and if two programmers are jointly hunting a bug in their code, they're probably not discussing transistors.

Since childhood, I've wondered how tangible physical stuff such as flesh and blood can give rise to something that feels as intangible, abstract, and ethereal as intelligence and consciousness. We've now arrived at the answer: These phenomena feel so nonphysical because they're substrate-independent, taking on a life of their own that doesn't depend on or reflect the physical details. We still don't understand intelligence to the point of building machines that can match all human abilities, but AI researchers are striking ever more abilities from their can't-do list—from image classification to Go-playing, speech recognition, translation, and driving.

But what about consciousness, by which I mean simply "subjective experience"? When you're driving a car, you're having a *conscious experience* of colors, sounds, emotions, etc. But why are you experiencing anything at all? Does it feel like anything to be a self-driving car? This is what philosopher David Chalmers calls the "hard problem," and it's distinct from merely asking how intelligence works.

I've been arguing for decades that consciousness is the way information feels when being processed in certain complex ways. This leads to a radical idea I really like: If consciousness is the way information feels when it's processed in certain ways, then it must be substrate-independent; *it's only the structure of the information-processing that matters, not the structure of the matter doing the information-processing.* In other words, consciousness is substrate-independent twice over!

We know that when particles move around in spacetime in patterns obeying certain principles, they give rise to substrate-independent phenomena—e.g., waves and computations. We've now taken this idea to another level: *If the information-processing itself obeys certain principles, it can give rise to the higher-level substrate-independent phenomenon we call consciousness.* This places your conscious experience not one but two levels up from the matter. No wonder your mind feels nonphysical! We don't yet know what principles information-processing needs to obey to be conscious, but concrete proposals have been made that neuroscientists are trying to test experimentally.

However, one lesson from substrate independence is already clear: We should reject carbon chauvinism and the common view that our intelligent machines will always be our unconscious slaves. Computation, intelligence, and consciousness are patterns in the spacetime arrangement of particles—patterns that take on a life of their own. It's not the particles but the patterns that really matter! Matter doesn't matter.

PT SYMMETRY

DANIEL HOOK
Theoretical physicist; Managing Director, Digital Science; Visiting Professor, Department of Physics, Washington University in St. Louis

When Paul Dirac formulated the postulates of quantum theory, he required hermiticity to be the fundamental symmetry for his equations. For Dirac, it was the mathematical device he needed to ensure that all predictions for the outcomes of real-world measurements of quantum systems resulted in a real number. This is important, because we observe only real outcomes in actual experimental observations. Dirac's choice of hermiticity as the fundamental symmetry of quantum theory was not seriously challenged for around seventy years.

Hermiticity is a subtle and abstract symmetry that is mathematical in its origin. Broadly speaking, the requirement of hermiticity imposes a boundary on a system. This is an idealization in which a system is isolated from any surrounding environment (and hence cannot be measured). While this gives a tractable mathematical framework for quantum theory, it's an unphysical requirement, since all systems interact with their environment—and if we wish to measure a system, then such an interaction is necessary.

In 1998, Carl Bender and Stefan Boettcher wrote a paper exploring the replacement of hermiticity with another symmetry. They showed they could replace the mathematically motivated symmetry of Dirac by a physically motivated symmetry preserving the reality of experimental outcomes. Their new theory had interesting new features. It was not a like-for-like replacement.

The underlying symmetry that Bender and Boettcher found was what they called PT symmetry. The symmetry here is geometric in nature and is hence closer to physics than hermiticity is. The "P" stands for "parity" symmetry, sometimes called mirror symmetry. If a system respects "P" symmetry, then the evolution of the system wouldn't change for a spatially reflected version of the system. The "T" stands for "time-reversal." Time-reversal symmetry is just as it sounds—a physical system respecting this symmetry would evolve in the same way regardless of whether time runs forward or backward. Some systems individually exhibit P or T symmetry, but it's the combination of the two symmetries that seems to be fundamental to quantum theory.

Instead of describing a system in isolation, PT symmetry describes a system in balance with its environment. Energy may flow in and out of the system; hence, measurements can be made within the theoretical framework of a system described by a PT symmetry. The requirement is that the same amount of energy flowing in must also flow out of the system.

This subtler definition of a system's relationship with its environment, provided by PT symmetry, lets us describe a much wider class of systems in mathematical terms—leading not only to an enhanced understanding of these systems but also to experimental results supporting the choice of PT as the underlying symmetry in quantum mechanics. Several physical models for specific systems which had been studied and rejected because they didn't respect hermiticity have been reexamined and found to be PT symmetric.

It's remarkable that the study of PT symmetry has progressed so rapidly. For many areas of theoretical physics, the time lag between theory and experiment is on the order of several decades. We may never be able to fully test string theory, and experimental verification of the fifty-year-old theory of supersymmetry remains elusive.

In the nineteen years since Bender and Boettcher's 1998 paper, experimentalists have created PT lasers, PT superconducting wires, PT NMR, and PT diffusion experiments—to mention just a few validations of their theory. As PT symmetry has matured, it has inspired the creation of exotic metamaterials that have properties allowing us to control light in new ways. The academic community, initially skeptical of such a fundamental change in quantum theory, has warmed to the idea of PT symmetry. Over 200 researchers from around the world have published scholarly papers on PT symmetry. The literature now extends to more than 2,000 articles, many in top journals such as *Nature, Science,* and *Physical Review Letters.*

The future is bright for PT-symmetric quantum mechanics, but there's still work to be done. Many of the experiments mentioned have quantum-mechanical aspects but aren't full verifications of PT quantum mechanics. Nevertheless, existing experiments are already leading to exciting results. PT is a hot topic in the optics and graphene communities, and the idea of creating a computer based on optical rather than electronic principles has recently been suggested. At the beginning of the 21st century, we're finding a new understanding of quantum theory with the potential to unlock new technologies, just as semiconductor physics was unlocked by the rise of quantum mechanics 100 years ago.

GRAVITATIONAL LENSING

PRIYAMVADA NATARAJAN

Theoretical astrophysicist, Yale University; author, *Mapping the Heavens*

> Every day you play with the light of the universe.
> Subtle visitor, you arrive in the flower and the water . . .

These lines from one of Pablo Neruda's poems capture the essence of light bending—gravitational lensing—which is ubiquitous in the cosmos. Reconceptualizing gravity in his general theory of relativity, Einstein postulated the existence of spacetime, a four-dimensional sheet to describe the universe. This is a beautiful marriage between geometry and physics, wherein all matter, both ordinary and exotic, in the universe would cause divots in the fabric of spacetime. So matter would dictate how spacetime curved, and spacetime, in turn, would determine how matter moves.

One important consequence of this formulation is the effect that the rumpled fabric of spacetime has on light propagation in the universe. Light emitted by distant galaxies is deflected by the divots generated by the mass it encounters en route. This phenomenon of the bending of light is referred to as gravitational lensing.

The consequence of lensing is that we see the shapes of distant galaxies systematically distorted; they appear more elongated than their true shapes. The strength of the lensing distortion is directly proportional to the number and depth of divots encountered—namely, the detailed distribution of matter along the line of sight across cosmic distances. Also, location matters; the ideal configuration for gravitational lensing is when the distant galaxy emitting the light is perfectly lined up behind the foreground cluster of galaxies causing the deflection of light rays. The strength of the light bending therefore also depends on the geometrical properties of spacetime, which are encapsulated in a set of cosmological parameters characterizing our universe.

Gravitational lensing is somewhat similar to the optical focusing and de-focusing produced by convex or concave glass lenses, which we're all familiar with from high school science experiments. Unlike the light bending produced by glass lenses, though, in gravitational lensing—occasionally, when the alignment is perfect—a single light beam might be cleaved in two, causing the appearance of a pair of images, when in reality the distant source of light is a single object. Production of multiple images of the same object is referred to as strong gravitational lensing. Most of the time, however, what we see are weak distortions in the shapes of distant galaxies.

Lensing was one of the key predictions of the general theory of relativity, proposed in 1915. It was proved in 1919, when light bending was detected during a solar eclipse. During such an eclipse, the divots generated by the sun and the Earth in spacetime line up, causing a measurable deflection in light from stars in the field. In this instance, there are no distortions or multiple images produced, since stars are point sources. But stars in the field have a displaced apparent position because of the curvature of spacetime during the line-up.

General relativity accurately predicted the displacement between the real and apparent positions. It was the verification of this prediction by the British astronomer Arthur Eddington that made Einstein a celebrity and a household name. The more dramatic predictions of gravitational lensing—the distortions in the shapes of lensed distant galaxies and the production of multiple images of a distant object—have been verified by data from ground-based telescopes and the Hubble Space Telescope. In fact, the most convincing evidence for the existence of copious amounts of dark matter in the universe comes from the lensing effects on the shapes of distant galaxies produced by the unseen matter.

THE COSMOLOGICAL CONSTANT, OR VACUUM ENERGY

RAPHAEL BOUSSO

Professor, Berkeley Center for Theoretical Physics, UC Berkeley

A hundred years ago, in 1917, Albert Einstein had a problem. He had just come up with a beautiful new theory of gravity called general relativity. But the theory predicted that the universe should either expand or contract. Even for Einstein, this was a bridge too far. The universe, though it might have had a beginning, certainly did not appear to be changing.

General relativity is a rigid theory; little of it can be changed without destroying its elegant mathematical structure. The only wiggle room was a single quantity, which Einstein called the *cosmological constant*. If he assumed this quantity was zero, the equations looked particularly simple, but they required the universe to be dynamical.

The gravity of ordinary matter (such as galaxies) is attractive. Introducing a positive cosmological constant adds a repulsive counterforce. By setting the cosmological constant to a particular nonzero value at which the two tendencies cancel each other, Einstein thought he could get his theory to spit out a static, unchanging universe.

Einstein later called this idea his "biggest blunder," an accurate assessment. First, his move doesn't actually accomplish the task: Einstein's static universe is unstable, like a pencil balanced on its tip. Because matter is distributed unevenly, the opposing forces couldn't possibly be arranged to balance out everywhere. Individual regions would soon begin to expand or contract. Worse, Einstein missed out on a spectacular prediction. Had he believed his own equations, he could have anticipated the 1929 discovery that galaxies are in fact receding from one another.

This dramatic turn of events proved that the universe was not static. But an expanding universe didn't imply that the cosmological constant was necessarily zero! As quantum field theory triumphed in the second half of the 20th century as a description of elementary particles, physicists recognized that the cosmological constant "wants to be there." The very theories predicting with unprecedented accuracy the behavior of small particles also implied that empty space should have some weight, or "vacuum energy." This kind of energy happens to be indistinguishable from a cosmological constant, as far as the equations of general relativity are concerned.

So it was no longer an option to set the cosmological constant to zero. Rather, it had to be calculated. Estimates indicated an enormous value—save for some unlikely precise cancellation between large positive and negative contributions from different particles. But a huge cosmological constant would show up as a repulsive force that would blow up the entire universe in a split second (or, alternatively, would cause it to collapse instantly, if the constant happened to come out negative). Evidently this was not what the universe was doing.

This drastic conflict between theory and observation is the "cosmological constant problem." It remains the most serious problem in theoretical physics. It has contributed to the development of revolutionary ideas, particularly the multiverse and the "landscape" of string theory.

The string landscape solves the cosmological constant problem by a strategy similar to throwing many darts randomly: Some will hit the bullseye by accident. In string theory there are many different ways of making empty space, and they would all get realized as vast regions in different parts of the universe. In some regions, the universe "hits the bullseye"—the cosmological constant is accidentally small. There, spacetime does not quickly explode or collapse, so structure and observers are more likely to evolve in these lucky regions.

A crucial prediction of this approach was pointed out by Steven Weinberg in 1987: If an accidental near-cancellation is the reason the cosmological constant is so small, then there's no particular reason for it to be exactly zero. Rather, it should be large enough to have a just-noticeable effect.

In 1998, astronomers did find such an effect. By observing distant supernovae, we can tell that the universe is not just expanding but accelerating (expanding ever more rapidly). Other observations, such as the history of galaxy formation and the present rate of expansion, have since provided independent evidence for the same conclusion. Empty space is filled with vacuum energy. In other words, there's a positive cosmological constant of a particular value, which we have now measured.

The cause of the acceleration is sometimes described more dramatically as a "mysterious dark energy." But in science we shouldn't embrace mystery where there is none. If it walks like a duck and quacks like a duck, we call it a duck. In this case, it accelerates the expansion and affects galaxy formation precisely like a cosmological constant, so we should call it by its name.

One of the most fascinating consequences of a positive cosmological constant is that we'll never see much more than the present visible universe. In fact, billions of years from today the most distant galaxies will begin to disappear, accelerated out of sight too far for light from them to reach us. Eventually our local group of galaxies will hover alone in a vast emptiness filled with nothing but vacuum energy.

INVARIANCE

JIM HOLT
Philosopher and essayist, *New York Times, The New York Review of Books, The New Yorker, Slate*; author, *Why Does the World Exist?*

Science is supposed to be about an objective world. Yet our observations are inherently subjective, made from a particular frame of reference, a point of view. How do we get beyond this subjectivity to see the world as it truly is?

Through the idea of *invariance*. To have a chance of being objective, our theory of the world must at least be intersubjectively valid: It must take the same form for all observers, regardless of where they happen to be located, or when they happen to be doing their experiment, or how they happen to be moving, or how their lab happens to be oriented in space (or whether they are male or female, or Martian or Earthling, or . . .). Only by comparing observations from all possible perspectives can we distinguish what is real from what is mere appearance or projection.

Invariance is an idea of enormous power. In mathematics, it gives rise to the beauties of group theory and Galois theory, since the shifts in perspective that leave something invariant form an algebraic structure known as a "group."

In physics, as Emmy Noether showed us with her beautiful theorem, invariance turns out to entail the conservation of energy and other bedrock conservation principles—"a fact," noted Richard Feynman, "that most physicists still find somewhat staggering."

And in the mind of Albert Einstein, the idea of invariance led first to $e = mc^2$ and then to the geometrization of gravity.

So why aren't we hearing constantly about Einstein's theory of invariance? Well, "invariant theory" is what he later said he wished he had called it. And that's what it should have been called, since invariance is its very essence. The speed of light, the laws of physics are the same for all observers. They're objective, absolute—invariant. Simultaneity is relative, unreal.

But no, Einstein had to go and talk about the "principle of relativity." So relativity—and not its opposite, invariance—is what his revolutionary theory ended up getting labeled. Einstein's "greatest blunder" was not (as he believed) the cosmological constant after all. Rather, it was a blunder of branding—one that has confused the public for over a century now and empowered a rum lot of moral relativists and lit-crit Nietzscheans.

Thanks, Einstein.

UNRUH RADIATION

JEREMY BERNSTEIN

Emeritus Professor of Physics, Stevens Institute of Technology; former staff writer, *The New Yorker*; author, *Quantum Leaps*

Many people have heard of Hawking radiation, which is a form of radiation emitted by a black hole. Less familiar is Unruh radiation, named after Canadian physicist W. G. Unruh, who first described it. It, too, is emitted by black holes.

Close to a black hole, the radiation is predominantly Unruh; farther away, it is predominantly Hawking. Unruh radiation is observed by a detector placed in a state of uniform acceleration, whereas if the same detector is at rest or in a state of uniform motion, no radiation is observed. The Unruh radiation, in the case of uniform acceleration, is like a black body with a temperature proportional to the acceleration.

The relevance to black holes is that close to a black hole, the spacetime geometry of a spherical black hole can be transformed to that of a detector moving with uniform acceleration. Hence Unruh radiation is detected. The theorist John Bell suggested that Unruh radiation might be observed in an electron storage ring. So far, the experiment has not been carried out. Unruh radiation is a novel prediction of the quantum theory of fields, which has always been a source of surprises.

DETERMINISM

JERRY A. COYNE

Professor Emeritus, Department of Ecology and Evolution, University of Chicago; author, *Faith Versus Fact: Why Science and Religion Are Incompatible*

A concept everyone should understand and appreciate is the idea of *physical determinism*—that all matter and energy in the universe, including what's in our brains, obey the laws of physics. The most important implication is that we have no "free will." At a given moment, all living creatures, including ourselves, are constrained by their genes and environment to behave in only one way—and could not have behaved differently. We *feel* as though we make choices, but we don't. In that sense, "dualistic" free will is an illusion.

This must be true from the first principles of physics. The brain, after all, is simply a collection of molecules following the laws of physics; it's a computer made of meat. That in turn means that given the brain's constitution and inputs, its output—our thoughts, behaviors, and "choices"—must obey those laws. There's no way we can step outside our minds to tinker with those outputs. And even molecular quantum effects, which probably don't affect our acts, can't possibly give us conscious control over our behavior.

Physical determinism of behavior is also supported by experiments that trick people into thinking they're exercising choice when they're really being manipulated. Brain stimulation, for instance, can produce involuntary movements, like arm-waving, that patients claim are really willed gestures. Or we can feel we're not being agents when we are, as with Ouija boards. Further, one can use fMRI brain scans to predict, with substantial accuracy, people's binary decisions up to ten seconds before they're even conscious of having made them.

Yet our feeling of volition—that we can choose freely, for instance, among several dishes at a restaurant—is strong. So strong that I find it harder to convince atheists that they don't have free will than to convince religious believers that God doesn't exist. Not everyone is religious, but all of us feel we could have made different choices.

Why is it important that people grasp determinism? Because realizing that we can't "choose otherwise" has profound implications for how we punish and reward people, especially criminals. It can also have salubrious effects on our thoughts and actions.

First, if we can't choose freely but are puppets manipulated by the laws of physics, then all criminals or transgressors should be treated as products of genes and environments that made them behave badly. The armed robber had no choice about whether to get a gun and pull the trigger. In that sense, *every* criminal is impaired. All of them, whether or not they know the difference between right and wrong, have the same excuse as those deemed "not guilty by reason of insanity."

Now, this doesn't mean we shouldn't punish criminals. We should—in order to remove them from society when they're dangerous, reform them so they can rejoin us, and deter others from apeing bad behavior. But we shouldn't imprison people as *retribution*—for making a "bad choice." And of course we should still reward people, because that rewires their brains, and those of onlookers, in a way that promotes good behavior. We can therefore retain the concept of *personal responsibility* for actions but reject the idea of *moral responsibility* which presumes that people can choose to do good or bad.

Beyond crime and punishment, how should the idea of determinism transform us? Well, understanding that we have no choices should create more empathy and less hostility toward others, once we grasp that everyone is the victim of circumstances over which they had no control. Welfare recipients couldn't have gotten jobs, and jerks had no choice about becoming jerks. In politics, this should give us more empathy for the underprivileged. And realizing we had no real choices should stave off festering regrets about things we wish we'd done differently. We couldn't have.

Many religions also depend critically on this illusory notion of free will. It's the basis, for instance, for the Christian belief that God sends people to Heaven or Hell based on whether they "choose" to accept Jesus as their savior. Also out the window is the idea that evil exists because it's an unfortunate but necessary by-product of the free will God gave us. We have no such will, and without it the Abrahamic religions dissolve into insignificance.

We should accept the determinism of our behavior because, though it may make us uncomfortable, it happens to be true—just as we must accept our own inevitable but disturbing mortality. At the same time, we should dispel the misconceptions about determinism that keep many from embracing it: that it gives us license to behave how we want, that it promotes lassitude and nihilism, that it means we can't affect the behavior of others, that embracing determinism will destroy the fabric of society by making people immoral. The fact is that our feeling of having free will and our tendency to behave well are so strong—probably partly ingrained by evolution—that we'll never feel like the meat robots we are. Determinism is neither dangerous nor dolorous.

There are some philosophers who argue that while we do behave deterministically, we can still have a form of free will, simply redefining the concept to mean things like "our brains are very complex computers" or "we *feel* we are free." But those are intellectual carny tricks. The important thing is to realize that we don't have any choice about what we do, and we never did. We can come to terms with this, just as we come to terms with our mortality. Though we may not like such truths, accepting them is the beginning of wisdom.

STATE

SCOTT AARONSON

David J. Bruton Centennial Professor of Computer Science, University of Texas at Austin; author, *Quantum Computing Since Democritus*

In physics, math, and computer science, the *state* of a system is an encapsulation of all the information you'd ever need to predict what it will do, or at least its probabilities to do one thing versus another, in response to any possible prodding of it. In a sense, then, the state is a system's "hidden reality," which determines its behavior beneath surface appearances. But in another sense, there's *nothing* hidden about a state—because any part of the state that never mattered for observations could be sliced off with Occam's razor, to yield a simpler and better description.

When put that way, the notion of "state" seems obvious. So why did Einstein, Alan Turing, and others struggle for years with the notion, on the way to some of humankind's hardest-won intellectual triumphs?

Consider a few puzzles:

- To add two numbers, a computer clearly needs an adding unit, with instructions on how to add. But then it also needs instructions for how to interpret the instructions. And it needs instructions for interpreting *those* instructions, and so on, ad infinitum. We conclude that adding numbers is impossible for any finite machine.
- According to modern ideas about quantum gravity, space might not be fundamental but, rather, emergent from networks of qubits describing degrees of freedom at the Planck scale. I was once asked: If the universe is a network of qubits, then where *are* these qubits? Isn't it meaningless, for example, for two qubits to be "neighbors" if there's no preexisting space for the qubits to be neighboring in?
- According to special relativity, nothing can travel faster than light. But suppose I flip a coin, write the outcome in two identical envelopes, then put one envelope on Earth and the other on Pluto. I open the envelope on Earth. Instantaneously, I've changed the state of the envelope on Pluto from "equally likely to say heads or tails" to

"definitely heads" or "definitely tails." (This is normally phrased in terms of quantum entanglement, but as we see, there's even a puzzle classically.)

The puzzle about the computer is a stand-in for countless debates I've had with non-scientist intellectuals. The resolution, I think, is to specify a state for the computer, involving the numbers to be added (encoded, say, in binary), and a finite control unit that moves across the digits, adding and carrying, governed by Boolean logic operations and ultimately by the laws of physics. You might ask, "What underlies the laws of physics themselves?" And whatever the answer, what underlies *that*? But those are questions for us. In the meantime, the computer works; everything it needs is contained in its state.

The question about the qubits is a cousin of many others; for example, if the universe is expanding, then what is it expanding into? These aren't necessarily bad questions. But from a scientific standpoint, a perfectly justified response is "You're proposing we tack something new onto the state of the world, such as a second space for 'our' space to live in or expand into. So would this second space make a difference to observation? If it never would, why not cut it out?"

The question about the envelopes can be resolved by noticing that your decision on Earth to open your envelope or not doesn't affect the probability distribution over envelope contents that would be perceived by an observer on Pluto. One can prove a theorem stating that an analogous fact holds even in the quantum case, and even if there's quantum entanglement between the envelope contents on Earth and on Pluto: Nothing you choose to do here changes the local quantum state (the so-called *density matrix*) over there. This is why, contrary to Einstein's worries, quantum mechanics is consistent with special relativity.

The central insight here—of equal importance to relativity, quantum mechanics, gauge theory, cryptography, artificial intelligence, and probably 500 other fields—could be summarized as "a difference that makes no difference is not a difference at all." This slogan might remind some readers of the early 20th-century doctrine of logical positivism, or of Karl Popper's insistence that a theory that ventures no falsifiable prediction is unscientific. For our purposes, though, there's no need to venture into the complicated debates about what exactly the positivists or Popper got right or wrong (or whether positivism is itself positivistic, or falsifiability falsifiable).

It suffices to concentrate on a simpler lesson: that, yes, there's a real world, external to our sense impressions, but we don't get to dictate from our armchairs what its state consists of. Our job is to craft an ontology around our best scientific theories, rather than the opposite. That is, our conception of "what's actually out there" always has to be open to revision, both to include new things that we've discovered can be distinguished by observation and to exclude things that we've realized can't be.

Some people find it impoverishing to restrict their ontology to the *state*—to that which suffices to explain observations. But consider the alternatives. Charlatans, racists, and bigots of every persuasion are constantly urging us to look beyond a system's state to its hidden essences, to make distinctions where none are called for.

Lack of clarity about the notion of "state" is even behind confusion over free will. Many people stress the fact that according to physics, your future choices are "determined" by the current state of the universe. But this ignores the fact that *regardless* of what physics had to say on the subject, the universe's current state could always be defined in such a way as to secretly determine future choices; indeed, that's exactly what so-called hidden-variable interpretations of quantum mechanics, such as Bohmian mechanics, do. To me, this makes "determination" an almost vacuous concept in these discussions, and actual predictability much more important.

State is my choice for a scientific concept that should be more widely known, because buried inside it, I think, is the whole scientific worldview.

PARALLEL UNIVERSES OF QUANTUM MECHANICS

FRANK TIPLER

Professor of Mathematical Physics, Tulane University; author, *The Physics of Christianity*

In 1957, a Princeton physics graduate student named Hugh Everett showed that the consistency of quantum mechanics required the existence of an infinity of universes parallel to our own. That is, there has to be a person identical to you reading this identical article right now, in a universe identical to ours. Further, since there have to be an infinite number of universes, thus there are an infinite number of people identical to you in them.

Most physicists—at least, most physicists who apply quantum mechanics to cosmology—accept Everett's argument. So obvious is Everett's proof for the existence of these parallel universes that Stephen Hawking once told me he considered their existence to be "trivially true." Everett's insight is the greatest expansion of reality since Copernicus showed us that our star was just one of many. Yet few people have even heard of parallel universes or thought about the philosophical and ethical implications of their existence. Kepler and Galileo emphasized that the Copernican revolution implied that humans and their planet Earth are important in the cosmos rather than being merely the "dump heap of the filth and dregs of the universe," to use Galileo's description of our standing in the Ptolemaic universe.

I'll mention only two implications of the existence of parallel universes which should be of general interest: the implications for the free-will debate and the implications for answering the question of why there is evil in the world.

The free-will question arises because the equations of physics are deterministic. Everything that you do today was determined by the initial state of all the universes at the beginning of time. But the equations of quantum mechanics say that although the future behavior of all the universes are determined exactly, it is also determined that in the various universes the identical yous will make different choices at each instant, and thus the universes will differentiate over time. Say you are in an ice-cream shop trying to choose between vanilla and strawberry. What is determined is that in one world you will choose vanilla and in another you will choose strawberry. But before the two yous make the choice, you two are exactly identical. The laws of physics assert that it makes no sense to say which one of you will choose vanilla and which strawberry. So before the choice is made, which universe you will be in after the choice is unknowable, in the sense that it is meaningless to ask.

To me, this analysis shows that we indeed have free will, even though the evolution of the universe is totally deterministic. If you think my analysis has been too facile—entire books can and have been written on the free-will problem—nevertheless, my simple analysis shows that these books are themselves too facile, because they never consider the implications of the existence of the parallel universes for the free-will question.

Another philosophical problem with ethical implications is the problem of evil: Why is there evil in the universe we see? We can imagine a universe in which we experienced nothing bad, so why is this evil-free universe not the universe we actually see? The German philosopher Gottfried Leibniz argued that we are actually in the best of all possible worlds, but this seems unlikely. If Hitler had never taken power in Germany, there would have been no Holocaust. Is it plausible that a universe with Hitler is better than a universe without him? The medieval philosopher Abelard claimed that existence was a good in itself, so that in order to maximize the good in reality, *all* universes, both those with evil in them and those without evil, have to be actualized. Remarkably, quantum mechanics says that the maximization of good as Abelard suggested is in fact realized.

Is this the solution of the problem of evil? I do know that many wonder "Why Hitler?" but no analysis considers the fact that—if quantum mechanics is correct—there is a universe out there in which he remained a house painter. No analysis of why evil exists can be considered reasonable unless it takes into account the existence of the parallel universes of quantum mechanics.

Everyone should know about the parallel universes of quantum mechanics!

THE COPERNICAN PRINCIPLE

MARIO LIVIO
Astrophysicist; author, *Why?: What Makes Us Curious*

Nicolaus Copernicus taught us in the 16th century that we are nothing special, in the sense that the Earth on which we live is not at the center of the solar system. This realization, which embodies a principle of mediocrity on the astrophysical scale, has become known as the *Copernican principle*.

In the centuries since Copernicus's discovery, it seems that his principle has significantly gained strength, through a series of steps demonstrating that our place in the cosmos is of lesser and lesser importance.

First, astronomer Harlow Shapley showed at the beginning of the 20th century that the solar system is not at the center of the Milky Way galaxy. It is in fact about two-thirds of the way out from the center. Second, astronomer Edwin Hubble showed that there exist galaxies other than the Milky Way. The most recent estimate of the number of galaxies in the observable universe gives the staggering number of 2 trillion. Third, recent estimates based on searches for extrasolar planets put the number of Earth-size planets in the Milky Way in the billions. A good fraction of those are even in the "Goldilocks" region (not too hot, not too cold) around their host stars—a region that allows liquid water to exist on a planetary surface. So we're not special in that respect, either.

Even the stuff we're made of—ordinary (that is, baryonic) matter—constitutes less than 5 percent of the cosmic energy budget. The rest is in the form of dark matter, which doesn't emit or absorb light (constituting about 25 percent) and dark energy, a smooth form of energy permeating all of space (about 70 percent). And if all that isn't enough, in recent years some theoretical physicists started speculating that our entire universe may be but one member in a huge ensemble of universes—a multiverse (another scientific concept we ought to get used to).

It seems, therefore, that the Copernican principle is operating on all scales, including the largest cosmological ones. Everybody should be aware of the Copernican principle, because it tells us that from a purely physical perspective we're just a speck of dust in the grand scheme of things.

This state of affairs may sound depressing, but from another point of view there's something extraordinarily uplifting about it. Notice that *every step along the way of the increasing validity of the Copernican principle also represents a major human discovery.* That is, each decrease in our physical significance was accompanied by a huge increase in our knowledge. The human mind expanded in this sense just as fast as the expansion of the known universe (or the multiverse). Copernican humility is therefore a good scientific principle to adopt, but at the same time we should keep our curiosity and passion for exploration alive and vibrant.

RHEOLOGY

MATTHEW PUTMAN
Applied physicist; Board Chairman, Pioneer Works; CEO, Nanotronics

The world is governed by scientific principles that are fairly well taught and understood. A concept that resonates throughout physical science is rheology, and yet there are so few rheologists in academia that an international symposium fills only a small conference room. *Rheo,* coming from the Greek "to flow," is primarily the study of how non-Newtonian matter flows. In practice, most rheologists look at how nanoparticles behave in complex compounds, often filled with graphene, silica, or carbon nanotubes. Rheology was key to the creation of tires and other polymer systems, but as new technologies incorporate flexible and stretchable devices for electronics, medical implants, and regenerative and haptic garments for virtual-reality experiences, rheologists will need to be as common as chemists.

The primary reason for the relative obscurity of such an important topic is the hidden complexity involved. The number of connections in the human brain has been discussed by neuroscientists and often invites a sense of awe and inspiration to reduce and unify. Despite some well-known and simple equations, a complex nano-filled elastomer is also nearly impossible to model. Before the dreams of today can be realized, experiments and theoretical models will need to converge in ways they don't now. Through rheology, van der Waals forces, semiconductivity, superconductivity, quantum tunneling, and other such properties of composite materials can be harnessed and exploited, rather than hindered. Rheology is the old, newly relevant transdisciplinary science.

THE PREMORTEM

RICHARD H. THALER

Father of behavioral economics; Director, Center for Decision Research, University of Chicago Graduate School of Business; author, *Misbehaving*

Before a major decision is taken—say, to launch a new line of business, write a book, or form a new alliance—those familiar with the proposal's details are given an assignment. Assume we are at some time in the future when the plan has been implemented and the outcome was a disaster. Write a brief history of that disaster.

Applied psychologist Gary Klein came up with "the Premortem," which was later written about by Daniel Kahneman. Of course, we're all too familiar with the more common "postmortem" that typically follows any disaster along with the accompanying finger-pointing. Such postmortems inevitably suffer from hindsight bias, also known as Monday-morning quarterbacking, in which everyone remembers thinking that the disaster was almost inevitable. As I often heard Amos Tversky say, "The handwriting may have been on the wall all along. The question is: Was the ink invisible?"

There are two reasons why premortems might help avert disasters. (I say "might" because I know of no systematic study of their use; organizations rarely allow such internal decision-making to be observed and recorded.) First, explicitly going through this exercise can overcome the natural organizational tendencies to groupthink and overconfidence. A devil's advocate is unpopular anywhere. The premortem procedure gives cover to a cowardly skeptic who otherwise might not speak up. After all, the entire point of the exercise is to think of reasons why the project failed. Who can be blamed for thinking of some unforeseen problem that would otherwise be overlooked in the excitement usually accompanying any new venture?

The second reason a premortem can work is subtle. Starting the exercise by assuming the project has failed, and then thinking of why that might have happened, creates the illusion of certainty, at least hypothetically. Laboratory research shows that by asking why it did fail rather than why it might fail gets the creative juices flowing. (The same principle can work in finding solutions to tough problems. Assume the problem has been solved and then ask how that happened. Try it!)

An example illustrates how this can work. Suppose a few years ago an airline CEO invited top management to conduct a premortem on this hypothetical disaster: "All of our airline's flights around the world have been canceled for two straight days. Why?" Of course, many will immediately think of some act of terrorism. But real progress will be made by thinking of much more mundane explanations. Suppose someone timidly suggests that the cause was that the reservation system crashed and the backup system didn't work properly.

Had this exercise been conducted, it might have prevented a disaster for a major airline that canceled nearly 2,000 flights over a three-day period. During much of that time, passengers couldn't get any information, because the reservation system was down. What caused this fiasco? A power surge blew a transformer and critical systems and network equipment didn't switch over to backups properly. This havoc was all initiated by the equivalent of blowing a fuse. Many companies that were once household names and now no longer exist might still be thriving if they had conducted a premortem, with the question being: "It's three years from now and we're on the verge of bankruptcy. How did this happen?"

And how many wars might not have started if someone had first asked, "We lost. How?"

IT'S ABOUT TIME

DUSTIN YELLIN
Artist; futurist; founder, Pioneer Works

Time is a scientific concept that deserves greater thought and study—though, despite advances in the mathematical behaviors of time over long horizons under what I'd call extreme conditions, it's a concept we will never be positioned to properly understand.

Time isn't simply a matter of duration; time is movement, motion, transition, and change. It's not static, to be ticked off, eliminated. If we consider one of the best known implications of $e = mc^2$, time is relative; it depends on perspective. Black holes, as hungry hippos of matter, could be considered a waste of time. Some say time flies when you're having fun, lending colloquial truth to the notion that time is relative.

There are boundaries to time in the temporal/spatial sphere, just as there are in the sphere of neurochemical subjectivity. Time varies according to experience, perception of pain, focus, thought or its absence. Meditation can make time meaningless. Mindlessness, in a sense, does away with time, just as enjoyment, like a TV movie, is often mindless, and over before we know it.

Self-help books talk confidently about "making time work for you." So do investment advisors. Physicists and gods laugh when we make plans. Some say all matter is basically fixed, others say bodies reconstitute molecularly, and most agree that neither energy nor matter are created or destroyed. Since 1662, we've invoked the immortal by the graveside prayer "Ashes to ashes, dust to dust"—so shall we all cycle in this eternal recurrence. Time plays the trump in every hand.

How does consciousness change with time? The serendipitous morphology of your world and your body is a brief flicker. We grapple with time, we fight gravity always seeking to bury us, we lean into the wind of prevailing wisdom; we rage against the uncaring forces of the cosmos, the amoral tyrant that is time itself.

Yet time is also the giver of life. It allows cell division, growth, love. Complex states are achieved through patterned development of eons of adaptive change; new states are reproduced through genetic programming and chromosomal mutations. Time gives our lives genetic high fidelity and brings the record player to the party, too.

What is matter but a cousin of time? Or a relative, at least. Matter is inextricably linked to time in a relationship of cause and effect, effect and cause; now and again, the quantum world behaves like a good boy should.

Life span is time-dependent, geology is time-dependent, volcanic belly-dancing and sculpted canyons are nothing if not the gorgeous supple movements of time, matter, and energy, dancing to a particular slo-mo rhythm. The canyons of Zion; the craters of our moon; the birth of the baby Krakatoa, the giant born from the belly of the sea—all miracles of time, no less marvelous for being partly understood.

Time is judge and jury, but perhaps not the best conversationalist. Time's a little slow on the uptake. Each of us is married to time; divorce is impossible. Thought, repeated over time, is part of the dowry the universe has given us. We have the ability to direct the actual arrangement and neurological formation of our brains by grappling regularly with the Sunday *New York Times* crossword puzzle—such mind-bending activities keep the mind elastic and supple.

Zen meditation, cyclical breathing, yoga, Tantra—these are ways of taking time for a walk, though in the end perhaps time is holding the leash and we are being led inexorably to an end most of us struggle mightily to frustrate. Time is the totalitarian ticket-seller of existence, the wizard behind the curtain, the puppeteer.

The plastic arts are ruled by time, too. If you can't keep time behind the drum kit, you're out of the band. The cathartic transcendence of music is rooted in meter, rhythm, repetition, leitmotif, climax—time is the key to what makes music particularly moving. Consider for a moment electronic music. Drumbeats are easily quantized, made to perfectly fix mathematical rhythm, but the effect is inhumane, robotic; the beat seems to slam you over the head like a hangover. It is the intersection between time and human frailty that makes everyone want to smile and dance, to feel the pocket of the beat.

Whatever may be the proper domain of time, clearly it's beyond human reckoning. Time is godlike, both completely outside our ken and inextricable from the best parts of life. Eternal life, arguably, would be less precious than a finite one, and it is because we lay our heads at the feet of time that we're able to find joy in life, to the extent we do. "All good things come to those who wait," goes another saying, and I argue that time—though we live by it every day and die by it in the great goodnight—is a concept both otherworldly and mundane, fresh and worn. Our understanding of it will never be complete.

MAXWELL'S DEMON

JIMENA CANALES

Thomas M. Siebel Chair in the History of Science, University of Illinois at Urbana-Champaign; author, *The Physicist and the Philosopher: Einstein, Bergson, and the Debate That Changed Our Understanding of Time*

Carl Sagan spoke too soon when he spoke about demons. Modern science, he told his numerous followers, banished witches, demons, and other such creatures from this world. Simply spread flour on the floor and check for suspicious footprints; this kind of reasoning, he claimed in *The Demon-Haunted World*, characterizes sound scientific thinking.

So why is Maxwell's demon still on the front lines of science? Since he was first conjured in 1874, modern inquisitors have valiantly chased after him with math and physics instead of holy water. But instead of going the way of phlogiston and the aether, he is a fixture in standard physics textbooks. Lest you think he's real, let me tell you he is not. But a demon he is, in spades. He sorts, and sorting is at the origin of sorcery.

Neat-fingered and vigilant, he can reverse time, momentarily violate the second law of thermodynamics, power a perpetual-motion machine, and generate pockets of hellish heat in substances that should otherwise reach temperature equilibrium. His smarts are debatable, matching those of a virtuoso piano player or humble switchman on the railway tracks. No offense is taken if he is compared to a simple valve; on the contrary, it means he can do a lot of work with minimum effort. Frequently portrayed as holding a cricket bat (to send molecules to-and-fro), manning a trapdoor (to let them pass or keep them out), and holding a torch, a flashlight, or a photocell (the better to see in the dark), this minuscule leviathan can wreak havoc.

He was named after the Scottish scientist James Clerk Maxwell, known for his theory of electromagnetism, by his colleague William Thomson (aka Lord Kelvin). Almost immediately after his public debut, he was exorcized. Sightings of his demonic activity were brushed away as statistical anomalies or insignificantly tiny. No need to worry, we were told.

But science can, and often does, turn imaginary beings into real things. Maxwell's demon is a case in point. Since he was first conjured, scientists have tried to bring him to life, re-purposing twisted metal, ratchets, and gears; molecules, enzymes, and cells; and even electronics and software. In 1929, the physicist Leo Szilard published a paper about him in *Annalen der Physik* (a paper that reached fame as "Szilard's exorcism") before pairing up with Einstein to patent a refrigerator that would momentarily and locally reverse entropy. Another milestone took place in 2007, when an article in *Nature* titled "A Demon of a Device" described some of the first successful molecular nanomotors. Sir James Fraser Stoddart considered it as auguring an entirely new approach to chemistry; he went on to win the 2016 Nobel Prize for chemistry. Artificial intelligence is another technology with a connection to Maxwell's being.

Maxwell's demon is neither the first nor the last of science's demons, but he is certainly the most well known. Descartes' demon preceded Maxwell's by more than two centuries. Like a master illusionist who can take over your sense of reality by throwing a cloak over your head, this genie can intercept your sense impressions and take over from there. He remains the patron saint of virtual reality.

Laplace's demon comes next: The master calculator who can relate every particle to the laws of motion and know the past and the future has inspired advances in supercomputing and Big Data. Yet another demon, this one called "the colleague of Maxwell's demon," appeared a few decades after the original; his arriviste career included traveling faster than light and helping to explain quantum entanglement. By sharing with Biblical *shedim* the power of instantaneous locomotion, he can intercept messages, mess with causality, and fiddle with time. These shape-shifting masters of disguise are getting smarter and more powerful. By taking on the name of the scientists who conjured them, they fulfill their patronymic destiny, reappearing regularly as spick-and-span newborns with the sagacity of old men.

At the turn of the 19th century, the French mathematician Henri Poincaré claimed he could almost see Maxwell's demon through his microscope, calculating that with a little patience (of millions of millions of centuries), his mischief would be evident to everyone. While some deemed him too small to matter for us, others noted that small causes were known to produce great effects, like the spark, or the pebble starting an avalanche. Other scientists argued that the universe as a whole was so large that he could confidently reign like a master in a confined territory. For Norbert Wiener, founder of cybernetics, it was clear that "there is no reason to suppose that metastable demons do not in fact exist." Richard Feynman wrote an entire article explaining why, when, and how Maxwell's demon would eventually tire. According to Isaac Asimov, everything that appears to us as if arriving by chance is really because "it is a drunken Maxwell's demon we are dealing with."

It is perhaps most ironic that Karl Popper, known for describing scientific progress as based on hypothesis creation and falsification, admired the demon's ability to survive: "Although innumerable attempts have been made on his life, almost from the day he was born, and although his non-existence has frequently been proved, . . . he will no doubt soon celebrate his hundredth birthday in perfect health and vigour."

Demons are here to stay. The reassuring predictions of scientists such as Carl Sagan have proved premature. In our hurry to make our world modern and base it on purely secular reason, we have failed to see the demons in our midst.

INCLUDED MIDDLE

MELANIE SWAN
Philosopher, economic theorist, New School for Social Research

"Included Middle" is an idea proposed by the Romanian philosopher Stéphane Lupasco (in *The Principle of Antagonism and the Logic of Energy*, 1951), further developed by Joseph E. Brenner and Basarab Nicolescu, and supported by Werner Heisenberg.

The notion pertains to physics and quantum mechanics and may have wider application in other domains, such as information theory and computing, epistemology, and theories of consciousness. The Included Middle is a theory proposing that logic has a three-part structure. The three parts are the positions of asserting something, the negation of this assertion, and a third position that is neither or both. Lupasco labeled these states A, not-A, and T. The Included Middle stands in opposition to classical logic, stemming from Aristotle. In classical logic, the principle of noncontradiction specifically proposes an Excluded Middle— that no middle position exists. *(Tertium non datur*—"There is no third option.")* In traditional logic, for any proposition, either that proposition is true or its negation is true; there's either A or not-A. While this could be true for circumscribed domains containing only A and not-A, there may also be a larger position not captured by these two claims—and which is articulated by the Included Middle.

Heisenberg noticed that there are cases where the straightforward classical logic of A and not-A doesn't hold. He pointed out how the traditional law of Excluded Middle has to be modified in quantum mechanics. In general cases at the macro scale, the law of Excluded Middle would seem to hold. Either there's a table here or there's not a table here; there's no third position. But in the quantum-mechanical realm there's the idea of superposition, in which both states could be true. Consider Schrödinger's cat being both dead and alive until an observer checks and possibility collapses into a reality state.

Thus a term of logic is needed to describe this third possible situation—hence, the Included Middle. It's not "middle" in the sense of being between A and not-A ("There's a partial table here") but in the sense that there's a third position, another state of reality, containing both A and not-A. This can be conceptualized by appealing to levels of reality. A and not-A exist at one level of reality, and the third position at another. At the level of A and not-A, there are no more than the two contradictory possibilities. At a higher level of reality, however, is a larger domain, where both elements could be possible; both elements are members of a larger set of possibilities.

Included Middle is a concept already deployed in a variety of scientific realms and we could benefit from a wider application if it's promoted to "meme" status. Beyond its uses in science, Included Middle is a model for thinking. The Included Middle is a conceptual model that overcomes dualism and opens a frame that is complex and multidimensional, not merely one of binary elements and simple linear causality. We now comprehend and address our world as one that is complex as opposed to basic; formal tools to support this investigation are crucial. The Included Middle helps expose how our thinking process unfolds. When we try grasping anything new, a basic "A, not-A" logic is the first step in understanding, but the idea is to progress to the next step. The Included Middle is a more robust model, with properties of both determinacy and indeterminacy—the universal and the particular, the part and the whole, actuality and possibility. The Included Middle is a position of greater complexity and possibility for addressing any situation. Conceiving of a third space that holds two apparent contradictions is what the Included Middle might bring to contemporary challenges in consciousness, artificial intelligence, disease pathologies, and unified theories in physics and cosmology.

RELATIVE DEPRIVATION

KURT GRAY

Assistant Professor of Psychology, University of North Carolina, Chapel Hill; co-author (with Daniel M. Wegner), *The Mind Club*

Middle-class Americans don't live like kings, they live better than kings.

If you showed Henry VIII the average American's living conditions, he would be awestruck. While most Americans don't have giant castles or huge armies, we do have luxuries that the royalty of yesteryear could scarcely dream of: big-screen TVs and the Internet, fast cars and even faster planes, indoor plumbing and innerspring mattresses, and vastly improved medical care. Despite this high standard of living, most of us don't feel like kings. Instead we feel like paupers because of relative deprivation.

Relative deprivation is the idea that people feel disadvantaged when they lack the resources or opportunities of another person or social group. Americans living in trailer parks have an objectively decent standard of living compared with the rest of the world and the long trail of human history: They have creature comforts, substantial freedom of choice, and significant safety. Nevertheless they feel deprived, because they compare their lives with those of celebrities and super-rich businessmen. Relative deprivation tells us that social and financial status is more a feeling than a fact, spelling trouble for traditional economics.

Economists largely agree that economic growth is good for everyone. In lifting the profits of corporations and the salaries of CEOs, the engine of capitalism also pulls up the lifestyle of everyone else. Although this idea is objectively true—standards of living are generally higher when the free market reigns—it is subjectively false. When everyone gets richer, no one feels better off, because, well, everyone gets richer. What people really want is to feel richer than everyone else.

Consider an experiment done by economist Robert Frank. He asked people to choose between two worlds. In World 1, you make $110,000 per year and everyone else makes $200,000; in World 2, you make $100,000 per year and everyone else makes $85,000. Although people have more purchasing power in World 1, most people chose World 2 to feel relatively richer than others.

The yearning for relative status seems irrational, but it makes sense from an evolutionary perspective. We evolved in small groups where relative status determined everything, including how much you could eat and whether you could procreate. Although most Americans can now eat and procreate adequately, we haven't lost that gnawing sensitivity to status. If anything, our relative status is now more important. Because our basic needs are met, we have a hard time determining whether we're doing well, so we judge ourselves based on our place in the hierarchy.

Relative deprivation can make sense of many curious human behaviors, such as why exposure to the rich makes middle-class people get sick and take dangerous risks. It also helps to understand the election of 2016.

Society today is economically more powerful than it was in the 1950s, with our money buying much more. In 1950, a 13-inch color TV cost over $8,000 (adjusting for inflation), whereas today a 40-inch LCD TV costs less than $200. Despite this objective improvement, one demographic group—white men without a college diploma—has seen a substantial relative decrease in their economic position since the 1950s. It is this relatively deprived group who wanted to Make America Great Again—not to have expensive TVs but to relive the days when they had a greater status than other groups.

The real problem with relative deprivation is that although it can change, it can never be truly eliminated. When one group rises in relative richness, another group feels worse because of it. When your neighbor gets an addition or a new convertible, your house and your car inevitably look inadequate. When uneducated white men feel better, then women, professors, and people of color inevitably feel worse. Relative deprivation suggests that economic advancement is less like a rising tide and more like a see-saw.

Of course, one easy way around relative deprivation is to change your perspective; each of us can look for someone who's relatively less successful. Unfortunately there's always someone at the very bottom, and they're looking straight up, wishing they lived like a king.

ANTISOCIAL PREFERENCES

STEVEN R. QUARTZ
Neuroscientist; Professor of Philosophy, Caltech; co-author (with Anette Asp), *Cool*

For at least fifty years, the rational, self-interested agent of neoclassical economics, *Homo economicus*, has been questioned, rebutted, and in some cases disparaged as a model psychopath. A seminal critique appeared in 1968 with the publication of Garrett Hardin's article "The Tragedy of the Commons." Hardin invited the reader to consider a pasture open to all neighboring herdsmen. If those herdsmen pursued their own rational self-interest, he reasoned, they'd continue to add cows to their herds, ultimately leading to the pasture's destruction—a multiplayer form of the Prisoner's Dilemma wherein the individual pursuit of rational self-interest inevitably leads to social catastrophe.

Contrary to *Homo economicus*, people seem to care about more than their own material payoffs. They care about fairness and appear to care about the welfare of others. They have what economists refer to as social preferences. In Dictator games, for example, a person is given some money they can share with a stranger. They'll often transfer some of it to the other player, even though the other player is merely a passive participant and cannot punish them for not transferring any money. Such behavior has been interpreted as evidence for strong reciprocity, a form of altruism on which human cooperation may depend. Emotions emerge at the level of psychological mechanism, such as reported links between empathy and sharing in Dictator games, which mark the distinction between real people and *Homo economicus*.

As much as we may celebrate the fall of *Homo economicus*, he'd never cut off his nose to spite his face. Far less well known is recent research probing the darker side of departures from rational self-interest. What emerges is a creature fueled by antisocial preferences who creates a whole variety of social dilemmas. The common feature of antisocial preferences is a willingness to make others worse off even at a cost to oneself. Such behaviors are distinct from more prosocial ones, such as altruistic punishment, where we may punish someone for violating social norms. It's more like basic spite, envy, or malice. An emerging class of economic games, such as money-burning games and vendetta games, is illustrative. In a basic Joy of Destruction game, for example, two players are each given $10 and asked whether they want to pay $1 to burn $5 of their partner's income.

Why would someone pay money to inflict harm on another person who has done nothing against them? The expression and intensity of antisocial preferences appears linked to resource scarcity and competition pressures. The German economist Sebastian Prediger and colleagues found, for example, that 40 percent of pastoralists from low-yield rangelands in southern Namibia burned their partner's money, compared to about 23 percent of pastoralists from high-yield areas.

Antisocial preferences thus follow an evolutionary logic found across nature and rooted in such rudimentary behaviors as bacteria that release toxins to kill closely related species. Harming behaviors reduce competition and should thus vary with competition intensity. In humans, such variations show up in, for example, the rate of "witch" murders in rural Tanzania. As Berkeley economist Edward Miguel found, these murders double during periods of crop failure. The so-called witches, killed by their relatives, are typically elderly women who are blamed for causing crop failure. As the most unproductive members of their households, their deaths help alleviate economic hardship in times of extremely scarce resources.

Why should the concept of antisocial preferences be more widely known? There are two main reasons: Although we still tend to blame *Homo economicus* for various social dilemmas, many of those are better explained by antisocial preferences. Consider, for example, attitudes toward income redistribution. If these were based on rational self-interest, anyone earning less than the mean income should favor redistribution, since they stand to benefit from that policy. Since income inequality skews income distribution rightward, making median income less than mean income, then, with increasing inequality, a larger share of the population has income below the mean, so support for redistribution should rise. Yet empirically this is not the case. One reason is antisocial preferences. As Princeton's Ilyana Kuziemko and her colleagues found, people exhibit "last-place aversion," both in the lab and in everyday social contexts. That is, individuals near the bottom of the income distribution oppose redistribution because they fear it might result in people below them either catching up to them or overtaking them, leaving them at the bottom of the status hierarchy.

The second reason has to do with long-run trends in resource scarcity and competition pressures. A nearly forty-year trend of broad-based wage stagnation and projections of anemic long-term economic growth mean increasing resource scarcity and competition pressures for the foreseeable future. Thus we should expect antisocial preferences to increasingly dominate prosocial ones as primary social attitudes. In the United States, for example, the poorest and unhealthiest states are the ones most opposed to federal programs aimed at helping the poorest and unhealthiest. We can make sense of such apparent paradoxical human behavior only through a broader understanding of the irrational, spiteful, and self-destructive behaviors rooted in antisocial preferences and the contexts that trigger them.

RECIPROCAL ALTRUISM

MARGARET LEVI
Director, Center for Advanced Study in the Behavioral Sciences;
Professor of Political Science, Stanford University; Jere L. Bacharach
Professor Emerita of International Studies, University of Washington

For societies to survive and thrive, some significant proportion of their members must engage in reciprocal altruism. All sorts of animals, including humans, will pay high individual costs to provide benefits for a non-intimate other. Indeed, this kind of altruism plays a critical role in producing cooperative cultures that improve a group's welfare, survival, and fitness.

The initial formulation of reciprocal altruism focused on a tit-for-tat strategy in which the altruist expected a cooperative response from the recipient. Game theorists posit an almost immediate return (albeit iterated), but evolutionary biologists, economists, anthropologists, and psychologists tend to be more concerned with returns over time to the individual or, more interestingly, to the collective.

Evidence is strong that for many human reciprocal altruists the anticipated repayment isn't necessarily for the person making the initial sacrifice or even for her family members. By creating a culture of cooperation, the expectation is that sufficient others will engage in altruistic acts as needed to ensure the well being of those within the given community. The return for such far-sighted reciprocal altruists is the establishment of norms of cooperation that endure beyond the lifetime of any particular altruist. Gift-exchange relationships documented by anthropologists are mechanisms for redistribution to ensure group stability; so are institutionalized philanthropy and welfare systems in modern economies.

At issue is how norms of giving evolve and help preserve a group. Reciprocal altruism—be it with immediate or long-term expectations—offers a model of appropriate behavior, but, equally important, it sets in motion a process of reciprocity that defines expectations of those in the society. If the norms become strong enough, those who deviate will be subject to punishment—internal in the form of shame and external in the form of penalties ranging from verbal reprimand, torture, or confinement to banishment from the group.

Reciprocal altruism helps us understand the creation of ethics and norms in a society, but we still need to better understand what initiates and sustains altruistic cooperation over time. Why would anyone be altruistic in the first place? Without some individually based motivations, far too few would engage in cooperative action. It may be that a few highly moralistic individuals are key; once there's a first mover willing to pay the price, others will follow, as the advantages become clearer or the costs are lowered.

Other accounts suggest that giving can be motivated by a reasonable expectation of reciprocity or rewards over time. Other factors also support reciprocal altruism, such as the positive emotions surrounding the act of giving, a lesson Scrooge learned. Most likely there's a combination of complementary motivations: Both Gandhi and Martin Luther King were undoubtedly driven by moral principles and outrage at the injustices they perceived, but they also gained adulation and fame. Less famous examples abound of sacrifice, charity, and costly cooperation, some demanding recognition, some not.

For far-sighted reciprocal altruism to be sustained in a society ultimately requires an enduring framework for establishing principles and ethics. Reciprocal altruism is reinforced by a culture that has norms and rules for behavior, makes punishment legitimate for deviations, and teaches its members its particular ethics of responsibility and fairness. If the organizational framework of the culture designs appropriate incentives and evokes relevant motivations, it will ensure a sufficient number of reciprocal altruists for survival of the culture and its ethics.

Far-sighted reciprocal altruism is key to human cooperation and the development of societies in which people take care of one another. However, there's huge variation in who counts in the relevant population and what they should receive as gifts and when. The existence of reciprocal altruism doesn't arbitrate these questions. Indeed, the expectation of reciprocity can both reduce and even undermine altruism. It may limit gift giving only to the in-group where such obligations exist. Perhaps if we stay only in the realm of group fitness (or, for that matter, tribalism), such behavior might still be considered ethical. But if we're trying to build an enduring, encompassing ethical society, tight boundaries around deserving beneficiaries of altruistic acts become problematic. If we accept such boundaries, we're in the realm of wars and terrorism, in which some populations are considered nonhuman or at least non-deserving of beneficence.

The concept of reciprocal altruism allows us to explore what it means to be human and to live in a humane society. Recognition of the significance of reciprocal altruism for the survival of a culture makes us aware of how dependent we are on one another. Sacrifices and giving, the stuff of altruism, are necessary ingredients for human cooperation, which itself is the basis of effective and thriving societies.

ISOLATION MISMATCH

DAVID C. QUELLER
Evolutionary biologist; Spencer T. Olin Professor, Washington University in St. Louis

You won't find the term "isolation mismatch" in any scientific dictionary. Isolation mismatches occur when two complex adaptive systems cannot be merged after evolving in isolation from each other. It's a generalization of a concept you might find in a scientific dictionary: "Dobzhansky-Muller incompatibilities," which cause isolated biological populations to become separate species. When ported into the realm of culture, isolation mismatches might explain how cultures can diverge and become incompatible. They might also account for our disconcerting human tendency toward xenophobia, or fear of other human cultures. But let's consider biology first, then the cultural analogy.

Formation of new biological species usually involves isolation and independent evolution of the two populations. As one toad population evolves via natural selection, each of its new genes is tried out in many toads and will necessarily be selected to work together with the other genes in its population. Any genes causing within-population mismatches are weeded out. But novel genes in one toad population are never tested with novel genes in another isolated toad population. Between-population mismatches won't be weeded out and will gradually accumulate, becoming apparent only when the populations later come into contact.

This is the dominant model (though not the only one) of how new species form. Interestingly, whereas much of evolution consists of fine-tuned adaptation, the formation of species in this manner is not directly adaptive. It's an accident of isolation, though such accidents will always happen, given enough time. There is, however, an add-on mechanism called reinforcement that's not accidental. When two partly incompatible populations come together, selection may directly favor reduced interbreeding. Individuals will have fewer mismatches and more viable offspring if they preferentially mate with their own type. For example, females of two closely related species of spadefoot toads have evolved to prefer the male call of their own type only in areas where the two species overlap.

Similar isolation mismatches may occur in other complex adaptive systems. Let two initially compatible systems evolve separately for long enough and they'll accrue mismatches. This includes cultural systems where what changes is the cultural equivalent of genes, which Richard Dawkins calls memes. For example, languages split, evolve independently, and become mutually unintelligible. Mixing diverged norms of social behavior can also cause mismatches. Acceptable behavior by a man toward another man's wife in one's own culture might prove fatal in another culture.

We're all familiar with isolation mismatches in technology. We see them when we travel between countries that drive on opposite sides of the road or have railroads of different gauges or electrical systems with different voltages and outlets. There's no logical reason for the mismatches; they just evolved incompatibly in different locations. We're also familiar with isolation mismatches resulting from computer-system upgrades, causing some programs that worked fine with the old system to crash.

Of course, software engineers can usually catch and correct such mismatches before a program's release. But they can be defeated if they don't know or don't care about a dependent program, or if they don't completely grasp the possible interactions in complex code. I suggest that throughout most of human cultural history, these two conditions often apply; cultures changed without concern about compatibility with other cultures, and people didn't understand how new cultural traits would interact.

Consider the Europeans' adoption of maize from the Amerindians. The crop spread widely because of its high yield, but it also caused pellagra, eventually traced to a deficiency of the vitamin niacin. The Amerindians suffered no such problem, because they de-hulled their corn by nixtamalization, a process involving soaking and cooking in alkaline solution. Nixtamalization also happens to prevent pellagra, possibly by releasing the niacin in maize from indigestible complexes. An 18th-century

Italian maize farmer who de-hulled by his own culture's supposedly superior mechanical methods got a very harmful isolation mismatch. If he too had adopted nixtamalization, he would have been fine, but he had no way of knowing that. Only being more conservative about adopting things foreign would have saved him.

So imagine a history of thousands of years of semi-isolated human bands, each evolving its own numerous cultural adaptations. Individuals or groups allowing too much cultural exchange would experience cultural mismatches and have decreased fitness. So a process parallel to reinforcement might be expected to occur. Selection could favor individuals or groups that avoid and perhaps even hate other cultures, much as the female spadefoot toads evolved to shun males of the other species. Xenophobic individuals or groups would be successful and propagate any genes or memes underlying xenophobia. For the same reason, selection could favor the adoption of cultural or ethnic markers that make it clear who belongs to your group, as suggested by anthropologist Richard McElreath and colleagues.

Many questions remain to be addressed. Is xenophobia selected by genic or cultural selection? Is it individuals or groups that are selected? How much isolation is required? Did prehistoric human groups frequently encounter other groups with sufficiently different cultures? When might selection favor the opposite of xenophobia, given that xenophobia can also lead to rejecting traits that would have been advantageous?

It should be stressed that no explanation of xenophobia, including this one, provides any moral justification for it. Indeed, understanding the roots of xenophobia might provide ways to mitigate it. The mismatch explanation is a relatively optimistic one, compared to the hypothesis that xenophobia is a genetic adaptation based on competition between groups for resources.

If correct, it tells us that the true objects of our evolutionary ire are certain cultural traits, not people. Moreover, even those traits work fine in their own context and, like software engineers, we may be able to figure out how to get them to work together.

MYSTERIANISM

NICHOLAS G. CARR
Author, *Utopia Is Creepy*

By leaps, steps, and stumbles, science progresses. Its seemingly inexorable advance promotes a sense that everything can be known and will be known. Through observation and experiment and lots of hard thinking, we will come to explain even the murkiest and most complicated of nature's secrets: consciousness, dark matter, time, the full story of the universe.

But what if our faith in nature's knowability is just an illusion, a trick of the overconfident human mind? That's the working assumption behind a school of thought known as mysterianism. Situated at the fruitful if sometimes fraught intersection of scientific and philosophic inquiry, the mysterianist view has been promulgated, in different ways, by many respected thinkers, from the philosopher Colin McGinn to the cognitive scientist Steven Pinker. The mysterians propose that human intellect has boundaries and that some of nature's mysteries may forever lie beyond our comprehension.

Mysterianism is most closely associated with the so-called hard problem of consciousness: How can the inanimate matter of the brain produce subjective feelings? The mysterians argue that the human mind may be incapable of understanding itself, that we'll never understand how consciousness works. But if mysterianism applies to the workings of the mind, there's no reason it shouldn't also apply to the workings of nature in general. As McGinn has suggested, "It may be that nothing in nature is fully intelligible to us."

The simplest and best argument for mysterianism is founded on evolutionary evidence. When we examine any other living creature, we understand immediately that its intellect is limited. Even the brightest, most curious dog is not going to master arithmetic. Even the wisest of owls knows nothing of the anatomy of the field mouse it devours. If all the minds that evolution has produced have bounded comprehension, then it's only logical that our own minds, also products of evolution, would have limits as well. As Pinker has observed, "The brain is a product of evolution, and just as animal brains have their limitations, we have ours." To assume there are no limits to human understanding is to believe in a level of human exceptionalism that seems miraculous, if not mystical.

Mysterianism, it's important to emphasize, is not inconsistent with materialism. The mysterians don't suggest that what's unknowable must be spiritual. They posit that matter itself has complexities lying beyond our ken. Like every other animal on Earth, we humans are just not smart enough to understand all of nature's laws and workings.

What's truly disconcerting about mysterianism is that if our intellect is bounded, we can never know how much of existence lies beyond our grasp. What we know or may know in the future might be trifling, compared with the unknowable unknowns. "As to myself," remarked Isaac Newton in his old age, "I seem to have been only like a boy playing on the seashore, and diverting myself in now and then finding a smoother pebble or a prettier shell than ordinary, whilst the great ocean of truth lay all undiscovered before me." It may be that we are all like that child on the strand, playing with the odd pebble or shell—and fated to remain so.

Mysterianism teaches us humility. Through science, we have come to understand much about nature, but much more may remain outside the scope of our perception and comprehension. If the mysterians are right, science's ultimate achievement may be to reveal to us its own limits.

RELATIVE INFORMATION

CARLO ROVELLI
Theoretical physicist, Centre de Physique Théorique, Marseille; author, *Reality Is Not What It Seems*

Everybody knows what "information" is. It's the stuff that overabounds online; it's what you ask for at the airport kiosk when you don't know how to get downtown; it's what's stored in your USB sticks. It carries meaning. Meaning is interpreted in our heads. So, is there anything out there that's just physical (i.e., independent from our heads) that is information?

Yes. It's called relative information. In nature, variables aren't independent. For instance, the two ends of a magnet have opposite polarities; knowing one amounts to knowing the other. So we can say that each end "has information" about the other. There is nothing mental in this; it's just a way of saying that there's a necessary relation between the polarities of the two ends. We say there's relative information between two systems anytime the state of one is constrained by the state of the other. In this precise sense, physical systems may be said to have information about one another, with no need for a mind to play a role.

Such relative information is ubiquitous in nature: The color of the light carries information about the object the light has bounced from; a virus has information about the cell it attaches to; neurons have information about one another. Since the world is a tangle of interacting events, it teems with relative information.

When this information is exploited for survival, extensively elaborated by our brain, and maybe coded in a language understood by a community, it becomes mental and acquires the semantic weight we commonly attribute to the notion of information. But the basic ingredient is down there, in the physical world—physical correlation between distinct variables. The physical world is not a set of self-absorbed entities that do their own thing; it's a tightly knit net of relative information, wherein each state reflects something else's state. We understand physical, chemical, biological, social, political, astrophysical, and cosmological systems in terms of these nets of relations, not in terms of individual behavior. Physical relative information is a powerful basic concept for describing the world. More so than "energy," "matter," or even "entity." Stating that the physical world is just a collection of elementary particles doesn't capture the full story; the constraints among them create the rich web of reciprocal information.

Twenty-four centuries ago, Democritus suggested that everything was made of atoms. But he also suggested that the atoms are like the letters of the alphabet: There are only twenty or so letters, but, as he put it, "It is possible for them to combine in diverse modes, in order to produce comedies or tragedies, ridiculous stories or epic poems." This is also true of nature: A few atoms combine to generate the phantasmagoric variety of reality.

But the analogy is deeper: The atoms are like an alphabet because how they're arranged is always correlated with how other atoms are arranged. Sets of atoms carry information.

The light arriving at our eyes carries information about the objects it has played across. The color of the sea has information on the color of the sky above it. A cell has information about the virus attacking it. A new living being has plenty of information, because it's correlated with its parents and its species. And you, dear reader, reading these lines, receive information about what I'm thinking while writing them—that is, about what's happening in my mind at the moment I write this text. What occurs in the atoms of your brain isn't independent of what is occurring in the atoms of mine: We communicate.

The world isn't just a mass of colliding atoms. It's also a web of correlations between sets of atoms, a network of reciprocal physical information between physical systems.

TIME WINDOW

ERNST PÖPPEL

Head of Research Group Systems, Neuroscience, and Cognitive Research, Ludwig-Maximilians-University Munich; Guest Professor, Peking University, China

Modern biology is guided by a principle summarized by Theodosius Dobzhansky in 1973 with the memorable sentence: "Nothing in biology makes sense, except in the light of evolution." On the basis of this conceptual frame, I both generalize and also suggest more specifically that nothing in neurobiology, psychology, the social sciences (or cognitive science in general) makes sense except in the light of synchronization—i.e., the creation of common time for temporally and spatially distributed sources of information or events. Without synchronization, neural information-processing, cognitive control, emotional relations, and social interactions would be either impossible or severely disrupted; without synchronization, we would be surrounded by informational chaos, desynchronized activities, unrelated events, or misunderstandings.

Synchronization as a fundamental principle is implemented on different operating levels by temporal windows with different time constants, from the sub-second range to seconds up to days, as reflected in circadian rhythms and even annual cycles. Temporal windows are the basis for the creation of perceptual, conceptual, and social identity, and they provide the necessary building blocks for the construction of experiential sequences or behavioral organization. Thus it may be said that "nothing in cognitive science makes sense except in the light of time windows."

Reference to time windows necessarily implies that information-processing is discontinuous, or discrete. Although of fundamental importance, the question of whether information-processing on the neural or cognitive level is of a continuous or discrete nature has been neglected in this scientific domain. Usually, and uncritically, continuous processing of information is taken for granted. The implicit assumption of a continuous processing mode of information may stem from an orientation of cognitive research and psychological reasoning to classical physics. This is how Isaac Newton in 1687 defined time: "Absolute, true, and mathematical time, of itself, and from its own nature, flows equably without relation to anything external." In this concept of an equal flow, time serves as a unidimensional and continuous "container" within which events are happening at one or another time. Does this theoretical concept of temporal continuity provide a solid conceptual background when we want to understand neural and cognitive processes? The answer is No. The concept of time windows speaks against such a frame of reference. Temporal processing has to be discrete to allow for efficient complexity reduction of information on different operating levels.

EFFECTIVE THEORY

LISA RANDALL
Theoretical physicist, Frank B. Baird, Jr., Professor of Science, Harvard University; author, *Dark Matter and the Dinosaurs: The Astounding Interconnectedness of the Universe*

People can disagree about many deep and fundamental questions, but we are all pretty confident that when we sit on a hard wooden chair it will support us, and that when we take a breath on the surface of the Earth we will take in the oxygen we need to survive.

Yet that chair is made of molecules, which are made of atoms composed of nuclei—protons and neutrons—with electrons orbiting them with probability functions in agreement with quantum-mechanical calculations. Those electrons are on average very far from the nuclei; from the perspective of matter, they're mostly empty space. And those protons and neutrons are made of quarks bound together by the dynamics of the strong force.

No one knew about quarks until the second half of the 20th century. And despite the wisdom of the ancient Greeks, no one really knew about atoms either, until at best a couple of hundred years ago. And, of course, air contains oxygen molecules and many others, too. Yet people had no trouble breathing air before this was known.

How is that possible? We all work in terms of effective theories. We find descriptions that match what we actually see, interact with, and measure. That a more fundamental description can underlie what we observe is pretty much irrelevant, until we have access to effects differentiating that description. A solid entity made from wood is a pretty good description of a chair as we go about our daily lives. It's only when we want to know more about wood's fundamental nature that we bother to change our description. It's only when we have the technological tools to study such questions that we can test whether our ideas about something's underlying nature are correct.

Effective theory is a valuable concept when we ask how scientific theories advance and what we mean when we say something is right or wrong. Newton's laws work extremely well. They're sufficient to devise the path that will send a satellite to the far reaches of the solar system, or to construct a bridge that won't collapse. Yet we know that quantum mechanics and relativity are the deeper underlying theories. Newton's laws are approximations that work at relatively low speeds and for macroscopic objects. Moreover, an effective theory tells us precisely its limitations—the conditions and values of parameters for which the theory breaks down. The laws of the effective theory succeed until we reach its limitations, when those assumptions are no longer true as our measurements or requirements become increasingly precise.

The notion of effective theory extends beyond the realm of science. It's really how we approach the world in all its aspects. We can't possibly keep track of all information simultaneously.

We focus on what's accessible and relate those quantities. We use a map that has the scale we need. It's pointless to know all the small streets around you when you're barreling down a highway.

Effective theory is practical and valuable. But we should be wary, since it also makes us miss things in the world—and in science. What lies beyond effective theory might be the more fundamental truth. Getting outside our comfort zone is how science and ideas advance. Sometimes only a little prodding will take us to this richer, more inclusive understanding of the world.

COARSE-GRAINING

JESSICA FLACK
Professor, Director, Collective Computation Group, Santa Fe Institute

In physics, a fine-grained description of a system is a detailed description of its microscopic behavior. A coarse-grained description is one in which some of this fine detail has been smoothed over.

Coarse-graining is at the core of the second law of thermodynamics, which states that the entropy of the universe is increasing. As entropy, or randomness, increases, there's a loss of structure. This simply means that some of the information we originally had about the system is no longer useful for making predictions about the system's behavior as a whole. To illustrate: Think about temperature.

Temperature is the average speed of particles in a system. It's a coarse-grained representation of the particles' behavior in aggregate. When we know the temperature, we can use it to predict the system's future state better than we could if we measured the speed of individual particles. This is why coarse-graining is so important: It's extremely useful, giving us what's called an effective theory. An effective theory lets us model a system's behavior without specifying all the underlying causes leading to system state changes.

A critical property of coarse-grained description is that it's "true" to the system, meaning that it's a reduction or simplification of the microscopic details. When we give a coarse-grained description, we don't introduce outside information; we don't add anything that isn't already in the details. This "lossy but true" property is one factor that distinguishes coarse-graining from other types of abstraction. A second property of coarse-graining is that it involves integrating over component behavior. An average is a simple example, but more complicated computations are also possible.

Normally when we talk of coarse-graining we mean coarse-grainings that scientists impose on a system to find compact descriptions of system behavior sufficient for good prediction—to help identify the relevant regularities for explaining that system's behavior. But we can also ask how adaptive systems *themselves* identify (in evolutionary, developmental, or learning time) regularities and build effective theories to guide their decision-making and behavior. Coarse-graining is one kind of inference mechanism that adaptive systems can use to build effective theories. To distinguish coarse-graining in nature from coarse-graining by scientists, we refer to coarse-graining in nature as endogenous coarse-graining.

Because adaptive systems are imperfect information processors, coarse-graining in nature is unlikely to be as perfect a simplification of the microscopic details as it is in the physics sense. Coarse-graining in nature is complicated by the fact that in adaptive systems it's often a collective process, performed by a large number of semi-independent components. One of many interesting questions is whether the subjectivity and error inherent in biological information-processing can be overcome through collective coarse-graining.

Two key questions for 21st-century biology are how nature coarse-grains and how that influences the quality of the effective theories that adaptive systems build to make predictions. The answers might give us traction on some slippery philosophical questions, such as "Is downward causation 'real'?" and "Are biological systems lawlike?"

COMMON SENSE

JARED DIAMOND
Professor of Geography, UCLA; author, *The World Until Yesterday*

You're much more likely to hear "common sense" invoked as a concept at a cocktail party than in a scientific discussion. It should be invoked more often in such discussions, where it is sometimes scorned. A scientist may string out a detailed argument that reaches an implausible conclusion contradicting common sense. But many of his colleagues may nevertheless accept the implausible conclusion because they get caught up in the details of the argument.

I was first exposed to this phenomenon in high school, when my plane-geometry teacher, Mr. Bridgess, gave us students a test consisting of a forty-nine-step proof on which we had to comment. The proof purported to demonstrate that all triangles are isosceles triangles—i.e., have two equal sides. Of course, that conclusion is wrong: Most triangles have unequal sides; only a tiny fraction are isosceles triangles. Yet there it was, a forty-nine-step proof, couched in the grammatically correct language of geometry, each step apparently impeccable and leading inexorably to the false conclusion that all triangles are isosceles triangles. How could that be?

None of us detected the reason. It turned out that somewhere around step thirty-seven the proof asked us to drop a perpendicular bisector from the triangle's apex to its base, then do further operations. The proof tacitly *assumed* that the perpendicular bisector did intersect the triangle's base, as is true for isosceles and nearly-isosceles triangles. But for triangles whose sides are very unequal in length, the perpendicular bisector doesn't intersect the base, and all of the proof's steps from step thirty-eight onward were fictitious. Conclusion: Don't get bogged down in the details of a proof if it leads to an implausible conclusion.

Distinguished scientists who should know better still fall into equivalents of Mr. Bridgess's trap. I'll tell you two examples. The first involves the famous Michelson–Morley experiment, one of the key experiments of modern physics. Beginning in 1881, the American physicists A. A. Michelson and E. W. Morley demonstrated that the speed of light in space did not depend on light's direction with respect to the Earth's direction of motion. The explanation for this came two decades later, with Albert Einstein's special theory of relativity, for which the Michelson–Morley experiment offered crucial support.

A further two decades later, though, another physicist carried out a complicated re-analysis of Michelson and Morley's experiment which showed that their conclusion had been wrong. If so, that would have shaken the validity of Einstein's formulation of relativity. Einstein was asked his assessment of the re-analysis. His answer, in effect, was: "I don't have to waste my time studying the details of that complex re-analysis to figure out what's wrong with it. Its conclusion is obviously wrong." Einstein was relying on his common sense. Eventually other physicists did waste their time studying the re-analysis and did discover where it had made a mistake.

My second example comes from the field of archaeology. Throughout most of human prehistory, human evolution was confined to the Old World; the Americas were uninhabited. Eventually, during the last Ice Age, humans did migrate from Siberia over the Bering Strait land bridge into Alaska. But for thousands of years thereafter, they were prevented from spreading farther south by the ice sheet that stretched uninterruptedly across Canada from the Pacific to the Atlantic. The first well-attested settlement of the Americas south of the Canada/U.S. border occurred around 13,000 years ago, as the ice sheets were melting. That settlement is attested to by the sudden appearance of stone tools of the radiocarbon-dated Clovis culture, named after the town of Clovis, New Mexico, where the tools and their significance were first recognized. Clovis tools have now been found over all of the lower forty-eight, south into Mexico. That sudden appearance of a culture abundantly filling the entire landscape is what one expects and observes whenever humans first colonize fertile, empty lands.

But any claim by an archaeologist to have discovered "the first X" is taken as a challenge by other archaeologists to discover an earlier X. In this case, archaeologists are bent on discovering pre-Clovis sites—sites with different stone tools and predating 13,000 years ago. Annually, new claims of pre-Clovis sites in the U.S. and South America are advanced and subjected to detailed scrutiny. Most are eventually invalidated by the equivalent of technical errors at step thirty-seven: For example, the radiocarbon sample was contaminated with older carbon, or the radiocarbon-dated material wasn't associated with the stone tools. But even after complicated analyses and objections and rebuttals, a few pre-Clovis claims still stand. The most widely discussed such claims are for Chile's Monte Verde site, Pennsylvania's Meadowcroft site, and one site each in Texas and Oregon. As a result, most American archaeologists currently believe in the validity of pre-Clovis settlement.

To me, it seems instead that pre-Clovis believers have fallen into the archaeological equivalent of Mr. Bridgess's trap. It's absurd to suppose that the first human settlers south of the Canada/U.S. border could have been airlifted by nonstop flights to Chile, Pennsylvania, Oregon, and Texas, leaving no unequivocal signs of their presence at intermediate sites. If there really had been pre-Clovis settlement, we'd already know it and no longer be arguing about it. That's because there would now be hundreds of undisputed pre-Clovis sites distributed everywhere from the Canada/U.S. border south to Chile.

As Mr. Bridgess told his plane-geometry students, "Use common sense, and don't be seduced by the details. Eventually, someone will discover the errors in those details." That advice is as true in modern science as it is in plane geometry.

"EVOLVE" AS METAPHOR

VICTORIA WYATT

Associate Professor of History in Art, University of Victoria, British Columbia

When scientific concepts become metaphors, nuances of meaning often get lost. Some such errors are benign. In the vernacular, "quantum leap" has come to mean tremendous change. This misrepresents the physics but usually without serious implications.

Yet errors in translation are not always neutral. As a popular metaphor, the phrase "to evolve" gets conflated with progress. "Evolved" means "better," as if natural law normally dictated constant improvement over time. In this interpretation, the significance of a species' dynamic relationship to a specific environment gets lost. Through natural selection, species become more equipped to survive in their particular environment. In a different environment, they may find themselves vulnerable. The popular metaphor of "evolve" misses this crucial point, often connoting progress without reference to context. People playfully tease that one or another friend needs to evolve. Businesses boast that they've evolved. In such usage, "evolved" means *essentially* better, more sophisticated, more developed. Since evolving occurs over time, the past is by definition inferior, a lower rung on the linear ladder to the future. Rapid changes in technology magnify the disconnect between present and past. Even the recent past may appear primitive, bearing little relevance to contemporary life.

A keen awareness of history is vital to intelligent decision-making. Used accurately, in its scientific sense, the evolution metaphor encourages historical acumen by emphasizing the significance of specific context. "Evolve" in its common usage actually obscures the importance of context, undermining interest in connections with the past. This erasure of relationships has serious implications. Species evolve in complex, nonlinear ecosystems; this use of the metaphor extracts the measure of progress from a multifaceted historical environment, linking it instead to a simple position in linear time. Recognizing current global challenges demands a different vision, one that acknowledges history and context. Climate-change denial does not rise from a deep appreciation of complex dynamic relationships. In simplistic linear paradigms, it thrives.

The popular usage of "evolve" reflects a symptom rather than a cause. The metaphor itself did not create disinterest in specific complicated conditions; rather, a widespread preference for simplicity and essentialism over complexity and connections shaped the metaphor. Now it perpetuates that outlook, while all signs point to an urgent need for a more accurate apprehension of reality.

Today we face problems much more daunting than the distortion of scientific concepts in popular metaphors. It's tempting to consider such misappropriations as mere annoyances. However, they reflect an issue informing many of our greater challenges: a failure to educate about how the sciences relate to the humanities, social sciences, and fine arts. Without such integrated education, scientific concepts dangerously mutate, as nonspecialists apply them outside the sciences. When these misunderstandings infiltrate popular language and thought, realistic approaches to global problem-solving suffer.

THE REYNOLDS NUMBER

GEORGE DYSON
Boat designer; author, *Turing's Cathedral; Darwin Among the Machines*

"The internal motion of water assumes one or other of two broadly distinguishable forms," Osborne Reynolds reported to the Royal Society in 1883. "Either the elements of the fluid follow one another along lines of motion which lead in the most direct manner to their destination, or they eddy about in sinuous paths the most indirect possible."

Reynolds puzzled over how to define the point at which a moving fluid (or a fluid with an object moving through it) makes the transition from stable to unstable flow. He noted that the transition depends on the ratio between inertial forces (characterized by mass, distance, and velocity) and viscous forces (characterized by the "stickiness" of the fluid). When this ratio, now termed the Reynolds number, reaches a certain critical value, the fluid shifts from orderly, deterministic behavior to disorderly, probabilistic behavior resistant to description in full detail. The two regimes are known as laminar and turbulent flow.

The Reynolds number is both nondimensional and universal, appearing consistently across a range of phenomena as diverse as blood pumping through a heart, a fish swimming through the sea, a missile flying through the air, burning gas flowing through a jet turbine, or weather systems flowing around the Earth. It is both descriptive, in the sense of capturing the characteristics of an existing flow, and predictive, in the sense that it gives a reliable indication of which regime will dominate a projected flow. Thanks to the Reynolds number, we can tackle otherwise intractable problems in fluid dynamics with scale models in a scaled flow and make predictions that hold up.

The notion of a Reynolds number and critical velocity can also be applied to nontraditional domains: for instance, the flow of information in a computer network, where velocity is represented by bandwidth and viscosity is represented by the processing speed of the individual nodes; or the flow of money in an economy, where velocity is represented by how fast funds are moving and viscosity is represented by transaction costs.

Wherever things (including ideas) are either moving through a surrounding medium or flowing past a boundary of some kind, the notion of a Reynolds number can help characterize the relation between inertial forces and viscosity, giving a sense for what may happen next.

Why do things go smoothly for a while, and then suddenly they don't?

METAMATERIALS

JOHN MARKOFF

Pulitzer Prize winner, former *New York Times* reporter; author, *Machines of Loving Grace*

First demonstrated in 1999 by a group of researchers led by physicist David R. Smith, metamaterials are now on the cusp of transforming entire industries. The term generally refers to synthetic composite materials that exhibit properties not found in nature. What Smith demonstrated was the ability to bend light waves in directions—described as "left-handed"—not observed in natural materials.

As a field of engineering, metamaterials are perhaps the clearest evidence that we're in the midst of a materials revolution that goes far beyond that of computing and communications.

The concept has been speculated about for more than a century, particularly in the work of Russian physicist Victor Veselago in the 1960s. However, the results of Smith's group touched off a new wave of excitement and experimentation in the scientific community. The notion captured the popular imagination briefly some years ago with the discussion of the possibility of invisibility cloaks; today, it's having more practical effects across the entire electromagnetic spectrum.

It may soon transform markets, such as the automotive industry, where there's a need for less expensive and more precise radars for self-driving vehicles. Metamaterial design has also begun to yield new classes of antennas that will be smaller and more powerful, tunable, and directional.

Some of the uses are novel, such as a transparent coating for cockpit windshields to protect pilots from harassing lasers beamed from the ground. This is already a commercial reality. Other applications are more speculative, yet promising. Several years ago, scientists at the French construction firm Menard published a paper describing a test of a new way to counteract the effects of an earthquake with a metamaterial grid of empty cylindrical columns bored into soil: The array produced a significant damping of a simulated earthquake.

While Harry Potter–style invisibility shields may not be possible, there is clear military interest in metamaterials for new stealth applications. A DARPA project is exploring the possibility of armor for soldiers which would make them less visible, and such technology might also be applied to vehicles such as tanks.

Another promising area is the creation of a "negative refractive index" (not found in nature), for so-called superlens microscopes that reach past the resolving power of today's scientific instruments. Metamaterials also promise to filter and control sound in new ways, enabling new kinds of ultrasound devices that will peer into the human body with enhanced 3D resolution.

Researchers note that the application of these synthetic materials far outstrips the imagination and that new applications will appear as engineers and scientists rethink existing technologies.

Perhaps Harry Potter invisibility is far-fetched, but several years ago Xiang Zhang, a UC Berkeley nanoscientist, speculated that it might be possible to make dust particles disappear in semiconductor manufacturing. Insertion of a metamaterial layer in the optical path of the exotic light waves now used to etch molecular-scale transistors might make the contamination effectively invisible. That, in turn, would lead to a significant jump in the yield of working chips—and conceivably put the computer industry back on the Moore's law curve of ever more powerful computing.

STIGLER'S LAW OF EPONYMY

WILLIAM POUNDSTONE
Journalist; author, *Head in the Cloud*

Stigler's law of eponymy says that no scientific discovery is named for its original discoverer. Notable examples include the Pythagorean theorem, Occam's razor, Halley's comet, Avogadro's number, Coriolis force, Gresham's law, Venn diagrams, Hubble's law. . . .

Statistician Stephen Stigler coined this law in a 1980 Festschrift honoring sociologist Robert K. Merton. It was Merton who had remarked that original discoverers never seem to get credit. Stigler playfully appropriated the rule, ensuring that Stigler's law would be self-referential.

The generalization is not limited to science. Elbridge Gerry did not invent gerrymandering nor Karl Baedeker the travel guide. Historians of rock music trace the lineage of the Bo Diddley beat, which didn't originate with that bluesman. The globe is filled with place names honoring explorers who "discovered" places already well known to indigenous peoples (Hudson

River, Hudson Bay, Columbia, the District of Columbia, and Columbus, Ohio, etc. Perhaps there are extraterrestrials who would consider the Magellanic Clouds a particularly egregious example).

Béchamel sauce is named for a once-famous gastronome, causing the rival Duke of Escars to complain: "That fellow Béchamel has all the luck! I was serving breast of chicken *à la crème* more than twenty years before he was born, but I have never had the chance of giving my name to even the most modest sauce."

Stigler's law is usually taken to be facetious, like Murphy's law (which predates Edward A. Murphy, Jr., by the way). It *is* facetious in its absolutism. But it says something nontrivial about the nature of discovery and originality.

The naïve take on Stigler's law is that the "wrong" people often get the credit. It's true that famous scientists, and others, sometimes get disproportionate credit relative to less famous colleagues. (That's actually a different law, the Matthew effect.)

What Stigler's law really tells us is that priority isn't everything. Edmund Halley's contribution was not in observing the 1682 comet but in recognizing that observations going back to 1531 (and millennia earlier, we now know) were of the same periodic comet. This claim would have made little sense before Newton's law of universal gravitation. Halley's achievement was developing the right idea at the right time, when the tools were available and the ambient culture was able to appreciate the result. Timeliness can matter as much as being first.

COMPARATIVE ADVANTAGE

ROBERT KURZBAN

Psychologist, University of Pennsylvania; Director, Penn Laboratory for Experimental Evolutionary Psychology (PLEEP); author, *Why Everyone (Else) Is a Hypocrite*

The intuitively clear effects of a tariff—a tax on goods or services entering a country—are that it helps domestic producers of the good or service and harms foreign producers. An American tax on Chinese tires, say, transparently helps American tire producers because the prices of the tires of their foreign competitors will be higher, allowing American producers to compete more easily. These effects—advantages for domestic firms, disadvantages for foreign firms—are useful for politicians to emphasize when contemplating tariffs and other forms of protectionism, because they appeal to nationalistic, competitive intuitions.

However, not all ideas surrounding international trade are so intuitive. Consider these remarks by economist Paul Krugman:

> The idea of comparative advantage—with its implication that trade between two nations normally raises the real incomes of both—is, like evolution via natural selection, a concept that seems simple and compelling to those who understand it. Yet anyone who becomes involved in discussions of international trade beyond the narrow circle of academic economists quickly realizes that it must be, in some sense, a very difficult concept indeed.

Comparative advantage is an important idea but is indeed difficult to grasp. Consider the economist's device of simplifying matters to see the underlying point. You and I are on a desert island, harvesting coconuts and catching fish to survive. You need an hour to harvest one coconut and two hours to catch one fish. (For this example, I'll assume you can meaningfully divide one fish—or one coconut—into fractions.) I, being old and no fisherman, am less efficient than you at both activities; I require two hours per coconut and six hours per fish. It might seem that you need not trade with me; after all, you're better than I am at both fishing and harvesting coconuts. But the economist David Ricardo famously showed that this isn't so.

Let's examine two cases. In the first, you and I don't trade. You produce, say, four coconuts and two fish in an eight-hour workday. I produce just one of each. In the second case, we specialize and trade: You agree to give me a fish in exchange for two-and-a-half coconuts. This is a good deal for both of us. For you, catching one less fish gets you two hours or two coconuts; two and a half is better. (For me, that fish saves me six hours, or three coconuts.) Now you produce one extra fish and give it to me in trade, leaving you with two fish and four-and-a-half coconuts. For ease of exposition, let's say I harvest only coconuts, produce four and give you two and a half of them, leaving me with one-and-a-half coconuts and the fish I got from you.

In this second case, you have just as many fish and an extra half-coconut; I also have the same number of fish (one), plus the half-coconut. Through the magic of trade, the world is one coconut better off, split between the two of us.

The lesson is that even though I produce both goods less efficiently than you, we are still both made better off when we specialize in the good for which we have a comparative advantage and then trade. This is an argument for a role for governments in facilitating, rather than inhibiting, specialization and trade. It should be clear, for instance, that if the island government forced me to pay them an extra coconut every time I purchased one of your fish—driving the coconut price to me from two-and-a-half to three-and-a-half coconuts—I would wind up with only half a coconut and a fish after the trade and so would prefer to revert to the first case, in which I split my time. In turn, you lose your trading partner and, similarly, revert to the previous case.

These gains reaped from world trade are less intuitive than the patriotic gains reaped by those of domestic firms protected—and therefore helped—by tariffs and other trade barriers. Given the outsize role that questions surrounding world trade play in current political discourse, the notion of comparative advantage ought to be more widely known.

PREMATURE OPTIMIZATION

KEVIN KELLY
Senior Maverick, *Wired*; author, *The Inevitable*

Why do the successful often fail to repeat their success? Because success is often the source of failure.

Success is a form of optimization—a state of optimal profits, or optimal fitness, or optimal mastery. In this state, you can't do any better than you're doing. In biology, a highly evolved organism might reach a state of supreme adaptation to its environment and competitors to reach reproductive success. A camel has undergone millions of revisions in its design to perfect its compatibility with an arid climate. Or in business, a company may have spent many decades perfecting a device until it was the number-one bestselling brand; say it manufactured and designed a manual typewriter that was difficult to improve. Successful individuals, too, discover a skill they're uniquely fit to master—a punk rock star who sings in an inimitable way.

Scientists use a diagram of a mountainous landscape to illustrate this principle. The contours of the undulating landscape indicate adaptive success of a creature. The higher the elevation of an entity, the more successful it is. The lower, the less fit. The lowest elevation is zero adaptation—or, in other words, extinction. The evolutionary history of an organism can thus be mapped over time as its population begins in the foothills of low adaptation and gradually ascends higher and higher mountains of increased environmental adaptation. This is known in biology and in computer science as "hill climbing." If the species is lucky, it will climb until it reaches a peak of optimal adaptation. *Tyrannosaurus rex* achieved peak fitness. The industrial-age Olivetti Corporation reached the peak of optimal typewriter. The Sex Pistols reached the summit of punk rock.

Their stories might have ended there with ongoing success for ages, except for the fact that environments rarely remain stable. In periods of particularly rapid co-evolution, the metaphorical landscape shifts and steep mountains of new opportunities rise overnight. What for a long time seemed a monumental Mt. Everest can quickly be dwarfed by a new, neighboring mountain that shoots up many times higher. During one era, dinosaurs, typewriters, or punk rock are at the top; in the next turn, mammals, word processors, and hip-hop tower over them. The challenge for the formerly successful entity is to migrate over to the newer, higher peak. Without going extinct.

Picture a world crammed with nearly vertical peaks, separated by deep valleys, rising and falling in response to each other. This oscillating geography is what biologists describe as a "rugged landscape." It's a perfect image of today's churning world. In this description, in order for any entity to move from one peak to a higher one, it must first descend to the valley between them. The higher the two peaks, the deeper the gulf between them. But descent, in our definition, means the entity must *reduce* its success. Descent to a valley means an organism or organization must first become less fit, less optimal, less excellent before it can rise again. It must lower its mastery and its chance of survival and approach the valley of death.

This is difficult for any species, organization, or individual. But the more successful an entity is, the harder it is to descend. The more fit a butterfly is to its niche, the harder it is for it to devolve away from that fit. The more an organization has trained itself to pursue excellence, the harder it is to pursue non-excellence, to go downhill into chaos. The greater the mastery a musician gains for her distinctive style, the harder it is to let it all go and perform less well. Each of their successes binds them to their peaks. But as we have seen, sometimes that peak is only locally optimal. The greater global optimal is only a short distance away, but it might as well be forever away, because an entity has to overcome its success by being less successful. It must go down against the gain of its core ability, which is going uphill toward betterment. When your world rewards hill climbing, going downhill is almost impossible.

Computer science has borrowed the concept of hill climbing as a way of discovering optimal solutions to complex problems. This technique uses populations of algorithms to explore a wide space of possible solutions. The possibilities are mapped as a rugged landscape of mountains (better solutions) and valleys (worse). As long as the next answer lands a little "higher uphill" toward a better answer than the one before, the system will eventually climb to the peak and thus find the best solution. But as in biology, it is likely to converge onto a local "false" summit rather than the higher global optimal solution. Scientists have invented many tricks to shake off the premature optimization in order to get it to migrate to the globally optimal. Getting off a local peak and arriving at the very best, repeatedly, demands patience and surrender to imperfection, inefficiency, and disorder.

Over the long haul, the greatest source of failure is prior success. So, whenever you're pursuing optimization of any type, you want to put into place methods that prevent you from premature optimization on a local peak: Let go at the top.

SIMULATED ANNEALING

EMANUEL DERMAN

Professor of Financial Engineering, Columbia University; Principal, Prisma Capital Partners; former head, Quantitative Strategies Group, Goldman Sachs; author, *Models.Behaving.Badly.*

Imagine you're nearsighted, so that you can see the region locally, near you, but nothing globally, far away. Now imagine you're on top of a mountain and want to make your way down by foot to the lowest point in the valley. But you can't see beyond your feet, so your algorithm is simply to keep heading downhill, wherever gravity takes you fastest. You do that, and eventually, halfway down the mountain, you end up at the bottom of a small oval ditch or basin, from which all paths lead up. As far as you can nearsightedly tell, you've reached the lowest point. But it's a local minimum, not truly the bottom of the valley, and you're not where you need to be.

What would be a better algorithm for someone with purely local vision? *Simulated annealing*, inspired by swordmakers.

To make a metal sword, you have to first heat the metal until it's hot and soft enough to shape. The trouble is that when the metal subsequently cools and crystallizes, it doesn't do so uniformly. Different parts of it tend to crystallize in different orientations (each of them a small local low-energy basin), so that the entire body of the sword consists of a multitude of small crystal cells with defects between them. This makes the sword brittle. It would be much better if the metal were one giant crystal with no defects. That would be the true low-energy configuration. How to get there?

Swordmakers learned how to anneal the metal, a process in which they first heat it to a high temperature and then cool it very slowly. Throughout the slow cooling, the swordmaker continually taps the metal, imparting enough energy to the individual cells so that they can jump up from their temporary basin into a higher energy state and then drop down to realign with their neighbors into a communally more stable, lower energy state. The tap paradoxically increases the energy of the cell, moving it up and out of the basin and farther away from the true valley, which is ostensibly bad; but in so doing, the tap allows the cell to emerge from the basin and seek out the lower energy state.

As the metal cools, as the tapping continues, more and more of the cells align, and because the metal is cooler, the tapping is less able to disturb cells from their newer and more stable positions.

In physics and mathematics, one often has to find the lowest-energy or minimum state of some complicated function of many variables. For very complicated functions, this can't be done analytically via a formula; instead, it requires an algorithmic search. An algorithm that blindly tried to head downward could get stuck in a local minimum and never find the global minimum.

Simulated annealing is a metaphorical kind of annealing, carried out in the algorithmic search for the minima of such complicated functions. When the descent in the simulated-annealing algorithm takes you to some minimum, you sometimes take a chance and shake things up at random, hoping that by shaking yourself out of the local minimum and temporarily moving higher (which isn't where you ultimately want to be), you may then find your way to a lower and more stable global minimum. As time passes, the algorithm decreases the probability of the shake-up, which corresponds to the cooling of the metal.

Simulated annealing employs judicious volatility in the hope that it will be beneficial. In an impossibly complex world, we should perhaps shun temporary stability and instead be willing to tolerate a bit of volatility in order to find a greater stability thereafter.

ATTRACTORS

KATE JEFFERY
Professor of Behavioral Neuroscience, Division of Psychology and Language Sciences, University College London

Quickly cool a piece of superheated liquid glass and a strange thing happens. The glass becomes hard but brittle—so brittle that it may abruptly shatter without warning. This is because the bonds between the molecules are under strain, and the cool temperature and low velocity means they cannot escape, as if caught in a negative equity trap with the neighbors from Hell. And, like warring neighbors, eventually something gives way and the strain relieves itself catastrophically. Glass makers avoid such catastrophes by annealing the glass, which means holding it for a long time at a high enough temperature that the molecules can move past each other but not too fast. In this way, the glass can find its way into a minimum-energy state, in which each molecule has had a chance to settle itself comfortably next to its neighbors with as little strain as possible, after which it can be completely cooled without problems.

Systems in which elements interact with their neighbors and settle into stable states are called *attractors*, and the stable states they settle into are called attractor states, or local minima. The term "attractor" arises from the system's tendency, when it finds itself near one of these states, to be attracted toward it, like a marble rolling downhill into a hollow. If there are multiple hollows—multiple local minima—then the marble may settle into a nearby one that's not necessarily the lowest point it can reach. To find the global minimum, the whole thing may need to be shaken up so the marble can jiggle itself out of its suboptimal local minimum and find a better one, including (one hopes) eventually the global one. This jiggling, or injection of energy, is what annealing accomplishes, and the process of moving into progressively lower energy states is called gradient descent.

Many natural systems show attractor-like dynamics. A murmuration of starlings, for example, produces aerial performances of such extraordinary, balletic synchrony that it seems like a vast, amorphous, purposeful organism, and yet the synchronized movements arise simply from the interactions between each bird and its nearest neighbors. Each flow of the flock in a given direction is a transient stable state, and periodic perturbations cause the flock to ruffle up and re-form in a new state, swooping and swirling across the sky. At a finer scale, brain scientists frequently recruit attractor dynamics to explain stable states in brain activity, such as the persistent firing of the neurons that signal which way you're facing or where you are. Unlike glass particles or starlings, neurons don't physically move, but they express states of activity that influence the activity of their "neighbors"—neurons they're connected to—such that the activity of the whole network eventually stabilizes. Some theoreticians even think that memories might be attractor states—presenting a reminder of a memory is akin to placing the network near a local minimum, and the evolution of the system's activity toward that minimum, via gradient descent, is analogous to retrieving the memory.

Attractors also characterize aspects of human social organization. The problem of pairing everybody off so that the species can reproduce successfully is a problem of annealing. Each individual is trying to optimize constraints: They want the most attractive, productive partner, but so do all their competitors, thus compromises must be made. Bonds are made and broken, made again and broken again, until each person (approximately speaking) has found a mate. Matching people to jobs is another annealing problem and one we haven't solved yet: How to find a low-strain social organization in which each individual is matched to their ideal job? If this is done badly, and society settles into a strained local minimum in which some people are happy but large numbers are trapped in jobs they dislike with little chance of escape, then the only solution may be an annealing one—to inject energy into the system and shake it up so it can find a better local minimum. This need to destabilize a system in order to obtain a more stable one might be why populations sometimes vote for seemingly destructive social change. The alternative is to maintain a strained status quo in which tensions fail to dissipate and society eventually ruptures, like shattered glass.

Attractors are all around us, and we should pay more attention to them.

ANTHROPOMORPHISM

DIANA REISS

Professor, Department of Psychology, Hunter College; author, *The Dolphin in the Mirror*

Anthropomorphism is the attribution of human characteristics, qualities, motivations, thoughts, emotions, and intentions to nonhuman beings and even nonliving objects. Writers and poets have freely used anthropomorphism in fictional and nonfictional narratives. Although this attribution has been a powerful and effective artistic device, in science it has been largely abandoned.

The successes of the reductionist approach in physics and chemistry motivated similar methodologies in biology and psychology. Anthropomorphism was considered an error in the context of scientific reductionism. However, it has remained curiously effective in certain areas of biology and psychology despite being controversial. For example, Richard Dawkins' invention of the selfish-gene meme was a brilliant use of anthropomorphism to introduce a crucial concept in evolutionary theory.

The view that the tendency to anthropomorphize is a source of error needs to be reconsidered. In his 1872 book *The Expression of the Emotions in Man and Animals,* Charles Darwin proposed evolutionary continuity in the animal world that extended beyond that of morphology into the realms of behavior and the expression of emotion. He argued that emotions evolved via natural selection in animals as well as in human beings and may be a substrate of their behavioral experiences.

Darwin's colleague George Romanes, a Canadian-English evolutionary biologist and physiologist who is considered the father of the field of comparative psychology, held similar opinions. Darwin's and Romanes' views on animal behavior and emotions were criticized as anthropomorphic and anecdotal and resulted in a scientific backlash that led to the rise of behaviorism. Although not denying the existence of cognitive processes in humans and other animals, behaviorist epistemology denied the ability to study them, focusing instead on observable behavior. But behaviorism ultimately failed to account for the complexity and richness seen in both human and animal behavior, and by the mid 1950s the cognitive revolution was underway— leading to increased research on animal cognitive and emotional processes and their underpinnings.

In 1976, a small book entitled *The Question of Animal Awareness*, by Rockefeller University zoologist Donald Griffin, was published. In his book, Griffin compares human brain processes with those of other animals. He claims that other vertebrate animals also have complicated brains—and in some cases brains that appear to be physically very much like our own. This suggests that what goes on in animal brains has a good deal in common with what goes on in human brains; laboratory experiments on animal behavior provide some measure of support for this suggestion. Griffin's small book seeded a new field of cognitive ethology—the marriage of cognitive science and ethology, in which scientists ask questions about the mental states of animals based on their interactions with their environment.

It's refreshing to see the revival of anthropomorphic language as a tool toward understanding the cognitive life of other animals, in the context of systematic studies of social behavior and our knowledge of the structure and complexities of the brains of other species. And we shouldn't fear or be fooled by the "ism" at the end of the term *anthropomorphism*; it isn't a school of thought or an ideology. Rather, it provides an alternative model to help us to interpret behavior. In the spirit of George Box's famous dictum about all models being wrong but some being useful, so anthropomorphism remains surprisingly useful in animal cognition studies. It's useful because it helps us understand and widen our appreciation of the similarities between other animals and ourselves. Anthropomorphism provides a view of continuity between the mental life of humans and other species, in contrast with often touted discontinuities—those traits that divide us from the rest of the animal world.

Frans de Waal has suggested, "To endow animals with human emotions has long been a scientific taboo. But if we do not, we risk missing something fundamental, about both animals and us." An epistemology that lets scientists use anthropomorphism as a tool to investigate and interpret behavior can enable us to see what English anthropologist, systems thinker, and linguist Gregory Bateson called "the patterns that connect us." He suggested an anthropomorphic approach to understanding other species by posing the question: "What is the pattern which connects all the living creatures?" This question can be asked at the morphological, behavioral, and emotional level. By anthropomorphizing, we may see evolutionary patterns connecting us to the rest of the animal world. And of course, the opposite of anthropomorphism is dehumanization—and we all know where that can lead us.

COGNITIVE ETHOLOGY

IRENE PEPPERBERG

Research Associate, Lecturer, Harvard University; Adjunct Associate
Professor, Brandeis University; author, *Alex & Me*

The term "cognitive ethology" was coined and used by zoologist Donald Griffin in the late 1970s to describe a field he was among the first to champion—the study of versatile thinking by nonhumans and how the data obtained could be used to examine the evolution and origins of human cognition. His further emphasis on the study of animal consciousness, however, caused many of his colleagues to shun all his ideas, throwing the baby out with the bathwater. Griffin's term nevertheless deserves a closer look and a renaissance in influence. The case is strengthened by a historical examination of the subject.

In the 1970s and 1980s, researchers studying nonhuman abilities were slowly moving away from the two ideologies that had dominated psychology and much of behavioral biology for decades—respectively, behaviorism and fixed-action patterns—but progress was slow. Proponents of behaviorism may have argued that little difference existed between the responses of humans and nonhumans to external stimuli, but they tried explaining all such responses in terms of a shaping by reward and punishment, avoiding any discussion of mental representations, manipulation of information, intentionality, or the like. Biology students were taught that animals were basically creatures of instinct that, when exposed to particular stimuli or contexts, engaged in species-specific invariant sequences of actions that would persist even if the environment changed. These patterns were thought to be controlled by hard-wired neural mechanisms—and, interestingly, there were no references to any sort of information processing, mental representation, etc.

Then came two major paradigm shifts, one in psychology and one in biology, but not much understanding by scientists of their common underlying themes. One shift, the so-called cognitive revolution in psychology, with its emphasis on all the issues ignored by the behaviorists, was initially conceived as relevant only for the study of human behavior. Nevertheless, far-sighted researchers such as S. H. Hulse, H. Fowler, and W. K. Honig saw how the human experiments could be adapted to study similar processes in nonhumans. The other shift, the advent of long-term observational studies of groups of nonhumans in nature by researchers (the most familiar being Jane Goodall) who collected extensive examples of versatile behavior, showed that nonhumans reacted to unpredictable circumstances in their environment in ways often suggesting humanlike intelligence.

Cognitive ethology was meant to be a synthesis of what were at the time seen as innovative psychological and biological approaches. But disaffection with Griffin's arguments about animal consciousness unfortunately prevented the term—and the field—from taking hold; thus the likelihood of interdisciplinary research has not been as great as its promise. Psychologists often remain in the lab and prefer to describe their research as "comparative." The term suggests an openness to looking at a variety of species and testing for similarities and differences in behavior on numerous tasks requiring, at the least, advanced levels of learning. Unfortunately, such studies usually occur under conditions far removed from natural circumstances. Furthermore, relatively few of them actually examine the cognitive processes underlying the exhibited behavior patterns.

Similarly, cognitive biologists (a term more common in Europe than in the United States) tend to be reductionist—more likely to compare the neuroanatomy and neurophysiology of various species. Even when comparing behavior patterns in the field, they often simply argue for either homology or analogy when similarities are found or merely highlight any observed differences. The connection isn't always made between neuro similarities and differences and how those are expressed in specific types of behavior. Studies from these areas come close to the goals of cognitive ethology but generally (although not exclusively) with less emphasis on examining cognitive processes with respect to the whole animal than would be true in cognitive ethology.

For example, psychologists may test whether a songbird in a laboratory can distinguish between the vocalizations of a bird in a neighboring territory versus that of a stranger, and they may determine what bits of song are relevant for that discrimination. Lab biologists may determine which bits of brain are responsible for these discriminations, and field biologists may collect information testing whether the size of the repertoire of a bird of a given species correlates with the quality of its territory or its reproductive success.

A cognitive ethologist, however, will be interested not only in these data but in how and why a bird chooses to learn a particular song or set of songs, why it chooses to sing a particular song from its repertoire to defend its territory against a neighbor as opposed to a stranger, how that choice varies with the environmental context (e.g., the distance from the intruder, the type of foliage separating them, the song being sung by the intruder), how other males respond to the interaction, *and* how the females in the area choose their mate based on the outcome of such male-male interactions. It's this type of inclusive research that provides real knowledge of the use of the song system.

The time has come to focus on the advantages of looking at nonhumans through the lens of cognitive ethology. Cognitive ethology should again be considered a means of bringing new views and methodologies to bear on the study of animal behavior and of encouraging collaborative projects. Whether the topic is communication, numerical competence, inferential or probabilistic reasoning, or any of a number of other possibilities, studies using this approach will provide a deeper and broader understanding of the data. Furthermore, a renewed interest in nonhuman cognition and intelligence, and how such intelligence is used in the daily life of nonhumans, will provide exciting evolutionary insights, as Griffin proposed: By examining and comparing mental capacities of large numbers of species, we can surmise much about the origins of human abilities.

MATING OPPORTUNITY COSTS

DAVID M. BUSS

Professor of Psychology, University of Texas at Austin; co-author (with Cindy M. Meston), *Why Women Have Sex*

The concept of *opportunity costs*—the loss of potential gains from alternatives not chosen when a mutually exclusive choice must be made—is one of the most important concepts in the field of economics. But the concept is not well appreciated in the field of psychology. One reason for its absence is the sheer difficulty of calculating opportunity costs that occur in metrics other than money. Consider mate choice. Choosing one long-term mate means forgoing the benefits of choosing an available and interested alternative. But how are non-monetary benefits calculated psychologically?

The complexities are multiple. The benefit-bestowing qualities of passed-over mates are many in number and disparate in nature. And there are inevitable trade-offs among competing and incommensurate alternatives. Sometimes the choice is between a humorless mate with excellent future job prospects and a fun-loving mate destined for a low-status occupation; or between an attractive mate who carries the costs of incessant attention from others versus a mate who garners little external attention but with whom you have less sexual chemistry. Another intangible quality also factors into the equation—the degree to which competing alternatives appreciate your unique assets, which renders you more irreplaceably valuable to one than the other.

Uncertainty of assessment surrounds each benefit-conferring quality. It's difficult to determine how emotionally stable someone is without sustained observation through times bad and good—events experienced with a chosen mate but unknown with a forgone alternative. Another complication centers on infidelity and breakups. There's no guarantee that you'll receive the benefits of a chosen mate over the long run. Mates higher in desirability are more likely to defect. Whereas less desirable mates are sure bets, more desirable partners represent tempting gambles. How do these mating opportunity costs enter into the complex calculus of mating decisions?

Despite the difficulties involved in computing non-monetary opportunity costs, probabilistic cues to their recurrent reality over evolutionary time must have forged a psychology designed to assess them, however approximate those computations may be. Although mating decisions provide clear illustrations, the psychology of opportunity costs is more pervasive. Humans surely have evolved a complex multifaceted psychology of opportunity costs, since every behavioral decision at every moment precludes potential benefits from alternative courses of action.

Many of these are trivial—sipping a cappuccino precludes downing a latte. But some are profound and produce post-decision regret, such as missed sexual opportunities or lamenting a true love that got away. The penalties of incorrectly calculating mating opportunity costs can last a lifetime.

SEX

HELENA CRONIN

Codirector, London School of Economics' Centre for Philosophy of Natural and Social Science; author, *The Ant and the Peacock: Altruism and Sexual Selection from Darwin to Today*

The poet Philip Larkin famously proclaimed that sex began in 1963. He was inaccurate by 800 million years. What began in the 1960s instead was a campaign to oust sex—in particular, sex differences—in favor of gender.

Why? Because biological differences were thought to spell genetic determinism, immutability, anti-feminism, and, most egregiously, women's oppression. Gender, however, was the realm of societal forces; "male" and "female" were social constructs, the stuff of political struggle, so gender was safe sex.

The campaign triumphed. Sex now struggles to be heard over a clamor of misconceptions, fabrications, and denunciations. And gender is ubiquitous, dominating thinking far beyond popular culture and spreading even to science—such that a respected neuroscience journal recently felt the need to devote an entire issue to urging that sex should be treated as a biological variable.

And, most profoundly, gender has distorted social policy. The campaign has undergone baleful mission-creep. Its aim has morphed from ending discrimination against women into a deeply misguided quest for sameness of outcome for males and females in all fields—above all, 50:50 across the entire workplace. This stems from a fundamental error: the conflation of *equality* and *sameness*. And it's an error all too easily made, if your starting point is that the sexes are "really" the same and that apparent differences are mere artifacts of sexist socialization.

Consider that 50:50 gender-equal workplace. A stirring call. But what will it look like? (These figures are U.K., but ratios are almost identical in all advanced economies.) Nursing, for example, is currently 90-percent female. So 256,000 female nurses will have to move elsewhere. Fortunately, thanks to a concomitant male exodus, 570,000 more women will be needed in the construction and building trades. Fifteen thousand women window-cleaners; 120,000 women electricians; 143,000 women auto mechanics;130,000 women metal machinists; and 32,000 women telecom engineers.

What's more, the most dangerous and dirty occupations are almost entirely 100-percent male—at least half a million jobs. So that will require a mass exodus of a quarter of a million women from further "unbalanced" occupations. Perhaps women teachers could become tomorrow's gender-equal refuse collectors, quarry workers, roofers, water-and-sewage plant operators, scaffolders, stagers, and riggers?

And perhaps gender-balanced pigs could fly? At this point, the question becomes: If that's the solution, what on earth was the problem? Gender proponents seem to be blithely unaware that thanks to their conflation of equality and sameness, they're now answering an entirely different set of concerns—such as "diversity," "under-representation," "imbalance"—without asking what those have to do with the original problem, discrimination.

And the confusions ramify. Bear in mind that equality isn't sameness. Equality is about fair treatment, not about people or outcomes being identical; so fairness does not and should not require sameness. However, when sameness gets confused with equality—and equality is, of course, to do with fairness—then sameness ends up undeservedly sharing the moral high ground, and male/female discrepancies become a moral crusade. Why so few women CEOs or engineers? It becomes socially suspect to explain this as the result not of discrimination but of differential choice.

Well, it shouldn't be suspect. Because the sexes do differ—and in ways that, on average, make a notable difference to their distribution in today's workplace.

So we need to talk about sex.

Here's why the sexes differ. A sexual organism must divide its total reproductive investment in two—competing for mates and caring for offspring. Almost from the dawn of sexual re-production, one sex specialized slightly more in competing for mates and the other slightly more in caring for offspring. This was because only one sex was able to inherit the mitochon-dria (the powerhouse of cells). So, that sex started out with sex cells larger and more resource-rich than those of the other sex. And thus began the great divide into fat, resource-laden eggs already investing in "caring" (providing for offspring) and slim, streamlined sperm, already competing for that vital investment. Over evolutionary time, this divergence widened, proliferat-ing and amplifying in every sexually reproducing species that has ever existed. The differences go far beyond reproductive plumbing. They are distinct adaptations for the different life strategies of competers and carers. Wherever ancestral males and females faced different adaptive problems, we should expect sex differences—encompassing bodies, brains, and behavior. And we should expect that, reflecting those differences, competers and carers will have correspondingly different life priorities. From that initial asymmetry, the same characteristic differences between males and females have evolved across all sexually-reproducing animals—differences that pervade what constitutes being male or female.

As for different outcomes in the workplace, the causes are, above all, different interests and temperaments (and not that women are "less clever" than men). Women, on average, have a strong preference for working with people—hence the nurses and teachers. Compared with men, they care more about family and relationships and have broader interests and priorities—hence little appeal in becoming CEOs. Men have far more interest in "things"—hence the engineers. And they're vastly more competitive: more risk-taking, ambitious, status-seeking, single-minded, opportunistic—hence the CEOs. Men and women have, on average, different conceptions of what constitutes success, despite the gender quest to impose the same (male) conception on all.

And here's some intriguing evidence. "Gender" predicts that as discrimination diminishes, males and females will increasingly converge. But a study of fifty-five nations found that it was in the most liberal, democratic, equality-driven countries that divergence was greatest. The less the sexism, the greater the sex differences. Differences, this suggests, are evidence not of oppression but of choice; not of socialization, not of patriarchy, not even of pink T-shirts or personal pronouns . . . but of female choice.

An evolutionary understanding shows that you can't have

sex without sex differences. Only within that powerful scientific framework, in which ideological questions become empirical answers, can gender be properly understood. And, as the fluidity of "sexualities" enters public awareness, sex is again crucial for informed, enlightened discussion.

So for the sake of science, society, and sense, bring back sex.

SUPERNORMAL STIMULI

NANCY ETCOFF
Assistant Clinical Professor of Psychology, Harvard Medical School /
Massachusetts General Hospital; Director, MGH Program in Aesthetics
and Well Being

Humans and other animals fall for hyperbole. Exaggeration is persuasive; subtlety exists in its shadows. In a famous set of studies done in the 1950s, biologist and ornithologist Niko Tinbergen created "supernormal stimuli," simulacra of beaks and eggs and other biologically salient objects which were painted, primped, and blown up in size. In these studies, herring gull chicks pecked more at big red knitting needles than at adult herring gull beaks, presumably because they were redder and longer than the actual beaks. Plovers responded more to eggs with striking visual contrast (black spots on white surround) than to natural but drab eggs with dark brown spots on light brown surround. Oystercatchers were willing to roll huge eggs into their nests to incubate. Later studies, as well as recording in the wild, show supernormal stimuli hijacking a range of biologically driven responses. For example, female stickleback fish get swollen bellies when they're ripe with eggs. When Tinbergen's student Richard Dawkins made a dummy rounder and more pear-shaped, greater lust was inspired. He called these dummies "sex bombs." Outside the lab, male Australian jewel beetles have been observed trying to perform sex with beer bottles of brown glass whose surface bumps resemble the bumpy brown carapace of female beetles.

Research on the evolution of signaling shows that animals frequently alter or exaggerate features to attract, mimic, intimidate, or protect themselves from conspecifics, sometimes setting off an arms race between deception and the detection of such deception. But only humans engage in conscious manipulation of signals using cultural tools in real time rather than relying on slow genetic changes over evolutionary time. We live in Tinbergen's world now, surrounded by supernormal signals produced by increasingly sophisticated cultural tools. We need only compare photoshopped images to the unretouched originals—or compare, as my own studies have done, the perceptions of the same face with and without cosmetics—to see that relatively simple artificially created exaggerations can be quite effective in eliciting heightened positive responses that may be consequential. In my studies, the makeup merely exaggerated the contrast between the woman's features and the surrounding skin.

How do such signals get the brain's attention? Studies of the brain's reward pathways suggest that dopamine plays a fundamental role in encouraging basic biological behaviors that evolved in the service of natural rewards. Dopamine is involved in learning and responds to cues in the environment that suggest potential gains and losses. In the early studies of the 1950s, before the role of dopamine was known, scientists likened the effects of supernormal stimuli to addiction, a process we now know is mediated by dopamine.

Are superstimuli leading to behavioral addictions? At the

least, we can say that they often waste time and resources with false promises. We fall down rabbit holes when we pursue information we don't need or buy products that seem exciting but offer little of real value or gain. Less obviously, superstimuli can have negative effects on our responses to natural stimuli—to nutritious foods rather than fast foods, to ordinary-looking people rather than photoshopped models, to the slow pleasures of novel and nonfiction reading rather than games and entertainment, to the examined life rather than the unexamined and frenetic one.

Perhaps we can move away from the pursuit of "supernormal" to at least sometimes consider the subtle and the fine, to close examination and deeper appreciation of the beauties and benefits that lie hidden in the ordinary.

COSTLY SIGNALING

STEVE OMOHUNDRO
Mathematical physicist; Founder, Self-Aware Systems, Palo Alto;
Cofounder, Center for Complex Systems Research, University of Illinois

If something doesn't make sense, your go-to hypothesis should be "costly signaling." The core idea is more than a century old, but new wrinkles deserve wider exposure. Thorstein Veblen's "conspicuous consumption" explained why people lit their cigars with $100 bills as a costly signal of their wealth. Later economists showed that a signal of a hidden trait becomes reliable if the cost of faking it is more than the expected gain. For example, Michael Spence showed that college degrees (even in irrelevant subjects) can reliably signal good future employees, because they're too costly for bad employees to obtain.

Darwin said, "The sight of a feather in a peacock's tail, whenever I gaze at it, makes me sick!" because he couldn't see its adaptive benefit. It makes perfect sense as a costly signal, however, because the peacock has to be quite fit to survive with a tail like that! Why do strong gazelles waste time and energy stotting (jumping vertically) when they see a cheetah? It's a costly signal of their strength, and the cheetahs will chase other gazelles. Biologists came to accept the idea only in 1990 and now apply it to signaling between parents and offspring, predator and prey, males and females, siblings, and many other relationships.

Technology is just getting on the bandwagon. The integrity of the cryptocurrency bitcoin is maintained by bitcoin "miners," who get paid in bitcoin. The primary deception risk is "Sybil attacks," where a single participant pretends to be many miners in an attempt to subvert the network's integrity. Bitcoin counters this by requiring miners to solve costly cryptographic puzzles in order to add blocks to the blockchain. Bitcoin mining currently burns up a gigawatt of electricity, which is about a billion dollars a year at U.S. rates. Venezuela is in economic turmoil and some of its starving citizens are resorting to breaking into zoos to eat the animals. At the same time, enterprising Venezuelan bitcoin miners are using the cheap electricity there to earn $1,200 per day. Notice the strangeness of this: By proving they've uselessly burned up precious resources, they cause another country to send them food!

As a grad student at Berkeley, I used to wonder why a preacher would often preach on the main plaza. Each time, he'd be harassed by a large crowd, and I never saw him gain any converts. "Costly signaling" explains that preaching to that audience was a much better signal of his faith and commitment than preaching to a more receptive audience. The very antagonism of his audience increased the cost and therefore the reliability of his signal.

A similar idea is playing out today in social media. Rationalist blogger Scott Alexander points out that the animal-rights group PETA is much better known than the related group Vegan Outreach. PETA makes outrageous statements and performs outrageous acts which generate a lot of antagonism and are therefore costly signals. They have thrown red paint on women wearing furs and offered to pay Detroit water bills for families who agree to stop eating meat. They recently called for the firing of the Australian zookeeper who punched a kangaroo to rescue his dog. Members who promote ambiguous or controversial positions signal their commitment to their cause in a way that generally accepted positions would not. For example, if PETA campaigned to prevent the torture of kittens, everyone would agree, and members wouldn't be sending a strong signal of their commitment to animal rights.

This connects to meme propagation in an interesting way. Memes everyone agrees with typically don't spread very far, because they don't signal anything about the sender. Nobody's tweeting that 2 + 2 = 4. But controversial memes make a statement. They cause people with an opposing view to respond with opposing memes. As vlogger CGP Grey beautifully explained, opposing memes synergistically help each other to spread. They also create a cost for the senders in the form of antagonistic pushback from believers in the opposing meme. But from the view of costly signaling, this is good! If you have enemies attacking you for your beliefs, you'd better demonstrate your belief and commitment by spreading your beliefs even more! Both sides get this boost of reliable signaling and are motivated to intensify meme wars.

One problem with all this costly signaling is that it's costly! Peacocks would do much better if they didn't have to waste resources on their large tails. Bitcoin would be much more efficient if it didn't burn up the electricity of a small country. People could be more productive if they didn't have endless meme wars to demonstrate their commitments.

Technology may be able to help us with this. If hidden traits could be reliably communicated, there would be no need for costly signals. If a peacock could show a peahen his genetic sequence, she wouldn't have to choose him based on his tail. If wealthy people could reliably reveal their bank accounts, they wouldn't need luxury yachts or fancy cars. If bitcoin miners could show they weren't being duplicitous, we could forget all those wasteful cryptographic puzzles. With proper design, AI systems can reliably reveal what's actually on their minds. And as our understanding of biology improves, humans may be able to do the same. If we can create institutions and infrastructure to support truthful communication without costly signaling, the world will become a much more efficient place. Until then, it's good to be aware of costly signaling and to notice it acting everywhere.

SEXUAL SELECTION

RORY SUTHERLAND

Executive Creative Director and Vice Chairman, OgilvyOne, U.K.; columnist, *The Spectator*; author, *The Wiki Man*

Having been born in the tiny Welsh village of Llanbadoc 141 years after Alfred Russel Wallace, I've always had a sneaking affinity with people eager to give Wallace joint billing with Darwin for "The Best Idea Anyone Ever Had."

Having said which, I don't think evolution by natural selection was Darwin's best or most valuable idea.

Earlier thinkers, from Lucretius to Patrick Matthew, grasped that there was something inevitably true in the idea of natural selection. Had neither Darwin nor Wallace existed, someone else would have come up with a similar theory; many practical people, whether pigeon fanciers or dog breeders, already understood the practical principles quite well.

But for its time, sexual selection was a truly extraordinary, out-of-the-box idea. It still is. Once you understand sexual selection—along with costly signaling, Fisherian runaway, proxies, heuristics, satisficing, and so forth—a whole host of behaviors that were previously baffling or seemingly irrational suddenly make perfect sense.

The body of ideas falling out of sexual-selection theory explain not only natural anomalies such as the peacock's tail but also the extraordinary popularity of many seemingly insane human behaviors and tastes—from the existence of Veblen Goods, such as caviar, to more mundane absurdities, such as the typewriter. For almost a century during which few people knew how to type, the typewriter must have damaged business productivity to an astounding degree; it meant that every single business and government communication had to be written twice, once by the originator in longhand and then once again by the typist or typing pool. A series of simple amends could delay a letter or memo by a week. But the ownership and use of a typewriter was a necessary expense to signal that you were a serious business. Any provincial solicitor who persisted in writing letters by hand became a tailless peacock.

But take note: I have committed the same offense as everyone else writing about sexual selection. I have confined my examples of sexual selection to those occasions where it runs out of control and leads to costly inefficiencies. Typewriters, Ferraris, peacock's tails. Elks will make an appearance any moment now, you expect. But this is unfair.

You may have noticed that there are very few famous Belgians. This is because when you are a famous Belgian (Magritte, Simenon, Brel), everyone assumes you're French. In the same way, there are few commonly cited examples of successful sexual selection, because when sexual selection succeeds, people casually attribute the success to natural selection.

But the tension between sexual and natural selection—and the interplay between the two divergent forces—may be the really big story here. Many human innovations wouldn't have got off the ground without the human status-signaling instinct. (For a good decade or so, cars were inferior to horses as a mode of transport—it was human neophilia and status-seeking car races, not the pursuit of "utility," that gave birth to the Ford Motor Company.) So might it be the same in nature? That, in the words of Geoffrey Miller, sexual selection provides the "early-stage funding" for nature's best experiments? So the sexual-fitness advantages of displaying ever more plumage on the sides of a bird (rather than, like the peacock, senselessly overinvesting in the rear spoiler) may have made it possible for birds to fly. The human brain's ability to handle a vast vocabulary probably arose more for the purposes of seduction than anything else. But most people will avoid giving credit to sexual selection whenever they possibly can. When it works, sexual selection is called natural selection.

Why is this? Why the reluctance to accept that life is not just a narrow pursuit of greater efficiency, that there is room for opulence and display as well? Yes, costly signaling can lead to economic inefficiency, but such inefficiencies are necessary to establish valuable social qualities such as trustworthiness and commitment—and perhaps altruism. Politeness and good manners are, after all, simply costly signaling in face-to-face form.

Why are people happy with the idea that nature has an accounting function but much less comfortable with the idea that nature necessarily has a marketing function as well? Should we despise flowers because they're less efficient than grasses?

If you're looking for underrated, underpromulgated ideas, a good place to start is always with those ideas that, for whatever reason, somehow discomfort people on both the political left and the political right. Sexual selection is one such idea. Marxists hate the idea. Neoliberals don't like it. Yet when the concepts underlying it—and the effects it has wrought on our evolved psychology—are better understood, it could be the basis for a new and better form of economics and politics. A society in which our signaling instincts were channeled toward positive-sum behavior could be far happier while consuming less.

But even Russel Wallace hated the idea of sexual selection. For some reason, it sits in that important category of ideas that most people—and intellectuals especially—simply don't want to believe. Which is why it's my candidate for the idea most in need of wider transmission.

PHYLOGENY

RICHARD PRUM
Evolutionary ornithologist; Director, Franke Program in Science and the Humanities, Yale University; author, *The Evolution of Beauty*

How are you related to the contents of your salad? Or to the ingredients of a slice of pepperoni pizza, or whatever your next meal might be? Consumption is an ecological relationship. Your body digests and absorbs the nutrients from your food which provide energy for your metabolism and material components for your cells. But another fundamental relationship is more cryptic and in many ways more profound.

The answer comes from one of Charles Darwin's least appreciated revolutionary ideas. Darwin is, of course, duly famous for his discovery of the process of natural selection, among the most successful concepts in the history of science. Darwin also discovered the process of sexual selection, which he viewed as an independent mechanism of evolution. But Darwin was the first person to imply that all of life came from a single or a few common origins and diversified over time through speciation and extinction to become the richness of the biotic world we know today. Darwin referred to the history of this diversification as "the great Tree of Life"; today, biologists refer to it as *phylogeny*. It may be Darwin's greatest empirical discovery.

Network science is an exploding field of study. Network analysis can be used to trace neural processes in the brain, uncover terrorist groups from cellphone metadata, or understand the social consequences of cigarette smoking and vaping among cliques of high school students. Biology is a network science. Ecology investigates the food web, while genetics explores the genealogy of variations in DNA sequences. But few understand that evolutionary biology is also a network science. Phylogeny is a rooted network, in which the edges are lineages of organisms propagating over time and the vertices are speciation events. The root of the phylogenetic network is the origin of life as we know it—diagnosable by the existence of RNA-/DNA-based genetic systems, left-handed amino acids, proteins, and sugars, and (likely) a lipid bilayer membrane. These are the features of the trunk of Darwin's great Tree of Life.

Thus you are related to the lettuce, the anchovies, the Parmesan, and the chicken eggs in your Caesar salad, through the historical network of shared common ancestry. Indeed, there's no food you can think of that cannot be placed on the Tree of Life. Being a member of this network is currently the most successful definition of life.

Darwin should be world-famous for his discovery of phylogeny. But, just as Einstein's discovery of the quantum nature of energy was eclipsed biographically by his discovery of the theory of relativity, Darwin had the mishap of discovering natural selection too. Despite its excellent intellectual roots, phylogeny remains underappreciated today because it was largely suppressed and ignored for most of the 20th century. The architects of the "New Synthesis" in evolutionary biology were eager to pursue an ahistorical science analyzing the sorting of genetic variations in populations. This required shelving the question of phylogeny for some decades. As population genetics became more successful, phylogeny came to be viewed merely as the residuum left behind by adaptive process. Phylogeny became uninteresting, not even worth knowing.

But the concept has come roaring back in recent decades. Today, discovering the full details of the phylogenetic relationships among the tens of millions of extant species and their myriad of extinct relatives is a major goal of evolutionary biology. Just as a basketball tournament with sixty-four teams has sixty-three games, the phylogeny of tens of millions of living species must have tens-of-millions-minus-1 branches. So biologists have a lot of work ahead. Luckily, genomic tools, computing power, and conceptual advances make our estimate of organismal phylogenies better all the time.

Despite a lot of empirical progress, the full implications of the concept of phylogeny have yet to be appreciated in evolutionary biology and the society at large. For example, the concept of homology—similarity relation among organisms and their parts due to common ancestry—can be understood only in terms of phylogeny. Infectious diseases are caused by various species from different branches on the tree of life. Defending against them requires understanding how to slow them down without hurting ourselves, which is greatly facilitated by understanding where they and we fit in the historical network of phylogeny.

Billions of dollars of biomedical research funds are invested in a few model organisms, such as *E. coli*, yeast, roundworms, fruit flies, and mice. But, like the diverse contents of your salad, the scientific results are usually unaccompanied by awareness of the complex hierarchical implications of the phylogenetic context.

Perhaps the most important implication of the singular phylogenetic history of life is its contingency. Given the pervasiveness of extinction pruning the network, our existence, or the existence of any other extant species, is possible only as a result of an unfathomable number of historically contingent events—speciation events, evolutionary changes within lineages, and survival. The history of any one branch connects to the whole individualized history of life.

NEOTENY

BRIAN CHRISTIAN
Author, *The Most Human Human*; co-author (with Tom Griffiths), *Algorithms to Live By*

The axolotl is a peculiar amphibian; it never undergoes metamorphosis, retaining its gills and living in water for its entire life, a kind of tadpole with feet. Studying the axolotl in the late 19th century, the German zoologist Julius Kollmann coined the term *neoteny* to describe this process—the retention of youthful traits into adulthood.

Neoteny has a provocative history within biology. Evolutionary biologists throughout the 20th century, including Stephen Jay Gould, discussed and debated neoteny as one of the mechanisms of evolution and one of the distinguishing features of *Homo sapiens* in particular. Compared to our fellow primates, we mature later, more slowly, and somewhat incompletely. We stay relatively hairless, with larger heads, flatter faces, bigger eyes. Human adults, that is, strongly resemble chimpanzee infants. It's interesting that our typical depiction of aliens (seen as even more highly evolved than ourselves) is one of enormous heads, huge eyes, tiny noses—that is, a species even more neotenous than we are.

Neoteny, depending on how far one wishes to extend the term beyond considerations of pure anatomy, also functions as a description of human cognitive development and human culture. A baby gazelle can outrun a predatory cheetah within several hours of being born. Humans don't even learn to crawl for six months. We're not cognitively mature (or allowed to operate heavy machinery) for decades.

Indeed, humans are, at the start of our lives, among the most useless creatures in the animal kingdom. Paradoxically, this may be part and parcel of the dominant position we hold today—via the so-called Baldwin effect, where we blend adaptation by genetic mutation with adaptation by learning. We are, in effect, tuning ourselves to our environment in software, rather than in hardware. The upside is that we can much more rapidly adapt (including genetically) to selective pressures. The downside: longer childhoods.

Human culture itself appears to progress by way of neoteny. Thirteen-year-olds used to be full-fledged adults, working the fields or joining the hunting parties. Now we grouse that "grad school is the new college," our careers beginning ever later, after ever-lengthening periods of study and specialization.

Computer scientists speak of the "explore/exploit" trade-off—between spending your energy experimenting with new possibilities and spending it on the surest bets you've found to date. One of the critical results in this area is that in problems of this type, few things matter so much as where you find yourself on the interval of time available to you.

The odds of making a great new discovery are highest the greener you are—and the value of a great discovery is highest when you've got the most time to savor it. Conversely, the value of playing to your strengths, going with the sure thing, only goes up over time, both as a function of your experience and as a function of time growing scarce. This naturally puts all of us on an inevitable trajectory—from play to excellence, from craving novelty to preferring what we know and love. The decision-making of the young—whether it's whom to spend time with, where to eat, or how to work and play—really should differ from the decision-making of the old.

Yet even here there's an argument to be made for neoteny, of a conscious and deliberate sort.

To imagine ourselves as making choices not only on our own behalf but on behalf of our peers, successors, and descendants is to place ourselves much more squarely at the beginning of the interval—an interval much vaster than our lifetime. The longer a future we feel ourselves to be stewarding, the more we place ourselves in the youth of the race.

This offers something of a virtuous circle. A host of results in neuroscience and psychology shows that the brain appears to mark time by measuring new events in particular. "Life is long," we think, and the effort of exploration is worthwhile. In turn, the explorer is immune to the feeling of time speeding up as they age. The mindset is self-fulfilling. To lengthen youthfulness is to lengthen life itself.

THE NEURAL CODE

JOHN HORGAN

Director, Center for Science Writings, Stevens Institute of Technology; author, *The End of War*

"Neural code" is by far the most underappreciated term, and concept, in science. It refers to the rules or algorithms that transform action potentials and other processes in the brain into perceptions, memories, meanings, emotions, intentions, and actions. Think of it as the brain's software.

The neural code is science's deepest, most consequential problem. If researchers crack the code, they might solve such ancient philosophical mysteries as the mind–body problem and the riddle of free will. A solution to the neural code could also give us unlimited power over our brains and hence minds. Science fiction—including mind control, mind reading, bionic enhancement, and even psychic uploading—will become reality. Those who yearn for the Singularity will get their wish.

More than a half-century ago, Francis Crick and others deciphered the genetic code, which underpins heredity and other biological functions. Crick spent his final decades seeking the neural code—in vain, because the most profound problem in science is also by far the hardest. The neural code is certainly not as simple, elegant, and universal as the genetic code. Neuroscientists have, if anything, too many candidate codes. There are rate codes, temporal codes, population codes, and grandmother-cell codes, quantum and chaotic and information codes, codes based on oscillations and synchronies.

But given the relentless pace of advances in optogenetics, computation, and other technologies for mapping, manipulating, and modeling brains, a breakthrough could be imminent. Question: Considering the enormous power that could be unleashed by a solution to the neural code, do we really want it solved?

SPONTANEOUS SYMMETRY BREAKING

GORDON KANE

Theoretical particle physicist, cosmologist; Victor Weisskopf Distinguished University Professor, University of Michigan; author, *String Theory and the Real World*

Spontaneous symmetry breaking is widespread and fundamental in physics and science. Most notably, it's the mechanism responsible for the importance of Higgs physics (the reason quarks and electrons are allowed to have mass) and for the vacuum of our universe not being nothing. The notion is widespread in condensed matter physics and, indeed, was first understood there. But it's much broader, potentially leading to confusion between theories and solutions in many areas.

The basic idea can be explained simply and generally. Suppose a theory is stated in terms of an equation, x times $y = 16$. For simplicity, consider only positive integer values of x, y as solutions. Then there are three solutions: $x = 1$ and $y = 16$, $x = 2$

and $y = 8$, and $x = y = 4$. What's important is that the theory ($xy = 16$) is symmetric if we interchange x and y, but some solutions are not. The most famous example is that the theory of the solar system has the sun at the center and is spherically symmetric, but the planetary orbits are ellipses—not symmetric. The spherical symmetry of the theory misled people to expect circular orbits for centuries. Whenever a symmetric theory has nonsymmetric solutions, which is common, it's called spontaneous symmetry breaking.

In the example above, as often in nature, there are three solutions, so more information is needed, either theoretical or experimental, to determine nature's solution. We could measure one of the values of x or y; then the other is determined by the equations. Improving the theory leads to an interesting case. Suppose there's an additional theory equation, $x + y = 10$, also symmetric if we interchange x and y, so the theory remains symmetric. But now there's a unique solution, $x = 2$, $y = 8$, and it is *not* symmetric. In fact, there are no symmetric solutions to the two theory equations.

Magnetism is a familiar real-world example. The equations describing individual iron atoms don't distinguish different directions in space. But when a piece of iron is cooled below about 770°C, it spontaneously develops a magnetic field in some direction. The original symmetry between different directions is broken. Describing this is how the name "spontaneous symmetry breaking" originated. In physics, what happens is understood: Known electromagnetic forces tend to make the spins of individual atoms parallel, and each spin is a little magnet.

Normally we expect all fields (such as electromagnetic fields) to be zero in the ground state or vacuum of the universe. Otherwise they add energy, and the universe will naturally settle in the state of minimum energy. Now we've learned that the universe is in a lower state of energy when the Higgs field is non-zero than when it's zero, a nonsymmetric result, and that's essential for understanding how electrons and quarks get mass. Nature's solution is a state of reduced symmetry.

In many fields we make theories to describe and explain phenomena. But the behavior of systems is described by the solutions to the theories, not by the theories alone. We saw here in this simple example that trying to deduce the properties of the solutions and the behavior of phenomena in sciences and social sciences, and the world in general, from the form of the theory can be completely misleading. Another way to view the situation is the reverse perspective: The properties of the theory (such as its symmetries) may be hidden when we observe only the nonsymmetric solutions. If the theory is described by equations such as those in our example it's easy to see this, but it's true more generally.

REGRESSION TO THE MEAN

JAMES J. O'DONNELL

Classics scholar; University Librarian, Arizona State University; author, *Pagans: The End of Traditional Religion and the Rise of Christianity*

In this somber time, asking what scientific term or concept ought to be more widely known sounds like the setup for a punchline, something like "2 + 2 = 4" or "To every action there is an equal and opposite reaction." We can make a joke like that, but the truth the joke reveals is that "science" is indeed very much a human conception and construction. Science is all in our minds, even as we see dramatic examples of the use of science all around us.

So this year's question is really a question about where to begin: What is there that we should all know that we don't know as well as we should, don't apply to our everyday and extraordinary challenges as tellingly as we could, and don't pass on to children in nursery rhymes and the like?

My candidate is an old, simple, and powerful one: the law of regression to the mean. It's a concept from the discipline of statistics, but in real life it means that anomalies are anomalies, coincidences happen (all the time, with stunning frequency), and the main thing they tell us is that the *next* thing to happen is *very* likely to be a lot more boring, ordinary, and predictable. Put in the simplest human terms, it teaches us not to be so excitable, not to be so worried, not to be so excited: Life really will be, for the most part, boring and predictable.

The ancient and late antique intellectuals whom I spend my life studying wouldn't talk so much about miracles and portents if they could calm down and think about the numbers. The baseball fans thrilled to see the guy on a hitting streak come to the plate wouldn't be so disappointed when he struck out. Even people reading election returns would see much more normality lurking inside shocking results than television reporters can admit.

Heeding the law of regression to the mean would help us slow down, calm down, pay attention to the long term and the big picture, and react with a more strategic patience to crises large and small. We'd all be better off. Now, if only I could think of a good nursery rhyme for it.

SCIENTIFIC REALISM

REBECCA NEWBERGER GOLDSTEIN
Philosopher, novelist; recipient, 2014 National Humanities Medal;
author, *Plato at the Googleplex: Why Philosophy Won't Go Away*

Has science discovered the existence of protons and proteins, neurons and neutrinos? Have we learned that particles are excitations of underlying quantum fields and that the transmission of inherited characteristics is accomplished by way of information-encoding genes? Those who answer *no* (as opposed to *dunno*) probably aren't unsophisticated science deniers. More likely they're sophisticated deniers of scientific realism.

Scientific realism is the view that science expands upon—and sometimes radically confutes—the view of the world that we gain by means of our sense organs. Scientific theories, according to this view, extend our grasp of reality beyond what we can see and touch, pulling the curtain of our corporeal limitations aside to reveal the existence of whole orders of unobserved and perhaps unobservable things, hypothesized in order to explain observations and having their reference fixed by the laws governing their behavior. In order for theories to be true (or at any rate, approximations of the truth), these things must actually exist. Scientific theories are ontologically committed.

Those who oppose scientific realism are sometimes called scientific non-realists and sometimes, more descriptively, instrumentalists. Their view is that scientific theories are instruments for predictions that don't extend our knowledge of what exists beyond what is already granted to us by way of observation.

Sure, theories seem to make reference to new and exotic entities, but bosons and fermions don't exist the way raindrops on roses and whiskers on kittens do. Quantum mechanics no more commits us to the existence of quantum fields than the phrase "for our country's sake" commits us to the existence of sakes. The content of a theory is cashed out in observable terms. A theory is a way of correlating observable input with observable output, the latter known as predictions. Yes, between the report of *what has been observed* and the prediction of *what will be observed* there's a whole lot of theory, complete with theoretical terms that function grammatically as nouns. But don't, as Wittgenstein warned, let language "go on holiday." These theoretical nouns should be understood as convenient fictions, to be spelled out in operational definitions. Science leaves ontology exactly as it finds it.

Instrumentalism is so deflationary a view of science that one might think it was conceived in the bitter bowels of some humanities department determined to take science down a notch. But in fact, in the 20th century, instrumentalism became standard in physics for a variety of reasons, including the difficulties in solving the stubborn measurement problem in quantum mechanics. Then, too, there was strong influence wafting from the direction of logical positivism—the program that, in an effort to keep meaningless metaphysical terms from infiltrating our discourse and turning it into fine-sounding gibberish, had proposed a criterion of meaningfulness that pared the meaning of a proposition down to its mode of verification.

The thrust of these pressures drove many of the most prominent scientists toward instrumentalism, by which scientists could both wash their hands of an unruly quantum reality, rife with seeming paradox, while also toeing the strict positivist line (as evidenced by the frequent use of the word "meaningless"). The Copenhagen interpretation, which was accepted as standard, dismissed the question of whether the electron was really a particle or a wave as meaningless and asserted that to ask where the electron was in between measurements was likewise meaningless.

There were, of course, scientists who resisted—Einstein, Schrödinger, Planck, de Broglie, and, later on, David Bohm and John Stewart Bell, staunch realists all. Said Einstein: "Reality is the business of physics," which is about as simple and direct a statement of scientific realism as can be. But Einstein's realism marginalized him.

The Copenhagen interpretation is no longer the only game in town, and the main competition—for example, the many-worlds interpretation (in which what quantum mechanics describes is a plethora of equally realized possibilities, albeit existing in other worlds) and Bohmian mechanics (in which the unobserved particles are nonetheless real particles having actual positions and actual trajectories)—are realist interpretations. Though they differ in their descriptions of what reality is like, they unflinchingly commit themselves to there being a reality they are attempting to describe. So far as their empirical content goes, all these interpretations are equivalent. They are, from the standpoint of instrumentalism, indistinguishable but from the standpoint of scientific realism vastly different.

The questions that press up behind the concept of scientific realism are still very much in play, and how we answer them makes a world of difference as to what we see ourselves doing when we're doing science. Are we employing a device that churns out predictions, or are we satisfying, in the most reliable way we have, our basic ontological urge to figure out where we are and what we are? Are we never carried, by way of science, beyond the contents of experience, or does science permit us to extend our reach beyond our meager sensory apparatus, en-

abling us to grasp aspects of reality—the elusive thing in itself—
that would be otherwise inaccessible?

What a different picture of science—and of us—these two
viewpoints yield. What then could be more central to the scientific mindset than the questions that swirl around scientific
realism, since without confronting these questions we can't even
begin to say what the scientific mindset amounts to.

FUNDAMENTAL ATTRIBUTION ERROR

RICHARD NISBETT

Professor of Psychology, University of Michigan; author, *Mindware*

Aristotle taught that a stone sinks when dropped into water because it has the property of gravity. Of course, not everything
sinks when dropped into water. A piece of wood floats, because
it has the property of levity. People who behave morally do so
because they have the property of virtue; people who don't
behave morally lack that property.

Molière lampoons this way of thinking by having a team of
physicians in *The Imaginary Invalid* explain why opium induces
sleep: namely, because of its dormitive power.

Lampoon or not, most of us think about the behavior of objects and people much of the time in purely dispositional terms.
It is properties of the object or person which explain its behavior. Modern physics replaced Aristotle's dispositional thinking
by describing all motion as due to the properties of an object
interacting in particular ways with the field in which it's situated.

Modern scientific psychology insists that explanation of the behavior of humans always requires reference to the situation the person is in. The failure to do so sufficiently is known as the "fundamental attribution error." In Stanley Milgram's famous obedience experiment, two-thirds of his subjects proved willing to deliver a great deal of electric shock to a pleasant-faced middle-aged man, well beyond the point where he became silent after begging them to stop on account of his heart condition. When I teach undergraduates about this experiment, I'm quite sure I'm not convincing a single one that their best friend might have delivered that amount of shock to the kindly gentleman, let alone that they themselves might have done so. They're protected by their armor of virtue from such wicked behavior. No amount of explanation about the power of the unique situation into which Milgram's subjects were placed is sufficient to convince them that their armor could have been breached.

My students, and everyone else in Western society, are confident that people behave honestly because they have the virtue of honesty, conscientiously because they have the virtue of conscientiousness. (In general, non-Westerners are less susceptible to the fundamental attribution error, lacking as they do sufficient knowledge of Aristotle!) People are believed to behave in an open and friendly way because they have the trait of extroversion, in an aggressive way because they have the trait of hostility. When people observe a single instance of honest or extroverted behavior, they're confident that in a different situation the person will behave in a similarly honest or extroverted way.

In actual fact, when large numbers of people are observed in a wide range of situations, the correlation between degree of honesty or extroversion or conscientiousness displayed in one situation and that displayed in another is 20 percent or less. Converting a 20-percent correlation to odds form, we can say that knowing that John was more honest than Carl in Situation A increases the odds that he'll be more honest than Carl in Situation B from the chance level of 50–50 to only about 55 percent. But people think there's an 80-percent chance that John will be more honest than Carl given that he was more honest in Situation A.

How can our opinions be so hopelessly miscalibrated with the facts? There are many reasons, but one of the most important is that we don't normally get trait-related information in a form that facilitates comparison and calculation. I observe John in one situation when he might display honesty or the lack of it, and then not in another for perhaps a few weeks or months. I observe Carl in a different situation tapping honesty and then not in another for many months.

This implies that if people could observe a large number of people in a large number of fixed situations they'd see that predictability from one situation to another is poor, and their calibration might improve. Our susceptibility to the fundamental attribution error—overestimating the role of traits and underestimating the importance of situations—has implications for everything from how to select employees to how to teach moral behavior.

HABITUATION

KATHERINE D. KINZLER
Associate Professor of Psychology and Human Development, Cornell University

In 1964, Robert Fantz published a brief paper in *Science* that revolutionized the study of cognitive development. Building on the idea that infants' gaze can tell you something about their processing of visual stimuli, he demonstrated that babies respond to familiarity and novelty differently. When infants see the same thing again and again, they look for less and less time—they habituate. When infants next see a new stimulus, they regain visual interest and look longer. Habituation establishes the status quo—the reality you no longer notice or attend to.

Developmental psychologists have expanded on this methodological insight to probe the fundamentals of human thinking. Applying the idea that babies get bored by the familiar and start to look to the new, researchers can test how infants categorize many aspects of the world as same or different. From this, scientists are able to investigate humans' early perceptual and conceptual discriminations of the world. Such studies of early thinking help reveal signatures of human thinking that can persist into adulthood.

The basic idea of habituation is exceedingly simple at its outset. And humans aren't the only species to habituate with familiarity; around the same time as Fantz's work, papers studying habituation in other species of infant animals were published. An associated literature on the neural mechanisms of learning and memory similarly finds that neural responses decrease after repeated exposure to the same stimulus. The punch line is clear: Organisms, and their neural responses, get bored. This intuitive boredom is etched in our brains and seen in babies' first visual responses. But the concept of habituation can also scale up to explain a range of people's behaviors, pleasures, and failures. In many domains of life, adults habituate too.

If you think about eating a whole chocolate cake, the first slice will almost certainly be more pleasurable than the last. It isn't hard to imagine being satiated. Indeed, the economic law of diminishing marginal utility describes a related idea. The first slice has a high utility or value to the consumer. The last one doesn't (and may even have negative utility, if it makes you sick). Adults' responses to pleasing stimuli habituate.

People are often at first unaware of how much they habituate. A seminal observation of lottery winners by psychologists Philip Brickman, Dan Coates, and Ronnie Janoff-Bulman found that after a while the happiness of lottery winners returned to baseline. The thrill of winning—and the pleasure associated with new possessions—wore off. Even among non-lottery winners, people overestimate the positive impact that acquiring new possessions will have on their lives. Instead, they habituate to a new status quo of having more things, and those new things become familiar and no longer bring them joy.

Behavioral economists such as Shane Frederick and George Loewenstein have shown that this "hedonic adaptation," or reduction in the intensity of an emotional response over time, can occur for both positive and negative life events. In addition to shifting their baseline of what they see as normal, people start responding with less intensity to circumstances they're habituated to. Over time, highs become less exhilarating, but lows also become less distressing.

Habituation may serve a protective function by helping people cope with difficult circumstances, but it can also carry a moral cost. People get used to situations, including those that (without prior experience) would be considered morally repugnant. Think of the frog in boiling water—it's only because the temperature is raised gradually that it doesn't jump out. In the (in)famous Milgram studies, participants were asked to shock a confederate by increasing the voltage incrementally; they weren't asked to administer a potentially lethal shock at the outset. If you've already administered many smaller shocks, adding just one more may not overwhelm the moral compass.

Future research exploring our propensity to habituation may help explain the situations that lead to our moral failures—to Hannah Arendt's "banality of evil." Reports of workplace misconduct find that large transgressions in business contexts often start with small wrongdoings—subtle moral breaches that grow over time. New studies are testing the ways our minds and brains habituate to dishonesty.

From the gaze of babies to the actions of adults, habituation can illuminate how people navigate their worlds, interpret familiar and new events, and make both beneficial and immoral choices. Many human tendencies, good and bad, are composed of smaller components of familiarity—slippery slopes people become habituated to.

GENERAL STANDARDIZATION THEORY

ROLF DOBELLI
Founder, Zurich.Minds Foundation; author, *The Art of Thinking Clearly*

Try building a tower by piling irregular stones on top of each other. It can be done, eight, nine, sometimes ten stones high. You need a steady hand and a good eye to spot each rock's surface features. You find such man-made "Zen stone towers" on riverbanks and mountaintops. They last for a while until the wind blows them over. What's the relationship here between skill and height? Take relatively round stones from a riverbank. A child of two can build a tower two stones high. A child of three with improved hand-eye coordination can manage three stones. You need experience to get to eight stones. And you need tremendous skill and a lot of trial and error to go higher than ten. Dexterity, patience, and experience can get you only so far.

Now try with a set of interlocking toy bricks as your stones. You can build much higher. More important, your three-year-old can build as high as you can. Why? Standardization. The stability comes from the standardized geometry of the parts. The advantage of skill is vastly diminished. The geometry of the interlocking bricks corrects the errors in hand movement. But structural stability is standardization's least impressive feat. Its advantages for collaboration are much more significant.

We've long appreciated the advantages of standardization in business. In 1840, the U.S. had more than 300 railroad companies, many with different gauges. Many companies refused to agree on a standard gauge because of heavy sunk costs and the need for barriers to competition. Where two rail lines connected, men had to offload the cargo, sometimes store it, and then load it onto new cars. In a series of steps, some by top-down enactment but mostly by bottom-up coordination, the industry finally standardized gauges by 1886. Other countries saw similar "gauge wars." England ended them by legislation in 1856.

In the last 100 years, every national government and supranational organization and virtually every industry has created bodies to deal with standardization. They range from the International Organization for Standardization (ISO) to the World Wide Web Consortium (W3C) to bodies like the Bluetooth Special Interest Group. Their goals are always a combination of improved product quality, reputation, safety, and interoperability.

What's the best way to achieve optimal standards? While game theory (coordination games) offers a vast body of knowledge, setting standards in the real world isn't easy. However, the advantages are huge. Thus, landing at a relatively low local peak is vastly preferable to no coordination. Let's call the sum of this theoretical and practical knowledge from management and game theory the special theory of standardization—akin to Einstein's special theory of relativity. However, standardization is a vastly more powerful concept, one that might lead to a general theory of standardization. Let's look at a few domains that are undergirded by standardization.

Take matter, which ranges from the elementary particles to the periodic table with its standardized atoms to an endless number of discrete molecules. Simple chunkiness doesn't seem to be enough to build a universe. Apparently that requires *standardized* chunks. From a general-standardization-theory point of view: Is this the optimal standard or just a local peak? Or take living matter. A cell can work only with standardized building blocks (amino acids, carbohydrates, DNA, RNA, etc.). Could something as complex as a cell ever work outside of standards? A general standardization theory might provide answers on the limits of complexity that can be achieved without standards.

Further up the chain, in biology, the question is how to get huge numbers of unrelated individuals to cooperate flexibly. Some anthropologists name the invention of religion as the solution. Others suggest the evolution of moral sentiments, the invention of written law, or Adam Smith's Invisible Hand. I suggest that standardization is at least part of the solution. People can cooperate in ample numbers without standards, through all the known mechanisms. But eventually groups that use standards outpace groups that don't. Is there a threshold where cooperation breaks down without the injection of standards?

My hypothesis: Yes, but it's much higher than Dunbar's number of approximately 150 individuals—possibly in the tens of thousands. Interestingly, only *Homo sapiens* and no other animal devised standardization. Then again, this advance took even humans a long time—until the fifth millennium BC, which brought the standardization of language (writing), the standardization of value (money), and standardized weights.

SCALING

ASHVIN CHHABRA

President, Euclidean Capital; author, *The Aspirational Investor*

The word "scale," when it refers to an object, may refer to a simple ruler. However, the act of measurement is the start of a deep relationship between geometry, physics, and many important human endeavors.

Scaling, in the geometrical context, is the act of re-sizing an object while preserving certain essential characteristics, such as its shape. It indicates a relationship that is robust under certain transformations.

One uncovers scaling relationships by measuring and plotting key variables against each other. The simplest scaling relation is a linear one—yielding a straight line on an x-y plot denoting proportionality. The number of miles you can travel before refueling scales roughly linearly with the amount of fuel left in your gas tank. Twice the gas, double the miles you can drive.

More complicated scaling functions include power laws, exponentials, and so on, each of which reflects the underlying spatial geometry or dynamics of a system. The circumference of a sphere scales proportionally with its radius. The surface area and volume, on the other hand, scale as the square and cube of the radius, respectively. These power-law relationships show up as straight lines on a log-log plot. The slopes reflect the dimension of the object being measured, demonstrating that length is one-dimensional, area is two-dimensional, and volume is three-dimensional. This is true regardless of the shape of these solids, so plots using other solids will show the same robust scaling relationships and scaling exponents.

But sometimes length doesn't scale with an exponent of.1. What then?

Lewis Fry Richardson's pioneering measurements of the length of coastlines across the world showed that measuring with increasing precision gave ever-increasing answers for their lengths. The various data sets (measurements with different precision) for each coastline scaled using an exponent greater than 1.

Benoit Mandelbrot, in his classic paper "How Long is the Coast of Britain?," provided the insight needed to understand this result. As one measures the coastline with ever-increasing precision, one measures and adds the lengths of ever smaller fjords, nooks, and crannies. The coastline is not a smooth object; rather, it is self-similar. Large fjords within it contain many small fjords, which in turn contain ever smaller fjords. It is jagged and undulating enough to be mathematically more than a line but less than a surface. This insight led to the development of fractal geometry—the geometry that describes irregular, self-similar objects, ranging from coastlines and mountains to broccoli. Systems that have this property of self-similarity show power-law scaling—alluding to a beautiful link between geometry and physics. Incorporating fractal geometry is also important for computer algorithms to generate the realistic-looking artificial worlds in movies and video games which we now take for granted.

More generally, a variety of natural phenomena appear scale-invariant (or look the same) when an important underlying physical parameter is changed (or re-scaled) in a specific way.

This physical parameter in many cases is a carefully constructed dimensionless quantity—e.g., the ratio of two key length scales in the system being studied. The equations describing the phenomena must then obey and thus reflect the observed scale invariance. Richardson's delightful ditty captures the scale invariance of the Navier-Stokes equations of fluid turbulence, which remain unsolved to this day.

Big whirls have little whirls that feed on their velocity, and little whirls have lesser whirls and so on to viscosity.

The laws of physics embody scaling. Newton's law of gravitation states that the attractive gravitational force between two bodies scales linearly with each of their masses and inversely with the square of the distance between them.

If you get large enough or small enough, almost any scaling law observed in the real world breaks down. This makes us think about the range of applicability and what causes the breakdown. The scaling relations implied by Newton's laws break down at very small distances—giving way to quantum mechanics. The same happens at very large velocities, giving way to Einstein's special theory of relativity. Thus it's often useful to refer to a scale, in order to specify the range over which a theory or observed phenomenon is applicable: the quantum (subatomic) scale, the human scale, the astronomical scale, and so on.

Scaling concepts have found broad applicability in unexpected areas, such as the study of social networks. Technology companies are often unconstrained by geography and can easily scale up their user base. Often, early-stage technology startups show exponential increases in usage that are then reflected in exploding valuations. Supply/demand problems requiring resources to scale exponentially are simply unsustainable. As companies mature, the exponential scaling and valuation must level off. On the other hand, power-law relationships and solutions indicate a business that may be scalable for a sustained period and are sought after by investors with a longer time horizon.

During the first Internet bubble (circa 2000), companies rushed to get users, and valuations were based on the number of people using their Web sites. Today, social-network companies value users on the strength and quality of their interactions with other users. One can argue that valuations of these companies should scale quadratically with the number of users. It remains to be seen if such scaling arguments used to justify higher valuations hold up, compared with more traditional measures, such as revenue and profitability of a company.

Scaling relationships inevitably break down beyond a certain range—an important clue that other effects, either ignored or not yet considered, are now important. One should not ignore these deviations. While solving for persistent social problems that appear at all scales, one should remember: What works for a family may not work for a business. What works for a business may not work for a nation.

THE MENGER SPONGE

CLIFFORD PICKOVER
Author, trilogy: *The Math Book, The Physics Book, The Medical Book*

In the 2006 Taiwanese thriller *Silk*, a scientist creates a Menger Sponge—a special kind of hole-filled cube—to capture the spirit of a child. The sponge not only functions as an antigravity device but seems to open a door into a new world. As fanciful as this film concept is, the Menger Sponge considered by mathematicians today is certainly beautiful to behold when rendered using computer graphics, and it's a concept that ought to be more widely known. Certainly it provides a wonderful gateway to fractals, mathematics, and reasoning beyond the limits of our intuition.

The Menger Sponge is a fractal object with an infinite number of cavities—a nightmarish object for any dentist to contemplate. The object was first described by the Austrian mathematician Karl Menger in 1926. To construct the sponge, we begin with a mother cube and subdivide it into twenty-seven identical smaller cubes. Next, we remove the cube in the center and the six cubes that share faces with it. This leaves behind twenty cubes. We continue to repeat the process forever with smaller and smaller cubes. The number of cubes increases as $20n$, where n is the number of iterations performed on the mother cube. The second iteration gives us 400 cubes, and by the time we get to the sixth iteration we have 64,000,000 cubes.

Each face of the Menger Sponge is called a Sierpiński carpet. Fractal antennae based on the Sierpiński carpet are sometimes used as efficient receivers of electromagnetic signals. Both the carpets and the entire cube have fascinating geometrical properties. For example, the sponge has infinite surface area while enclosing zero volume. Imagine the skeletal remains of an ancient dinosaur that has turned into the finest of dust through the gentle acid of time. What remains seems to occupy our world in a ghostlike fashion but no longer "fills" it.

The Menger Sponge has a fractional dimension (technically referred to as the Hausdorff dimension) between a plane and a solid, approximately 2.73, and it has been used to visualize certain models of a foamlike spacetime. The origami artist Dr. Jeannine Mosely has constructed a Menger Sponge model from over 65,000 business cards; it weighs about 150 pounds.

It's important for the general public to become familiar with the Menger Sponge, partly because it reaffirms the idea that the line between mathematics and art can be a fuzzy one; the two are fraternal philosophies, formalized by ancient Greeks like Pythagoras and Ictinus and dwelled on by such greats as Fra Luca Bartolomeo de Pacioli (c.1447–1517), the Italian mathematician and Franciscan friar who published the first printed illustration of Leonardo's rhombicuboctahedron, in *De divina proportione*. The rhombicuboctahedron, like the Menger Sponge, is a beauty to behold when rendered graphically—an Archimedean solid with eight triangular faces and eighteen square faces, with twenty-four identical vertices and one triangle and three squares meeting at each.

Fractals such as the Menger Sponge often exhibit self-similarity, which suggests that various exact or inexact copies of an object can be found in the original object at smaller scales. The detail continues for many magnifications—like an endless nesting of Russian dolls. Some of these shapes exist only in abstract geometric space, but others can be used as models for complex natural objects such as coastlines and blood-vessel branching. The dazzling computer-generated images can be intoxicating, perhaps motivating students' interest in math as much as any other mathematical discovery in the last century.

The Menger Sponge reminds students, educators, and mathematicians of the need for computer graphics. As Caltech computer scientist Peter Schroeder once wrote: "Some people can read a musical score and in their minds hear the music. . . . Others can see, in their mind's eye, great beauty and structure in certain mathematical functions. . . . Lesser folk, like me, need to hear music played and see numbers rendered to appreciate their structures."

THE HOLOGRAPHIC PRINCIPLE

DONALD D. HOFFMAN

Professor of Cognitive Science, UC Irvine; author, *Visual Intelligence*

The most famous case study in science, before Freud, was published in 1728 in the *Philosophical Transactions of the Royal Society* by the English surgeon William Cheselden, who attended Isaac Newton in his final illness. It bore a snappy title: "An Account of Some Observations Made by a Young Gentleman, Who Was Born Blind, or Lost His Sight so Early, That He Had no Remembrance of Ever Having Seen, and Was Couch'd between 13 and 14 Years of Age."

The poor boy "was couch'd"—his cataracts removed—without anesthesia. Cheselden reported:

> When he first saw, he was so far from making any Judgment about Distances, that he thought all Objects whatever touch'd his Eyes, (as he express'd it) as what he felt, did his Skin; . . . We thought he soon knew what Pictures represented, which were shew'd to him, but we found afterwards we were mistaken; for about two Months after he was couch'd, he discovered at once, they represented solid Bodies.

The boy saw, at first, patterns and colors pressed flat upon his eyes. Only weeks later did he learn to perform the magic that we daily take for granted: to inflate a flat pattern at the eye into a three-dimensional world.

The image at the eye has but two dimensions. Our visual

world, vividly extending in three dimensions, is our holographic construction. We can catch ourselves in the act of holography each time we view a drawing of a Necker cube—a few lines on paper which we see as a cube, enclosing a volume, in three dimensions. That cubic volume in visual space is, of course, virtual. No one tries to use it for storage. But most of us—both layman and vision-science expert—believe that volumes in visual space usually depict, with high fidelity, the real volumes of physical space, volumes that can properly be used for storage.

But physics has a surprise. How much information can you store in a volume of physical space? We learn from the pioneering work of physicists such as Gerard 't Hooft, Leonard Susskind, Jacob Bekenstein, and Stephen Hawking that the answer depends not on volume but on area. For instance, the amount of information you can store in a sphere of physical space depends only on the area of the sphere. Physical space, like visual space, is holographic.

Consider one implication. Take a sphere that is, say, a meter across. Pack it with six identical spheres that just fit inside. Those six spheres, taken together, have about half the volume of the big sphere, but about 3 percent more area. This means that you can cram more information into six smaller spheres than into one larger sphere that has twice their volume. Now, repeat this procedure with each of the smaller spheres, packing it with six smaller spheres that just fit. And then do this, recursively, a few hundred times. The many tiny spheres that result have an infinitesimal volume but can hold far more information than the original sphere.

This shatters our intuitions about space and its contents. It's natural to assume that spacetime is a fundamental reality. But the holographic principle and other recent discoveries in physics tell us that spacetime is doomed—along with the objects it contains and their appearance of physical causality—and must be replaced by something more fundamental if we're to succeed, for instance, in the quest for a theory of quantum gravity.

If spacetime isn't fundamental, then our perception of visual space, and of objects in that space, is not a high-fidelity reconstruction of fundamental reality. What, then, is it? From the theory of evolution, we can conclude that our sensory systems have been shaped by natural selection to inform us about the fitness contingencies relevant to us in our niche. We've assumed that this meant that our senses inform us of fitness-relevant aspects of fundamental reality. Apparently they don't. They simply inform us about fitness, not fundamental reality.

In this case, the holographic principle points to a different conception of our perception of visual space. It isn't a reconstruction of an objective and fundamental physical space. It's simply a communication channel for messages about fitness and should be understood in terms of concepts that are standard for any communication channel—concepts such as data compression and error correction. If our visual space is simply the format of an error-correcting code for fitness, this would explain its holographic nature. Error-correcting codes introduce redundancy to permit correction of errors. If I wish to send you a bit that's either 0 or 1, but there's a chance that noise might flip a 0 to a 1 or vice versa, then we might agree that I'll send you that bit three times instead of just once. This is a simple Hamming code. If you receive a 000 or a 111, you will interpret this, respectively, as 0 and 1. If you receive a 110 or 001, you will interpret this, respectively, as 1 and 0, correcting an error in transmission. In this case we use a redundant three-dimensional format to convey a lower-dimensional signal. The holographic redundancy in our perception of visual space might be a clue that this space, likewise, is simply an error-correcting code—for fitness.

What about physical space? Research by Fernando Pastawski, Beni Yoshida, Daniel Harlow, John Preskill, and others indicates that spacetime itself is an error-correcting code—a holographic, quantum, secret-sharing code. Why this should be so is, for now, unclear, and tantalizing.

But it's clear that the holographic principle can shatter false convictions, stir dogmatic slumbers, and push on where intuitions fear to tread. That's why we do science.

THE NAVIER-STOKES EQUATIONS

IAN McEWAN
Novelist; author, *Sweet Tooth; Solar; On Chesil Beach; Nutshell*

The 2017 *Edge* Question puts us in danger of resembling the man who looks for his dropped watch only under the light of a street lamp: The scientific concept that has the widest impact on our lives may not necessarily be the simplest. The Navier-Stokes equations date from 1822 and apply Newton's second law of motion to viscous fluids. The range of applications is vast—in weather prediction, aircraft and car design, pollution and flood control, hydroelectric architecture, in the study of climate change, blood flow, ocean currents, tides, turbulence, shock waves, and the representation of water in video games or animations.

The name of Claude-Louis Navier is to be found inscribed on the Eiffel Tower, whereas the Irishman, George Stokes, once president of the Royal Society, is not well known outside of math and physics. Among many other achievements, Stokes laid the foundations of spectroscopy. It needs a John Milton of mathematics to come down among us and metamorphose the equations into lyrical English (or French) so that we can properly celebrate their ingenuity and enduring use and revive the reputations of these two giants of 19th-century science.

THE SCIENTIST

STUART FIRESTEIN

Neuroscientist; Professor, Department of the Biological Sciences, Columbia University; author, *Failure: Why Science Is So Succcessful*

It's said that Charles Darwin left on the *Beagle* as a "natural philosopher" and returned as a scientist. Not because of anything he did while on the voyage, although he did plenty, but because in 1833 the polymath William Whewell, Cambridge professor and Master of Trinity College, invented the word "scientist." It wasn't the only word he coined (he also came up with "ion," "cathode," and "anode" for Michael Faraday), but it's perhaps the most influential. Until Whewell invented the word, all those people we would today call scientists—beginning with Aristotle and including Newton, Galileo, Mendel, Galen—were known as "natural philosophers." The distinction is revealing. Among the purposes of natural philosophers was to understand the mind of the Creator through the study of the natural world. The study of science was an intellectual pursuit not distinct from theological examination. But that was changing.

Whewell's suggestion of the term "scientist" was in response to a challenge from the poet Samuel Taylor Coleridge at a meeting of the British Association for the Advancement of Science in Cambridge. Coleridge, old and frail, had dragged himself to

Cambridge and was determined to make his point. He stood and insisted that men of science in the modern day should not be referred to as philosophers, since they were typically digging, observing, mixing, or electrifying—that is, they were empirical men of experimentation and not philosophers of ideas. The remark was intended to be both a compliment and a slight. Science was everyday labor, and philosophy was lofty thought. There was much grumbling among those in attendance, when Whewell masterfully suggested that in "analogy with 'artist' we form 'scientist.'" Curiously, this almost perfect linguistic accommodation of workmanship and inspiration—of the artisanal and the contemplative, the everyday and the universal—was not readily accepted. The term "scientist" came into popular use in America before it was generally adopted in England; and, indeed, for a time it was erroneously thought to have originated among those crass Americans then ruining the English language. It took some thirty years for it to come into general usage.

The root word "science" has now been regularly co-opted to mean the supposed rigorous practice of any area that used to be considered "just" scholarship. Thus we have library science, political science, linguistic science, etc., etc. Of course there's nothing wrong with rigor per se, only that appending the word "science" doesn't necessarily make it so. On the other hand, the word "scientist," the person who stands behind the concept, has not been thus co-opted—yet. The scientist is still recognizable as someone who does experiments, observes data, theorizes, and does her best to explain phenomena. She is still someone who tries hard not to be fooled, knowing scientists are the easiest people to be fooled (to paraphrase Richard Feynman). Most important, she knows that she knows far less than she doesn't know.

The objections of so many 19th-century scientists to the word "scientist" is instructive, because we can now see that its coinage was the beginning of a revolution in scientific practice no less disruptive than the first scientific revolution had been. Those natural philosophers were, with few exceptions, wealthy men who dabbled in scientific exploration because that was considered the highest form of intellectual pursuit. They were not workers or laborers, and they would never have described their scientific enterprise as something so pedestrian as a job. None were paid for their work, there were no grants, and no one would have thought to patent their work. But that was about to change. Science was indeed to become a career, a position in society and in the academy. At least in theory, anyone could become a scientist, given sufficient training and intellect. Science was professionalized, and the scientist was a professional.

But how unfortunate it is that we've lost Whewell's brilliant consilience (also a word he invented) between art and science—that "in analogy with 'artist' we form 'scientist.'" The professionalism of science has overtaken, in the public mind and the mindset of many scientific institutions, the importance of values like creativity, intuition, imagination, and inspiration in the scientific process. Too often believing there's a simple recipe for producing cures and gadgets—the so-called scientific method (invented by Whewell's contemporary and sometime sparring partner, Francis Bacon)—we're disappointed when scientists say they're uncertain or that there are changing opinions—i.e., new ideas—about this or that supposedly settled fact.

Coleridge, the poet whose quarrels goaded Whewell into inventing the scientist, was actually quite attracted to science (claiming it provided him with some of his best metaphors) and was a close friend and confidant of the famed chemist and head of the Royal Society, Humphry Davy. In an especially notable correspondence with Davy, Coleridge likened science to art because "being necessarily performed with the passion of Hope,

[it] is poetical." Perhaps the modern scientist meme should be updated to include more of the hopeful poet than the authoritarian demagogue.

BAYES' THEOREM

SEAN CARROLL
Research Professor of Physics, Caltech; author, *The Big Picture*

You're worried that your friend is mad at you. You threw a dinner party and didn't invite them; it's just the kind of thing they'd be annoyed about. But you're not really sure. So you send them a text: "Want to hang out tonight?" Twenty minutes later, you receive a reply: "Can't, busy." How are we to interpret this new information?

Part of the answer comes down to human psychology, of course. But part of it is a bedrock principle of statistical reasoning known as Bayes' theorem.

We turn to Bayes' theorem whenever we're uncertain about the truth of some proposition and new information comes to light affecting the probability of that proposition's being true. The proposition could be our friend's feelings, or the outcome of the World Cup, or a presidential election, or a particular theory about what happened in the early universe. In other words, we use Bayes' theorem all the time. We may or may not use it correctly, but it's everywhere.

The theorem itself isn't so hard: The probability that a proposition is true, given some new data, is proportional to the probability that it was true before that data came in times the likelihood of the new data if the proposition were true.

So there are two ingredients. First, the prior probability (or simply "the prior"), the probability we assign to an idea before we gather any new information. Then, the likelihood of some particular piece of data being collected if the idea is correct (or simply "the likelihood"). Bayes' theorem says that the relative probabilities for different propositions after we collect new data is just the prior probabilities times the likelihoods.

Scientists use Bayes' theorem in a precise, quantitative way all the time. But the theorem—or really, the idea of "Bayesian reasoning" that underlies it—is ubiquitous. Before you sent your friend a text, you had some idea of how likely it was that your friend was mad at you or not. You had, in other words, a prior for the proposition "mad" and another one for "not mad." When you received your friend's response, you implicitly did a Bayesian updating on those probabilities. What was the likelihood they would send that response if they were mad, and what was the likelihood if they weren't? Multiply by the appropriate priors, and you can now figure out how likely it is that they're annoyed with you, given your new information.

Behind this bit of dry statistical logic lurk two enormous, profound, worldview-shaping ideas.

One is the very notion of a prior probability. Whether you admit it or not, no matter what data you have, you implicitly have a prior probability for just about every proposition you can think of. If you say, "I have no idea whether that's true or not," you're really just saying, "My prior is fifty percent." And there's no objective, cut-and-dried procedure for setting your priors. Different people can sharply disagree. To one, a photograph showing what looks like a ghost is incontrovertible evidence for life after death; to another, the photograph is more likely to be a fake. Given an unlimited amount of evidence and perfect rationality, we should all converge to similar beliefs no matter what priors we start with—but neither evidence nor rationality is perfect or unlimited.

The other big idea is that your degree of belief in an idea should never go all the way to either zero or one. It's never absolutely impossible to gather a certain bit of data, no matter what the truth is. Even the most rigorous scientific experiment is prone to errors, and most of our daily data-collecting is far from rigorous. That's why science never "proves" anything; we just increase our credences in certain ideas until they're almost (but never exactly) 100 percent. Bayes' theorem reminds us that we should always be open to changing our minds in the face of new information, and tells us exactly what kind of new information we would need.

UNCERTAINTY

LAWRENCE M. KRAUSS
Theoretical physicist, cosmologist, Arizona State University; author, *The Greatest Story Ever Told—So Far*

Nothing feels better than being certain, but science has taught us over the years that certainty is largely an illusion. In science, we don't "believe" in things or claim to know the absolute truth. Something is either likely or unlikely, and we quantify *how* likely or unlikely. That's perhaps the greatest gift science can give.

That uncertainty is a gift may seem surprising, but it's precisely for this reason that the scientific concept of uncertainty needs to be better and more broadly understood.

Quantitatively estimating uncertainties—the hallmark of good science—can have a substantial effect on the conclusions we draw about the world, and it's the only way we can clearly counteract the human tendency to assume that whatever happens to us is significant.

The physicist Richard Feynman was reportedly fond of saying, "You won't believe what happened to me today!" and then adding, "Absolutely nothing!" We all have meaningless dreams night after night. Yet if we dream that someone breaks his leg and later hear that a cousin had an accident, it's easy to assume some correlation. But in a big and old universe, rare accidents will happen all the time. You need a healthy skepticism, because the easiest person to fool in this regard is yourself.

To avoid the natural tendency to impute spurious significance, all scientific experiments include an explicit quantitative characterization of how likely it is that results are as claimed. Experimental uncertainty is inherent and unremovable. It's not a weakness of the scientific method to recognize this, but a strength.

There are two different sorts of uncertainties attached to any observation. One is purely statistical. Because no measurement apparatus is free from random errors, any sequence of measurements will vary over some range, determined by the accuracy of the measurement apparatus and also by the size of the sample being measured. Say a million voters are asked to go to the polls a second time to check the results of an election and asked to vote for exactly the same candidate the second time around. Random measurement errors suggest that if the margin of difference was reported to be less than a few hundred votes the first time around, on the second round a different candidate might be declared the winner.

Take the recent "tentative" observation of a new particle at the Large Hadron Collider which would have revolutionized our picture of fundamental physics. After several runs, calculations suggested the statistical likelihood that the result was spurious was less than 1 percent. But in particle physics, we can usually amass enough data to reduce the uncertainty to a much smaller level—less than 1 in a million. (This isn't always possible in other areas of science before claiming a discovery.) And after more data was amassed, the signal disappeared.

There's a second kind of uncertainty, called *systematic uncertainty*, that's generally much harder to quantify. A scale, for example, might not be set to zero when no weight is on it. Experimenters can often test for systematic uncertainties by playing with their apparatus, readjusting the dials and knobs to see what the effect is, but this isn't always possible. In astronomy, one cannot fiddle with the universe. However, you can try to estimate systematic uncertainties in your conclusions by exploring their sensitivity to uncertainties in the underlying physics you use to interpret the data.

Systematic uncertainties are particularly important when considering unexpected and potentially unlikely discoveries. In the election example I offered earlier, an error was discovered in the design of the ballot, so that selecting one candidate sometimes ended up being recorded as a vote for two candidates, in which case the ballot would be voided. Even a very small systematic error of this type could affect the result in any close election.

In 2014, the BICEP2 experiment claimed to have observed gravitational waves from the earliest moments of the Big Bang. This would have been one of the most important scientific discoveries in recent times if it were true. However, a later analysis discovered an unexpected source of background noise—dust in our own galaxy. When all the dust settled, if you'll forgive the pun, it turned out that the observation had only a 92-percent likelihood of being correct. In many areas of human activity this would be sufficient to claim its validity. But extraordinary claims require extraordinary evidence. So the cosmology community has decided that no such claim can yet be made.

Over the past several decades we've been able to refine the probabilistic arguments associated with the determination of likelihood and uncertainty, developing an area of mathematics called Bayesian analysis which has turned the science of determining uncertainty into one of the most sophisticated areas of experimental analysis. Here we first fold in a-priori estimates of likelihood and then see how the evidence changes our estimates. This is science at its best: Evidence can change our minds, and it's better to be wrong rather than to be fooled.

In the public arena, scientists' inclusion of uncertainties has been used by some critics to discount otherwise important results. Consider the climate-change debate. The evidence for human-induced climate change is neither controversial nor surprising. Fundamental physics arguments have anticipated the observed changes. When the data show that the last sixteen years have been the warmest in recorded human history, and when measured CO_2 levels exceed those determined over the past 500,000 years, and when the West Antarctic ice sheet is observed to be melting at an unprecedented rate, the fact that responsible scientists report many small uncertainties associated with each of these measurements shouldn't discount the threat we face.

Louis Pasteur once said, "Fortune favors the prepared mind." Incorporating uncertainties prepares us to make better-informed decisions about the future. This doesn't obviate our ability to draw rational and quantitatively reliable conclusions on which to base our actions—especially when our health and security may depend on them.

302

EQUIPOISE

NICHOLAS A. CHRISTAKIS

Physician and social scientist, Yale University; co-author (with James H. Fowler), *Connected: The Surprising Power of Our Social Networks and How They Shape Our Lives*

There's an old word in our language, "equipoise," which has been around since at least the 16th century, when it meant something like "an equal distribution of weight." With respect to science, it's analogous to standing at the foot of a valley and not knowing the best way to climb to the top—poised, that is, between alternative theories and ideas about which, given current information, one is neutral. Use of the word peaked around 1840 and has declined roughly fivefold since then, according to Google Ngram, though it appears to be enjoying a resurgence in the last decade. But attention to equipoise ought to be greater.

The concept found a new application in the 1980s, when ethicists were searching for deep justifications to conduct randomized clinical trials in medicine. A trial was justified, they rightly argued, only when the doctors and researchers doing the trial considered (relying on their medical knowledge) the new drug and its alternative (a placebo, perhaps) as potentially *equally* good. If they felt otherwise, how could they justify the trial? Was it ethical, for the sake of research, to place patients at risk if the researchers suspected that one course of action might be materially better than another?

So *equipoise* is a state of equilibrium in which scientists cannot be sure which of the alternative theories they're contemplating might be true.

In my view, equipoise is related to that famous Popperian *sine qua non* of science itself: falsifiability. Something isn't science if it isn't capable of disproof. We can't even imagine an experiment that would disprove the existence of God—so that's what makes a belief in God religion. When Einstein famously conjectured that matter and energy warp the fabric of space and time, experiments to test the claim weren't possible but they were at least imaginable, so the theory was capable of disproof. Eventually he was proved right, based on astronomical observations of the orbit of Mercury, and also the bending of light from distant stars, observed during a 1919 solar eclipse—and most recently by the magnificent discovery by LIGO of gravitational waves from the collision of two black holes over a billion years ago. Yet even if he had been wrong, his conjecture would still have been scientific.

If falsifiability solves the "problem of demarcation" that Popper identified between science and non-science, equipoise addresses the problem of origin: Where ought scientists to start from? Thinking about where scientists do—and should—start from is often lacking. Too often, we simply begin from where we are. In some ways, therefore, equipoise is an antecedent condition to falsifiability. It is a state we can be in before we hazard a guess that we might test. It's not quite a state of ignorance but rather a state of quasi-neutrality, when glimmers of ideas enter our mind.

Scientific equipoise tends to characterize fields both early and late in their course, for different reasons. Early in a field or a new area of research, it's often true that little is known about anything, so any direction can seem promising and might actually be productive. An exciting neutrality prevails. Late in the exploration of a field, much is known, so it might be hard to head toward new things—or the new things, even if true, might be small or unimportant. An oppressive neutrality can rule.

Equipoise carries with it aspects of science which are sorely needed these days. It connotes judgment, for it asks what problems are worthy of consideration. It connotes humility, for we don't know what lies ahead. It connotes open vistas, because it looks out at the unknown. It connotes discovery, because, whatever way forward we choose, we will learn something. And it connotes risk, because there are dangers in embarking on unknown journeys.

Equipoise is a state of hopeful ignorance, the quiet before the storm of discovery.

ANSATZ

NEIL GERSHENFELD
Physicist; Director, MIT's Center for Bits and Atoms; author, *FAB*

"Ansatz" is a fancy way to say that scientists make stuff up.

The most common formulation of physics is based on what are called differential equations, which are formulas that relate the rate at which things change. Some of these are easy to solve, some are hard to solve, and some can't be solved. It turns out that there's a deep reason why there's no universal way to find these solutions, because if that existed it would let you answer questions we know to be uncomputable (thanks to Alan Turing).

But differential equations do have a handy property: Their solutions are unique. If you find *a* solution, it's *the* solution. You can guess a solution, try it out, and fiddle with it to see if you can make it work. If it does, your guess is justified after the fact. That's what an ansatz is—a guess that you test. It's a German word that could be translated as "initial placement," "starting point," "approach," or "attempt."

Hans Bethe famously demonstrated this technique in 1931, with an ansatz for the behavior of a chain of interacting particles. His solution has since been used to study such systems as electrons in a superconducting wire that can carry current without resistance and trapped atoms in a quantum computer.

There's a similar concept in probability, called a prior. This is a guess you make before you have any evidence. Once you do make observations, the prior gets updated to become what's called a posterior. It's initially equally plausible for the universe to be explained by a Flying Spaghetti Monster or the Feynman *Lectures on Physics*; the latter becomes more probable once its predictions are tested.

Finding an ansatz or a prior is a creative, rigorous process; they can come from hunches or whims or rumors. The rigor is in how you evaluate them. You could call this a hypothesis, but the way that term is taught misses both how these can start without justification and how you initially expect them to be wrong but then patch them up.

My favorite approach to research management is "ready, fire, aim." You have to get ready by doing your homework in an area, then do something without thinking too much about it, then think carefully about what you just did. The problem with the more familiar "ready, aim, fire" is that if you aim first, you can't hit anything unexpected. There's a sense in which everything I've ever done in the lab has failed at what I set out to do, but as a result something better has consistently happened.

Research progress is commonly expected to meet milestones. But a milestone is a marker that measures distance along a highway. To find something that's not already on the map, you need to leave the road and wander about in the woods beside it. The technical term for that is a biased random walk, which is how bacteria search for gradients in chemical concentrations. The historical lesson is just how reliable that random process of discovery is.

The essential misunderstanding between scientists and non-scientists is the perception that scientific knowledge emerges in a stately procession of received knowledge. As a result, ambiguity isn't tolerated and changing conclusions are seen as a sign of weakness. Conversely, scientists shouldn't defend their beliefs as privileged; what matters isn't where they come from but how they're tested. Science appears to be goal-directed only after the fact. While it's unfolding, it's more like a chaotic dance of improvisation than a victory march. Fire away with your guesses, then be sure to aim.

"ON THE AVERAGE"

ROBERT SAPOLSKY
John A. and Cynthia Fry Gunn Professor, Professor of Neurology and Neurosurgery, Stanford University; author, *Behave: The Biology of Humans at Our Best and Worst*

I'm voting for this concept, one so central to the scientific process, so much a given, that hardly any scientist ever utters those words.

Scientists present their work—say, "We manipulated variable X, and observed that this caused Z to happen," or "We measured this and found that it takes Z amount of time to happen." And when they do, most of the time what they're actually saying is, "We manipulated variable X, and observed that, on the average, this caused Z to happen." "We measured this and found that, on the average, it takes Z amount of time to happen." Everyone knows this.

Of course. Everyone in a population doesn't have the exact same levels of something-or-other in their bloodstream. A causes B to happen most of the time, but not every single time. There's variability.

When scientists present their data, they typically display the average—the mean, the X on a graph, the bar of a particular height in the figure. And it always comes with an error term—a measure of how much variability there was around that average, a measure of how much confidence there is in saying "on the average." Measure something-or-other in three people, observe values of 99, 100, and 101, producing an average of 100; measure that something in three other people and observe values of 50, 100, and 150, producing an average of 100. These are two very different circumstances; "The average was 100" tells you a lot more about how some sliver of the universe works in the first case than in the second.

This is how scientists in most disciplines go about their business, with the recognition that you're always seeing how things work on the average. So why is it important that this be more widely known? I can think of at least three reasons, of increasing importance.

First, this should constitute a big shout-out to scientists and the scientific process. Perhaps counter to the general perception, scientists don't pronounce upon some fact that they've discovered; they pronounce upon the temporary way station of statistical confidence that they've discovered. "On the average" means there's stuff you can't explain yet, and there's even the possibility that you're entirely wrong. It's a badge of the humility that defines science, when things are working right. And it sure couldn't hurt if that sort of humility became more commonplace in lots of other settings.

The second reason is that the variability around an average is usually much more interesting than the average itself. "More interesting" in the scientific sense. If variability means there's stuff you can't explain yet, it's also the guide to where to look to understand things more, to identify previously unappreciated factors giving rise to the variability.

For example, "On the average, having a particular variant of a gene produces a particular behavior in people . . . unless, as it then turned out with additional research, someone had a particular type of childhood." Pursuing the question of why there are exceptions reveals all sorts of things about environmental regulation of gene transcription, child development, gene/environment interactions, and so on.

Moreover, variability is often more interesting in the human sense as well. All things considered, it's not that exciting that humans average a score of 100 on this thing called an IQ test. It's the variation that interests us. Or that, on the average, adult human males can run 100 meters in, say, 25 seconds. It's Usain Bolt that gets our attention. Or that there's an average life expectancy; it's what you and your loved ones are destined for that matters. Crowds of protestors don't gather in some nation's capital because of the average income in that country; it's because of the magnitude of the variance, the extent of inequality.

The third reason is the most important and subtle and ultimately has little to do with science. Take a population of people. Figure out their average height. Their average weight. Average IQ, shoe size, number of friends, hip/waist ratio, radiance of smile, symmetry of face, athletic skill, sex drive, scores on psych instruments that measure perkiness or optimism or gumption. Define your average human across these parameters. And then good luck trying to find such a person. Because they don't exist. Even if someone seems to be, say, the average weight, they won't really be if you look closely enough, measuring things out to the level of grams, or milligrams, or micrograms, or. . . . Nothing and no one is precisely average, because "averageness" is an emergent property of populations, an artificial construct. It's like a strange attractor in chaotic systems, which oscillates around a singular point, a hypothetical average that, no matter how closely you look, is never actually achieved. Oh, it does "on the average," but never in reality.

This matters because psychologically we tend to morph "average" into "the norm" and then into "normal" or "ideal." And what that means is that we all always come up short in achieving what we've labeled as normal and ideal; we're all a little too heavy, or too tall, with a nose that's a little too much this, a personality that's a bit too little that. We all deviate from the norm, from something that's an artificial, statistical construct that doesn't really exist. We're all "abnormal," in a sense that's more pejorative than statistical. And thus we feel bad about who we are. What "on the average" truly means in populations is liberating.

BLIND ANALYSIS

SARAH DEMERS
Horace D. Taft Associate Professor of Physics, Yale University

When we measure a value consistent with our prediction, our tendency is to trust the result. When we get an unexpected answer, we apply greater scrutiny, trying to determine whether we've made an error before believing the surprise. This tilts us toward revealing errors in one category of measurements and leaving them unexposed in another. It's particularly dangerous when our assumptions are flawed—and if there's anything we should bet on, given the history of progress within physics, it's that our underlying assumptions are in some way flawed. Bias can creep into the scientific process in predictable and unpredictable ways.

Blind analyses are employed as a protection against bias. The idea is to fully establish procedures for a measurement before we look at the data, so we can't be swayed by intermediate results. Such analyses require rigorous tests along the way, to convince ourselves that the procedures we develop are robust and that we understand our equipment and techniques. We can't "unsee" the data once we've taken a look.

There are options when it comes to performing a blind analysis. If you're measuring a particular number, you can apply a random offset to the number which is stored but not revealed to the analyzers. You complete the full analysis and reveal the offset and true result only when the work is done. Another method is to designate a sensitive segment of the data, the "signal," as off limits. You don't look at the signal until you're convinced you understand the remaining data, the "background." You can fully develop your analysis using the background and a simulated fake signal. Only when the analysis is fully developed do you look at the signal and obtain the result, a process known as "opening the box." Another flavor of blind analysis is employed by LIGO in the search for gravitational waves. Fake signals are periodically inserted into the data, so that full analyses are undertaken without analyzers knowing whether the signals they're seeing are real. LIGO carries the analyses all the way to the point of preparing the corresponding publication before learning whether they've analyzed real or fake data.

Blind analyses force scientists to approach their work with humility, acknowledging the potential for bias to influence the process. They require both creativity and rigor as they establish an understanding of the data without direct access to it. They enforce good stewardship of the data, which can represent significant investments in experiments that aren't easily repeatable. They highlight the mystery and anticipation inherent in the discovery process, in which opening the box may well reveal a surprise. Humility, rigor, stewardship, and mystery are the essential ingredients of blind analyses and represent the best that science has to offer.

HOMOPHILY

MATTHEW O. JACKSON

William D. Eberle Professor of Economics, Stanford University; External Faculty, Santa Fe Institute; author, *Social and Economic Networks*

No, homophily has nothing to do with sexual orientation. In the 1950s, a pair of sociologists, Paul Lazarsfeld and Robert Merton, coined the term to refer to the pervasive tendency of humans to associate with others like themselves.

Even if you don't know homophily by name, you have experienced it throughout your life. In whatever elementary school you went to, in any part of the world, girls tended to be friends with girls and boys with boys. If you went to a high school that had people of more than one ethnicity, you saw homophily there. Yes, you may have been friends with someone of another ethnicity, but such friendships are the exception rather than the rule. We see strong homophily by age, ethnicity, language, religion, profession, caste, and income level.

Homophily is not only instinctual (just watch people mingle at any large social event where they're all strangers) but also makes sense, for many reasons. New parents learn from talking with other new parents and help care for one another's children. People of the same religion share beliefs, customs, holidays, and norms of behavior. By the very nature of any workplace, you'll spend most of your day interacting with people in the same profession and often in the same field. Homophily also helps us navigate our networks of connections. If you need to know a doctor's reputation, which one of your friends would you ask? The one who's in the healthcare industry, of course, as they would be the most likely to know the doctor or know someone who knows the doctor. Without homophily, you'd have no idea of whom to ask.

As simple and familiar as it is, homophily is very much a scientific concept: It's measurable and has predictable consequences. In fact, it's so ubiquitous that it should be thought of as a fundamental scientific concept. But it's the darker side of homophily that makes it such an important one.

As the world struggles with inequality and immobility, we can debate the importance of the role political regimes or capital accumulation plays, but we miss a primary constraint on upward mobility if we ignore homophily. To understand why many young Americans join gangs and so many end up shot or in jail before their twenty-fifth birthday, we've only to look at what they observe and experience from a young age. If we want to know why universities like Stanford, Harvard, and MIT have more than twenty times as many students from the top quarter of the income distribution as from the bottom quarter, homophily is a big part of the answer. High school students in poor neighborhoods often have little idea of the financial aid available or the benefits of higher education, or even what higher education is. By the time they talk to a high school counselor who might have a few answers, it's much too late. Homophily affects how their parents raise them, the culture they experience, the role models they see, their beliefs, the opportunities coming their way, and ultimately the expectations they have for their lives.

Although we're all familiar with homophily, thinking of it as a measurable phenomenon may add to the discourse on how to increase upward mobility and decrease inequality around the world. Achieving these goals requires recognizing that persistent segregation by income and ethnicity prevents information and opportunities from reaching those who need them most. Homophily lies at the root of many social and economic problems, and understanding it can help us better address them—from inequality and immobility to political polarization.

SOCIAL IDENTITY

ZIYAD MARAR

Global Publishing Director, SAGE Publications; author, *Intimacy: Understanding the Subtle Power of Human Connection*

We know we're ultrasocial animals, yet we have a consistent blind spot about how truly social we are. Our naïve realism fosters a self-image as individual, atomistic, rational agents experiencing life as though peering out on the world through a window. Like the fish unaware of the water in which it swims, we struggle to see the social reality in which our actions are meaningfully conducted.

Contrary to this view, psychology has shown repeatedly how deeply permeated each of us is by a social identity. This is an important corrective to our illusory self-image and gives us more insight into our social natures. But even when our social identity enters the picture, it's often crucially misunderstood.

Social identity has been explored in earnest ever since World War II, in an attempt to understand how ordinary people could have committed, or at least allowed, genocidal horrors. Much of this work, such as the Milgram experiments on obedience, has suggested that if you dial up the social, you dial down the thinking individual. The social psychologist Henri Tajfel's minimal-group experiments divided boys into two arbitrary groups, each group affiliated with a painter they'd never heard of, and showed how quickly they started discriminating against boys in the other group. All it took was the creation of a meaningless boundary to create an in-group, an out-group, and the conditions for favoritism and conflict. But this important insight, explored in many other contexts over the decades, has led to only a partial understanding of social identity and to unfortunate misinterpretations. Phrases like "bystander effect," "diminution of responsibility," "groupthink," "herd or mob mentality," and so on, suggest that when we join groups we lose our minds and become malleable, prone to irrational or regrettable actions.

But this view gets it backward to some extent. To introduce the social is not to add distortion to otherwise clear thinking. For good and for ill, our social identities are minded, not mindless. Two social psychologists, Stephen Reicher and Mark Levine, studied whether or not British football fans were willing to help an injured fan of the opposing team. They found that if the subjects thought of themselves as fans of Manchester United, they wouldn't be inclined to help a Liverpool fan, but if they saw themselves as football fans generally, they would stop and help.

Contrary to the stereotype of fans as mindless thugs, they are highly minded, their actions depending on which group they see themselves as belonging to.

The important point is that social identities can change, and as they do, the view of who is "one of us" changes, too. My sense of myself as a father, a publisher, a Londoner, a manager, or as someone with Arabic heritage and family, profoundly shapes the decision space around what's rational for me to think and do. My allegiances, self-esteem, prejudices, willingness to be led or influenced, sense of fairness, sense of solidarity, biases about "people like me," are all to an extent shaped by the collective self salient to me at the time. This isn't to deny my individuality but to recognize how it's irreducibly expressed through a social lens—and that my social identity changes how it makes sense for me to engage with the world.

This matters, because when we see ourselves purely as rational, individual actors, we miss the fact that the social is not just the context in which we act but also deeply constitutive of who we are. If we adopt the collective view and see only irrational actions—"mad" rioters, "crazy" extremists, "evil" people who have ideological commitments different from our own—we're condemned to judging others with no chance of understanding them. A better comprehension of our true social identities would not just enable us to better understand those we might ordinarily dismiss as irrational but also help us better understand our ultrasocial selves.

REFLECTIVE BELIEFS

HUGO MERCIER

Cognitive scientist, CNRS [Centre Nationale de la Recherche
Scientifique], Lyons; co-author (with Dan Sperber), *The Enigma of Reason*

Do Christians believe that God is omniscient in the same way
they believe there's a table in the middle of their living room?
We can refer to both attitudes using the same term, "belief," and
Christians would readily assent to both. But these two beliefs are
markedly different.

The belief about the living room table is, so to speak, free to
roam our minds, guiding our behavior (we must go around the
table; we can put dishes on it) and our inferences (a child might
use it as a hiding place; its size limits how many guests we can
have for dinner).

By contrast, the belief about God's omniscience is more constrained. It guides some behaviors—for instance, verbal behavior
when quizzed on the subject—but not others. Believers in God's
omniscience might still try to hide actions or thoughts from
God. They sometimes try to attract God's attention. Believers in
God's omnipotence may still imagine God attending to prayers
one after the other.

That people behave and draw a variety of inferences in a way
that ignores or contradicts some of their beliefs is obvious to
some extent, but it has also been experimentally demonstrated.
It's true of a variety of religious beliefs and for many scientific beliefs as well. You learned in school that the Earth revolves

around the sun, but you may still think of the sun as rising in the east and setting in the west.

To help explain these apparent contradictions, the cognitive scientist Dan Sperber has introduced a distinction between intuitive and reflective beliefs. Intuitive beliefs are formed through simple perceptual and inferential processes. They can also be acquired through communication, provided the information communicated is of a kind that could have been picked up by simple perception and inference. For instance, if someone tells you they have a table in their living room, you can form an intuitive belief about the table. Intuitive beliefs are the common stock of our minds, the basic data we rely on to guide our behavior and inference in everyday life—as do many other animals.

However, humans have an extraordinary ability to hold a variety of attitudes toward thoughts. You can believe that Bob is Canadian, but you can also doubt that Bob is Canadian, suppose that Bob is Canadian for the sake of an argument, attribute the belief that Bob is Canadian to someone else, and so on. Most of these attitudes toward a thought don't entail believing the thought—if you doubt that Bob is Canadian, you clearly don't believe that he is. Holding some of these attitudes toward a thought, however, amounts to treating this thought as a belief of yours: For instance, if you believe there's a document proving that Bob is Canadian, or if you believe that Susan, who told you Bob is Canadian, is trustworthy in this respect, then you have compelling reasons to accept as true the thought that Bob is Canadian. At least initially, this thought occurs in your mind not as a free-floating belief but embedded in a higher-order belief that justifies believing that Bob is Canadian. This makes your belief that Bob is Canadian *reflective,* in Sperber's sense. In such trivial cases, of course, you may dis-embed the thought that Bob is Canadian from the higher-order belief that justifies it and accept it as a plain intuitive belief free to roam your mind. You may even forget how you initially came to know that Bob is Canadian.

In the same vein: If you're told, by someone you trust in this respect, that God is omniscient, you should come to believe in a reflective way that God is omniscient. However, by contrast with the case of Bob's being Canadian, it's unclear how you could turn the belief in God's omniscience into an intuitive belief free to roam your mind. The very idea of omniscience isn't part of the standard furnishing of our minds; omniscience cannot be perceived, or inferred from anything we might perceive; there's nothing intuitive about it. When we think about agents, we think of them as having cognitive and sensory limitations—things they know and things they don't know, things they can see and things they can't see—because that's how normal agents are. Thus the belief in God's omniscience is stuck in its position of reflective belief; it cannot be dis-embedded and turned into an intuitive belief. As such, it's largely insulated from our ordinary inferences and from guiding our mundane behavior.

If the belief in God's omniscience is stuck in this reflective status, how can it still influence some of our actions? Through the intuitive belief it's embedded in—the belief that someone you trust in this respect believes God is omniscient. This higher-order belief has been acquired through intuitive processes that calibrate our trust in others, and it can be used in guiding inferences and behaviors—for instance by making a Christian affirm and agree that God is omniscient.

The word "belief" collapses together at least two functionally different attitudes: intuitive and reflective beliefs. That some of our most cherished beliefs are reflective helps solve some apparent paradoxes, such as how people can hold contradictory beliefs or ignore much of their beliefs in actual practice. By drawing attention to the differences in the cognitive mechanisms that interact with intuitive and reflective beliefs—and the intuitive beliefs in which reflective beliefs are embedded—this concept offers a more sophisticated and accurate picture of how our minds work.

ALLOPARENTING

ABIGAIL MARSH

Professor of Psychology; Director, Laboratory on Social and Affective Neuroscience, Georgetown University

To *alloparent* is to provide care for offspring that aren't your own. It's an unimaginable behavior for most species (few of which even care for their own offspring); rare even among relatively nurturant classes of animals, like birds and mammals; but central to the existence of humankind. The vigor and promiscuity with which humans in every culture around the world alloparent stands in stark contrast to widespread misconceptions about who we are and how we should raise our children.

Humans' survival as a species over the last 200,000 years has depended on our motivation and ability to care for one another's children. Our babies are born as helpless and needy as it's possible for a living creature to be. The actress Angelina Jolie was once derided for describing her newborn as a "blob," but she wasn't far off. Human infants arrive into the world unable to provide the smallest semblance of care for themselves. Worse, more than a decade will pass before a human child becomes self-sufficient—a period during which that child requires intensive, around-the-clock feeding, cleaning, transport, protection, and training in hundreds or thousands of skills. No other species is on the hook for anywhere near the amount of care we humans must provide our children.

Luckily, evolution never meant us to do it alone. As the anthropologist Sarah Hrdy has described, among foraging cultures that best approximate our ancestral conditions, human babies never rely on only one person, or even two people, for care. Instead they're played with, protected, cleaned, transported, and fed (even nursed) by a wide array of relatives and other group members—as many as twenty different individuals *every day*, in some cases. And the more alloparenting that children get, the more likely they are to survive and flourish.

You'd never know any of this from reading most modern books on child development or child rearing. Attachment to and responsive care from a single primary caregiver (invariably the mother) is nearly always portrayed as the critical ingredient for a child's optimal development. When fathers or other caregivers are mentioned at all, their effect is seen as neutral at best. The implicit message is that for a baby to spend significant time apart from its mother and in the care of other caregivers, like babysitters or daycare providers, is unnatural and potentially harmful.

But the opposite is more likely true. As the historian Stephanie Coontz has put it, human children "do best in societies where child rearing is considered too important to be left entirely to parents." When children receive care from a network of loving caregivers, not only are mothers relieved of the nearly impossible burden of caring for and rearing a needy human infant alone but also their children can learn from an array of supportive adults—form bonds with them and learn to love and trust widely, rather than narrowly.

Children aren't the only beneficiaries of humans' extensive alloparenting practices. Across primate species, the prevalence of alloparenting is also the single best predictor of a behavior that theories portraying human nature as motivated strictly by rational self-interest struggle to explain: altruism. Not reciprocal altruism or altruism toward close kin (which are self-interested) but costly acts of altruism for unrelated others, even strangers. This sort of altruism can seem inexplicable, according to dominant accounts of altruism like reciprocity and kin selection. But it's perfectly consistent with the idea that, as alloparents *sine qua non*, humans are designed to be attuned to, and motivated to care for, a wide array of needy and vulnerable others. Altruism toward one another is likely an exaptation of evolved neural mechanisms that equip us to alloparent.

Remember this if you're ever tempted to write off humanity as a lost cause. We have our flaws, without a doubt, but we can also claim to be the species shaped by evolution to possess the most open hearts and the greatest proclivity for caring on Earth.

CUMULATIVE CULTURE

CRISTINE H. LEGARE
Associate Professor, Department of Psychology, University of Texas at Austin; Director, Cognition, Culture, and Development Lab

In the 7 million years since humans and chimpanzees shared a common ancestor, the inventory of human tools has gone from a handful of stone implements to a technological repertoire capable of replicating DNA, splitting atoms, and interplanetary travel. In the same evolutionary time span, the chimpanzee toolkit has remained relatively rudimentary. It was tool innovation—constructing new tools or using old tools in new ways—that proved crucial in driving increasing technological complexity over the course of human history.

How can we explain this wide divergence in technological complexity between such closely related primate species? One possibility is that humans are unique among primate species in our ability to innovate. If so, we might expect that innovation develops early in childhood, like walking or language acquisition. Yet there's little evidence for precocious innovation in childhood. Although young children are inquisitive and keen to explore the world around them, they're astonishingly poor at solitary tool innovation. New Caledonian crows and great apes outperform young children in tool-innovation tasks. This is particularly striking, given the dazzling technological and social innovations associated with human culture. How does a species with offspring so bad at innovation become so good at it?

Technological complexity is the outcome of our species' remarkable capacity for cumulative culture. Innovations build on each other and are progressively incorporated into a population's stock of skills and knowledge, generating ever more sophisticated repertoires. Innovation is necessary to ensure cultural and individual adaptation to new and changing challenges, as humans spread to every corner of the planet. Cultural evolution makes individuals more innovative by allowing for the accumulation of prefabricated solutions to problems—solutions that can be recombined to create new technologies. The subcomponents of technology are typically too complex for individuals to develop from scratch. The cultural inheritance of the technologies of previous generations enables the explosive growth of cultural complexity.

Children are cultural novices. Much of their time is spent trying to become like those around them—to do what they do, speak like they do, play and reason like they do. Learning from and imitating others lets them benefit from and build on cumulative cultural transmission. Cumulative culture requires the high-fidelity transmission of two qualitatively different abilities—instrumental skills (e.g., how to keep warm in winter) and social conventions (e.g., how to perform a ceremonial dance). Children acquire these skills through high-fidelity imitation and behavioral conformity. These abilities afford the rapid acquisition of behavior more complex than could otherwise be learned exclusively through individual discovery or trial-and-error learning.

Children often copy when uncertain. This proclivity is useful, given that a vast amount of our behavior is opaque from the perspective of physical causality. High-fidelity imitation is an adaptive human strategy, fostering social learning of instrumental skills faster than would be possible if copying required a full causal representation of an event. It's so adaptive, in fact, that it's often employed at the expense of efficiency—as when kids "over-imitate" behavior causally irrelevant to accomplishing a particular task.

The unique demands of acquiring instrumental skills and social conventions like rituals provide insight into when children imitate or innovate and to what degree. Instrumental behavior is outcome-oriented. Innovation often improves the efficiency of solving defined problems. When learning instrumental skills, high-fidelity imitation decreases with experience. In contrast, conventional behavior is process-oriented. The goals are affiliation and group inclusion. When learning social conventions, imitative fidelity stays high regardless of experience, and innovation stays low. Indeed, innovation impedes learning well-prescribed social conventions. Imitation and innovation work in tandem, deployed at different times for different purposes, to support learning group-specific skills and practices. The distinct goals of instrumental skills and social conventions drive cumulative culture and illuminate human cognitive architecture.

Cumulative culture allows the collective insights of previous generations to be harnessed for future discoveries, in ways more powerful than the solitary brainpower of even the most intelligent individual. Our ability to build on the innovations of others, within and across generations, drives our technological success. The capacity for cumulative culture has set genus *Homo* on an evolutionary pathway quite distinct from that of all others.

LIFE HISTORY

ALISON GOPNIK
Professor of Psychology, Affiliate Professor of Philosophy, UC Berkeley; author, *The Gardener and the Carpenter*

Imagine that an Alpha Centauran scientist came to Earth 150,000 years ago. She might note in passing that the newly evolved *Homo sapiens* were just a little better at tool use, cooperation, and communication than were their primate relatives. But as a well-trained evolutionary biologist, she'd be far more impressed by their remarkable and unique life history.

"Life history" is the term biologists use to describe how organisms change over time: how long an animal lives, how long a childhood it has, how it nurtures its young, how it grows old. Human life history *is* weird. We have a much longer childhood than any other primate—twice as long as chimps—and that long childhood is related to our exceptional learning abilities. Fossil teeth suggest that this long childhood evolved in tandem with our big brains; we even had a longer childhood than Neanderthals. We also rapidly developed special adaptations to care for our helpless children—"pair-bonding" and "alloparents." Fathers and unrelated kin help take care of human children, which is not the case with our closest primate relatives.

And we developed another unusual life-history feature—post-menopausal grandmothers. The killer whale is the only other animal we know of that outlives its fertility. The human life span expanded at both ends—a longer childhood and a longer old age. In fact, anthropologists have argued that grandmothers were a key to the evolution of learning and culture. They were crucial for the survival of those helpless children, and they also could pass on two generations' worth of knowledge.

Natural selection often operates on life-history characteristics, and life history plays an important role in evolution in general. Biologists long distinguished between "K" species and "R" species. R species—most fish, for example—may produce thousands of offspring, but most of them die and the rest live only a short time. In contrast, K species—like primates and whales—have only a few babies, invest a great deal in their care, and live a long time. Generally speaking, a K life-history strategy is correlated with a larger brain and higher intelligence. We are the ultimate K species.

Life history is also important because it's especially responsive to information from the environment, not only over evolutionary time but also in the lifetime of a single animal. Tiny water fleas develop a helmet when they mature to protect them from certain predators. When the babies, or even their pregnant mothers, detect more predators in the environment, the developmental process speeds up—the helmets grow earlier and larger, even at a cost to other functions. In the same way, in other animals, including human beings, early stress triggers a "live fast, die young" life history. Young animals who detect a poor and risky environment grow up more quickly and die sooner.

Our unique human developmental trajectory has cumulatively led to much bigger differences in the way we live and behave. A hundred and fifty thousand years ago, the Alpha Centauran biologist wouldn't have seen much difference between adult humans and our closest primate relatives—art, trade, religious ritual, and complex tools were still far in the future, not to mention agriculture and technology. Our long childhood and our extended investment in our children allowed those changes to happen; think of all the grandmothers passing on the wisdom of the past to a new generation of children. Each human generation had a chance to learn a little more about the world from their caregivers, and to change the world a little more themselves.

Evolutionary psychologists have tended to focus on adult men; hunting and fighting got a lot more attention than caregiving. We've all seen the canonical museum diorama of the mighty early human hunters bringing down the mastodon. But the children and grandmothers in the background were just as important parts of the story.

You still often read psychological theories describing both the young and the old in terms of their deficiencies, as if they were just preparation for, or decline from, an ideal grown-up human. But new studies suggest that both the young and the old may be especially adapted to receive and transmit wisdom. We may have a wider focus and a greater openness to experience when we're young or old than we do in the hurly-burly of feeding, fighting, and reproduction that preoccupies our middle years.

"Life history" is an important idea in evolution, especially human evolution. But it also gives us a richer way of thinking about our lives. A human being isn't just a collection of fixed traits but part of an unfolding and dynamic story. And that isn't just the story of our own lives; caregiving and culture link us both to the grandparents who were there before we were born and to the grandchildren who will carry on after we die.

HALDANE'S RULE OF THE RIGHT SIZE

PAUL SAFFO

Futurist; technology forecaster; Consulting Associate Professor in Mechanical Engineering, Stanford University

Toss a mouse from a building. It will land, shake itself off, and scamper away. But if similarly dropped, "a rat is killed, a man is broken, a horse splashes." So wrote J. B. S. Haldane in his 1926 essay "On Being the Right Size." Size matters, but not the way a city-stomping Godzilla or King Kong might hope.

Every organism has an optimum size, and a change in size inevitably leads to a change in form. Tiny lizards dance weightlessly up walls, but grow one to Godzilla size and the poor creature would promptly collapse into a mush of fractured bones and crushed organs. This principle isn't just a matter of extremes: A hummingbird scaled to the size of a blue jay would be hopelessly earthbound, fluttering its wings in the dust.

If gravity is the enemy of the large, surface tension is the terror of the small. Flies laugh at gravity but dread water. As Haldane noted, "An insect going for a drink is in as great danger as a man leaning out over a precipice in search of food." No wonder most insects are either unwettable or do their drinking at a distance through a strawlike proboscis.

Thermoregulation is an issue for all organisms, which is why Arctic beasts tend to be large whereas tropical critters are small. Consider a sphere: As it grows, the interior volume increases faster than the surface area. Small animals have lots of surface area relative to their volume, an advantage in the Torrid Zone, where survival depends on efficient cooling. In the Arctic, the same ratio works in reverse: Large beasts rely on a lower surface-area ratio to help stay warm.

The power of Haldane's rule is that it applies to far more than just organisms. Hidden laws of scale stalk humankind everywhere we turn. Like birds, the minimum power an aircraft requires to stay in flight increases faster than its weight. This is why large birds soar instead of flapping their wings—and Howard Hughes' Spruce Goose never got more than a few feet off the water.

Size inevitably comes at a cost of ever greater complexity. In Haldane's words, "Comparative anatomy is largely the story of the struggle to increase surface in proportion to volume." Which is why intestines are coiled and human lungs pack in 100 square yards of surface area. Complexity is a Goldilocks tool for the large, widening the zone of "just right."

But complexity can expand the envelope of "just right" only so far before bad things happen. Like the engine on an underpowered aircraft, the cost can be catastrophic. Everything from airplanes to institutions has an intrinsic right size, which we ignore at our peril. The 2008 banking crisis taught us that companies and markets aren't exempt from Haldane's rule. But we got the lesson backwards: It wasn't a case of "too big to fail" but, rather, "too big to succeed." One cannot help but fret that in their globe-spanning success, megacompanies are flirting with the unforgiving limits of right size.

Our political institutions also cannot escape the logic of Haldane's rule. The Greeks concluded that their type of democracy worked best in a unit no larger than a small city. Haldane adds that the English invention of representative government made possible a scale-up to large, stable nation-states. Now it seems that the U.S. and other nations are growing beyond the right size for their political systems. Meanwhile, globalization is stalling in a cloud of conflict and confusion, precisely because no workable political structure right-sized for the entire planet exists. The turbulence of the past year very likely is mere prologue to more wrenching shifts ahead.

Haldane wrote decades before the advent of globalization and digital media, but his elegant rule of right size hovers behind our biggest challenges today. Can megacities be sustained? Have social networks scaled beyond the optimum size for sensible exchange? Has cyberspace become so large and complexly interdependent that it's at risk of catastrophic scale-driven failure? Is our human population outstripping its right size given the state of our planetary systems? As we face these and the myriad other surprises to come, we'd do well to remember that mice bounce, horses splash—and size truly matters.

PHENOTYPIC PLASTICITY

NICOLAS BAUMARD
CNRS Research Scientist, Department of Cognitive Science, École Normale Supérieure, Paris; author, *The Origins of Fairness*

Humans all over the world share the same genome, the same neural architecture, and the same behavioral niche (three-generational system of resource provisioning, long-term pair-bonding between men and women, high levels of cooperation between kin and non-kin). At the same time, human cultures are highly variable. Some societies see revenge as a duty, others as a sin; some regard sex as a pleasure, others as a danger; some reward innovations, whereas others prefer established practice. In many instances, these cultural differences are robust and can last for millennia, despite cultural contacts, political assimilation, or linguistic replacement.

How can we account for such variability? Traditionally, it's assumed that cultural variability can't be explained by species-specific evolved mechanisms—that it must be the product of socially transmitted norms in the form of religious beliefs, informal enforcement, or political conquest.

This assumption is based on a common misconception about natural selection, which is wrongly thought to select mechanisms that systematically produce universal, uniform, and unchanging behaviors. But all evolved mechanisms, physiological or psychological, come with a certain level of flexibility in response to local contexts. This is called *phenotypic plasticity*. The genotype codes for a mechanism that can express different phenotypes (organs, behaviors) in response to detectable and recurring changes in the environment.

Tanning is a case in point. While the mechanism of skin pigmentation is a universal adaptation to protect human cells from ultraviolet damage and to synthesize vitamin D, it responds differently to different contexts, making skin darker in low latitudes (and in the summer) and lighter in high latitudes (and in the winter).

Phenotypic plasticity has been shown to be evolutionarily advantageous when there are different optimal phenotypes in different environments. This is of course the case for tanning: The optimal level of tanning differs depending on whether an individual lives in high or low latitudes and whether it's winter or summer. Natural selection must therefore be able to maintain the right skin color despite variations in the environment. One solution is to select for a certain level of plasticity (skin pigmentation isn't completely plastic) and to build a mechanism that can detect the amount of light and adjust the level of melatonin accordingly.

The study of phenotypic plasticity has long been limited to physical traits (such as skin pigmentation). However, recent research in neuroscience, ecology, and psychology shows that phenotypic plasticity extends to behaviors. For instance, in a harsh and unpredictable environment where the future is dreary, organisms tend to adopt a short-term life strategy, maturing and reproducing earlier, investing less in offspring and pair-bonding, being more impulsive. In a more favorable and predictable environment, organisms switch to a long-term strategy, maturing and reproducing later, investing in offspring and pair-bonding, and being more patient. These switches between present- and future-oriented behaviors can affect all kinds of behaviors: reproduction and growth, of course, but also attitudes toward consumption, investment in learning or in health, trust in others, political opinions, technological innovation, etc. In fact, every behavior (and neural structure) for which time and risk are relevant dimensions is likely to involve a certain degree of plasticity.

Could phenotypic plasticity be relevant to explaining cultural differences?

When we observe cultural differences between two societies, we can't help but think that the difference has its roots in different cultural heritages (religious, legal, literary, etc.). This is because we have no alternative mechanism to explain such differences. Suggesting that the two groups differ in terms of psychological mechanisms seems a retreat to 19th-century "national character" studies or to mysterious "cultural mindsets," which redescribe the phenomenon rather than explaining it. By contrast, phenotypic plasticity offers a plausible (and not mutually incompatible) mechanism to explain why people have different mindsets in different societies: why they're more impulsive, why they trust others less, why they're afraid of innovation. Even better, it makes predictions about how the environment should affect people's psychology. For instance, a commonsense idea is that people should innovate when they're in danger and need an urgent solution. Evolutionary theory suggests the opposite: When resources are scarce and unpredictable, innovation is too risky: Better stick to what you know than jeopardize everything. And, indeed, this is what people do.

Finally, phenotypic plasticity may solve several limitations of the standard culture-as-transmission-of-information paradigm: Why does the same cultural background (language, religion, ethnicity) give rise to radically different behaviors according to whether an individual was born in a low vs. a high social class or in an old vs. recent generation? Or how does an old and apparently robust cultural phenomenon crumble in a few generations, sometimes in a few years, and without any cultural external input? This may be because different environments triggered different "behavioral strategies" in people, transforming a common cultural heritage into diverging new cultures.

To sum up, phenotypic plasticity is key in the study of human behavior. It provides a framework to account for the fact that the same genome and the same neural architecture can give rise to cultural variability in humans.

SLEEPER SENSITIVE PERIODS

LINDA WILBRECHT

Associate Professor, Department of Psychology and Helen Wills
Neuroscience Institute, UC Berkeley

If you moved from the United States to France as a child, you'd
likely become fluent in French in a short period, but if you
moved to France as an adult you might never become fluent.
This difference in the capacity for learning language exists because there are sensitive periods in development when the brain
is particularly plastic and able to receive and retain information
with greater efficacy.

A well-established field of sensitive-period biology seeks to
explain how people learn to speak, how birds learn to sing, and
how our sensory systems wire up, among other things. The field
has been particularly successful in explaining how the brain co-
ordinates the information streaming in from the two eyes to
allow binocular vision, useful for depth perception. In the last
century, it was discovered that when a person was born with a
"lazy" eye or had their vision clouded in one eye by a cataract,
their binocular vision would be impaired for life. However, if a
correction was made in early life, then the brain and binocular
vision could recover to develop normally. This human phenomenon can be modeled in rodents by closing one eye in early life.
Extensive study of this model provides the basis for our understanding of the cellular mechanisms regulating sensitive periods
across the cortical regions of the brain.

You might conclude that early experiences are simply the

most powerful; the juvenile brain is, in general, more plastic than the adult brain. However, the often glossed-over details show that the younger brain is not always more sensitive to experience than the older brain. When the biology can be studied in carefully controlled laboratory experiments, we find that periods of greater sensitivity are often delayed, perhaps even timed, until the incoming experience is appropriate to sculpt the brain. For example, the peak of sensitive-period plasticity for the development of binocular vision occurs about a month after birth in rodent brains, which is more than a week after eye opening. Scientists are still working on the why and how. Nonetheless, it's clear that the brain can and does hold highly sensitive plasticity under wraps and then unveils it when appropriate. It's thought that years of evolution have sculpted brain development to be not only experience-dependent but also carefully timed, such that it is experience-expectant—that is, dormant until needed.

What this means for the big picture is that human development probably involves a staggered sequence of undiscovered sensitive periods stretching late into the second or even third decade of life. Hence, we should be on the lookout for "sleeper" sensitive periods. For example, there may be teenage social-sensitive periods when we learn to interact with peers, or cognitive-sensitive periods when we sculpt our decision-making style. These sensitive periods might be timed to overlap with important transitions, as when we leave our parents' protection to explore the world, or go through puberty, or become a parent. The boundaries may be sharp, triggered by events like puberty onset, or gradual slopes that rise and fall with age and experience. We don't yet know when, where, and how these more subtle cognitive- and emotion-sensitive periods may work.

It may be easier to see evidence of complex sensitive periods in development in other species. Life-history ecologists have identified a wide array of nonhuman species that adapt their phenotype according to the sampled statistics of their particular environment. For example, if developing crickets are exposed to spiders in the environment, then the adult crickets are better at surviving where there are spiders. If food is scarce during development for a species of mite, then an alternate body type and foraging strategy may be used in adulthood. Less is known about the neurobiology of these phenomena in these non-mammalian species.

Sensitive-period biology may in future provide important insights into understanding and preventing mental illness. Sensitive-period plasticity enables adaptation to experience, but this adaptation doesn't ensure an optimal or even favorable outcome. For example, negative experience during a sensitive period could generate a persistent negative bias in the processing of events, potentially leading to mental illness. It's known that negative experiences do have different effects at different ages in humans and animal models, but we don't know exactly when it's better or worse to endure negative experience and why.

Sensitive-period biology may also influence behaviors commonly thought to make up a person's personality. Experience at different times might alter someone's appetite for risk or tolerance for delayed gratification, or kindle an interest in music. The experience of poverty, even during a brief window of development, could alter the brain and behavior for a lifetime. When money is available for educational or public-health intervention, knowledge of sensitive-period biology should become a central aspect of strategy. If sleeper sensitive periods exist in late childhood or teenage years, these periods may become more efficient target years.

ZONE OF PROXIMAL DEVELOPMENT

ATHENA VOULOUMANOS

Associate Professor of Psychology; Director, NYU Infant Cognition and
Communication Laboratory, New York University

Clear instruction is essential for learning. But even the clearest
instruction can be of limited use if the learner isn't at the right
place to receive it. Psychologist Lev Vygotsky had a remarkable
insight about how we learn. He coined the term *zone of proximal development* to describe a sweet spot for learning in the gap
between what a learner could do alone and what she could do
with help from someone providing knowledge or training just
beyond her level. With such guidance, learners can succeed at
tasks too difficult for them to master alone. Crucially, guidance
can then be taken away like scaffolding, and learners can perform the task on their own.

The zone of proximal development introduces three interesting twists to cognitive scientists' notions of learning. First, it
prompts us to reconsider notions of what a person "knows" and
"knows how to do." Instead, conceptualizing peak knowledge or
abilities as a learner's current maximal accomplishments under
guidance draws our attention to his learning potential and helps
us avoid reifying test scores and grades. Second, it introduces the
idea of socially constructed knowledge, created in the interstitial
space between learner and mentor. Seeing knowledge as an act
of dynamic creation empowers teachers and learners alike. Third,
it provides a nuanced caveat to findings showing that explicit

instruction can inhibit learning in some situations. Recent studies show that novices given instruction generated less creative solutions than novices engaged in unguided discovery-based exploration, but the zone of proximal development reminds us that the nature of the instruction relative to the learners' state of readiness matters.

This zone should be more widely recognized by parents, teachers, and anyone learning anything new (which includes all of us). Teachers who understand students' current level of knowledge can present new information that takes them just beyond it, to a new level of understanding. Subtraction, for example, might be introduced using a simpler term, like "taking away." Parents who understand their children's current abilities can give specific guidance by, say, suggesting they look for puzzle pieces with straight edges to create a jigsaw puzzle's frame, or showing how two puzzle pieces can interlock. Whereas encouraging children with generic praise can help them persevere, giving specific verbal or physical guidance in the child's zone of proximal development can help them learn to solve puzzles on their own.

LENGTH-BIASED SAMPLING

ELIZABETH WRIGLEY-FIELD
Assistant Professor, Department of Sociology and Minnesota Population Center, University of Minnesota–Twin Cities

Here are three puzzles:

- American fertility fluctuated dramatically in the decades surrounding World War II. Parents created the smallest families during the Great Depression and the largest families during the postwar baby boom. Yet children born during the Great Depression came from *larger* families than those born during the baby boom. How can this be?

- About half the prisoners released in any given year in the United States will end up back in prison within five years. Yet the proportion of prisoners ever released who will ever end up back in prison over their lifetime is just one third. How can this be?

- People whose cancers are caught early by random screening often live longer than those whose cancers are detected later, after they're symptomatic. Yet those same random screenings might not save any lives. How can this be?

And here's a twist: These are all the same puzzle.

The solution is adopting the right perspective. Consider the family puzzle. One side is about what parents do; the other side is about what kids experience. If families all had the same number of kids, those perspectives would coincide: The context parents create is the context kids live in. But when families aren't all the same size, it matters whose perspective you take.

Imagine trying to figure out the average family size in a particular neighborhood. You could ask the parents how many kids they have. Big families and small families will count equally. Or you could ask the children how many siblings they have. A family with five kids will show up in the data five times, and childless families won't show up at all. The question is the same: "How big is your family?" But when you ask kids instead of parents, the answers are weighted by the size of the family. This isn't a data error so much as a trick of reality: The average kid actually has a bigger family than the average parent does. And (as the great demographer Sam Preston has pointed out), during the Great Depression, when families were either very small or very large, this effect was magnified—so the average child came from a very large family, even though the average adult produced a small family.

The recidivism puzzle is the family puzzle on a slant. When we look at released prisoners at a moment in time, we see the ones who leave prison most often—who are also the ones who return most often. We see, as the econometrician William Rhodes and his colleagues recently pointed out, the repeat offenders. Meanwhile, the population that *ever* leaves prison has 2-to-1 odds of never going back.

Snapshots bias samples: When some people experience something—like a prison release—more often than others, looking at a random moment in time guarantees a non-random assortment of people.

And the cancer screenings? Screenings reveal cancer at an intermediary stage, when it's advanced enough to be detectable but not so advanced that the patient would have shown up for testing without being screened. And this intermediary, detectable stage generally lasts longer for cancers that spread slowly. The more time the cancer spends in the detectable stage, the more likely it is to be detected. So the screenings disproportionately *find* the slower growing, less lethal cancers whether or not early detection *does* anything to diminish their lethality. Assigning screenings randomly to people necessarily assigns screenings selectively to tumor types.

The twist in these puzzles is length-biased sampling: This works when we see clusters in proportion to their size. Length-biased sampling reveals how life spans—of people, of post-prison careers, of diseases—bundle time the way families bundle children.

All this may seem like a methodological point—and, indeed, researchers go awry when we ask about one level (such as, what parents do) but unwittingly answer about another (such as, what children experience). To be correct, an answer needs to be about the same population as the question it answers. But length-biased sampling also explains how our social positions can give us very different experiences of the world—as when, if a small group of men each harasses many women, few men know a harasser but many women are harassed.

Most fundamentally, length-biased sampling is the deep structure of nested categories. It's not just that the categories *can* play by different rules but that they *must*.

Consider again those different-size families, now stretching out over generations. If we each had the same number of children as our parents did, small families would beget a small number of new small families—and large families would beget larger and larger numbers of families with many children of their own. With each passing generation, the larger a family is, the more common families of its size would become. The mushrooming of families with many kids would sprout into wild, unchecked population growth.

This implies that, as Preston's analysis of family sizes showed us, population stability requires family instability: If the population is to stay roughly the same size, most children must grow up to have fewer kids than their parents did, each generation rejecting tradition anew. And indeed, most of us do. We make different choices than our parents did, and our culture evolves as each generation finds meaning in the families we create, which are smaller than the ones we came from. Adages about the rebelliousness of youth may have roots in culture or in developmental psychology, but their truth is also demographic: Between a country and the families it comprises, size can be stable at one level or the other, but never both.

Categories nestle inside one another: tumor inside person

inside family inside nation. They nestle not as Russian dolls, a regress of replicas, but rather like layers of rock and soil, each layer composing the world differently. Whether we see equality or divergence, stasis or change, depends in part on the level at which we look. Length-biased sampling is the mathematics that links the levels, and the tunnel that lets us walk between them.

CONSTRUAL

ELDAR SHAFIR

Class of 1987 Professor in Behavioral Science and Public Policy, Princeton University; co-author (with Sendhil Mullainathan), *Scarcity: The New Science of Having Less and How It Defines Our Lives*

Here's a trivial fact about our mental lives. So trivial it's rarely noticed—and also hard to talk about without sounding sophisticated. The postmodernists have explored versions of it, but the notion I mean to promote—the notion of *construal*—is painfully obvious. It refers to the fact that our attitudes and opinions and choices pertain to things not as they are in the world but as they're represented in our minds.

Economic theorizing presumes that people choose between options in the world: job A versus job B, or car A versus car B. From the point of view of a psychologist, however, that presumption is radical: When a person is given a choice between options A and B, she chooses not between A and B as they are in the world but rather as they're represented by the three-pound machine she carries between her ears. And that representation isn't a complete and neutral summary but a selective and constrained rendering—a construal.

There's no way around it. The behaviorists tried to avoid it by positing that behaviors were direct responses to stimuli, that mental life didn't interfere in relevant ways. But clearly that's not the case. We now know a lot about our rich mental lives, which shape what we experience, making construal not neutral. A food that's 10-percent fat is less appetizing than one that's 90-percent fat-free. A risky venture that entails some lives saved and others lost is a lot more appealing when our attention is directed toward the lives saved rather than those lost. Our attempts to elicit empathy for global catastrophes are ineffective—a phenomenon referred to as "psychic numbing"—partly because our construal processes cannot trigger differential indignation for outcomes as a function of their enormity.

Visual illusions provide a compelling illustration of how our experience of an object doesn't conform to the actual object in the world. Susan Sontag famously observed, "To photograph is to frame, and to frame is to exclude." In fact the mind is a lot messier than a camera. We don't merely choose where to look; our minds influence what we see. And they influence what we see both when we think fast and when we think slowly—both when we respond impulsively, without conscious thought, and when we deliberately choose what to take seriously and what to ignore.

Construal lies at the core of behavioral economics. Violations of standard rationality assumptions arise not from stupidity, computational limitations, or inattention, but from the simple fact that things in the world, depending on how they're described or interpreted, get construed differently, yielding inconsistent judgments and preferences.

Real-world options, like automobiles, houses, job offers, potential spouses, all come with multiple attributes. How much weight we give each attribute is largely a function of where our attention is directed, our pet theories, what we expect or wish to see, the associations that come to mind. One rule of construal is that things are judged in comparative rather than absolute terms. How water feels to the hand depends on whether the hand was previously in colder or warmer water. In the delivery room, a doctor's decision of whether to perform a Caesarian section depends on the gravity of immediately preceding cases.

Knowledge in the form of scripts, schemas, and heuristics serves to make sense of stimuli in ways that transcend what is given. What we experience is determined not simply by the objective circumstances of the situation but also by what we know, care about, attend to, understand, and remember. And what we care about, attend to, and remember is malleable. In one study, participants were invited to play a Prisoner's Dilemma game, which was referred to in the study as either the Wall Street or the Community game. While the payoffs and setup were identical, the mere label altered participants' construal, changing their tendency to cooperate or to defect.

Psychological costs and subsidies also enter people's construal and are different from the financial costs and subsidies that policymakers are typically concerned with. In one well-known study, when fines were introduced for picking up children late from daycare, parents were more likely to pick up their children late. Parents who had previously felt guilty (had incurred a psychic cost) for showing up late now construed the fine as a contract—paying a fee entitled them to a late pick-up.

Psychologists see construal as an integral feature of human cognition, but if your aim is to influence behavior, construal presents a challenge. The difference between success and failure often boils down to how things are construed. Although similar from an accounting point of view, the Earned Income Tax Credit (EITC)—compared with Temporary Assistance for

Needy Families (TANF) and other forms of welfare—has been an effective form of government assistance. This is attributed to the construal of EITC as just reward for labor, delivered in the form of a tax-refund check, rather than as a separate assistance payment. It's seen as an entitlement rather than welfare, designating beneficiaries as taxpaying workers rather than "on the dole."

Construal needs to be more widely appreciated, because so much thinking and intuition, in policy and in the social sciences, tends to focus on actual circumstances as opposed to how they're construed. The words making up this essay are just words. It's partly their construal that will make some readers think they're useful and others think they're of little use.

DOUBLE BLIND

SCOTT DRAVES
Software artist

In 1620, Sir Francis Bacon published *Novum Organum* and kicked off the Scientific Revolution by defining its basic method: hypothesis, experiment, and result. By 1687, we had Newton's *Principia*, and the rest is history. Today, public primary schools teach the scientific method. It's well known.

It turns out that following the method isn't so simple. People, including scientists, aren't perfectly rational. People have biases and even when we try to be good, sometimes, unconsciously, we do wrong. When the outcome of an experiment has career implications, things start to get complicated. And when the outcome has financial implications for a powerful institution, people have been known to actively game the system. For example, starting in 1953 the Tobacco Industry Research Committee waged a war on truth, until it was dissolved as part of a master settlement in 1998.

The stakes are high. Tobacco killed 100 million people in the 20th century. Climate change threatens our very way of life.

In the years since the 17th century, science has developed a much more detailed playbook for the scientific method, in order to defend itself against bias. Double-blind experiments are an essential part of the modern gold-standard scientific method.

What is a double-blind experiment? Consider this scenario:

A pharmaceutical company develops a new arthritis pill and hires you to prove its efficacy. The obvious experiment is to give the drug to a group of people and ask them if it relieved their pain. What's not so obvious is that you should also have a control group. These subjects get placebos and this is done without the subjects' knowledge of which kind of pill they get. That's a *single-blind* experiment, and the idea is that it keeps the subjects' expectations or desires that this pill will do something from influencing their reported results.

There's a remaining problem, however. You, the experimenter, also have expectations and desires, and those could be communicated to the subjects or influence how the data are recorded. Another layer of blindness can be introduced, so that the experimenter doesn't know which pills are which as they're administered, the subjects are surveyed, and the data are collected. Normally this is done by using a third party to randomly assign the subjects to the groups and keep the assignment secret from the researchers until after the experiment is complete. The result is a double-blind experiment: Both subjects and experimenters are unaware of who got what.

The first single-blind experiment was performed in 1784 by Benjamin Franklin and Antoine Lavoisier, who were commissioned by the French Academy of Sciences to investigate Franz Mesmer's claims of animal magnetism. The claims were debunked.

The first recorded double-blind experiment was done in 1835 in Nuremberg. Friedrich Wilhelm von Hoven, a public-health official and hospital administrator, got into a public dispute with Johann Jacob Reuter, who claimed the odds were 10 to 1 that a single grain of salt dissolved in 100 drops of snow-melt, and then diluted 30 times by a factor of 100 each time, would produce "extraordinary sensations" in someone who drank it. Twenty-five samples of homeopathic saltwater and twenty-five samples of plain distilled water were randomly assigned to the subjects. The assignment was sealed, and the water was administered. In the end, eight subjects did report feeling something, but three of those had actually had plain water, the placebo. Reuter lost the bet by the rules they'd agreed on in advance.

That was huge progress, and science and medicine have come a long way as a result.

Invaluable though double-blind experiments are, the process is still imperfect. The existence of one double-blind study cannot be considered conclusive. If the question is of consequence, and you do your research, you'll likely find many competing studies, authors, and institutions. Reputations and careers come into play. Research labs get long-term funding from corporations and governments who have skin in the game, and over time their influence works its magic. Truth can be lost in the hall of mirrors of the Internet.

There's a constant struggle in science to distinguish signal from noise, to discern a pattern in experimental data, and to come up with a theory that explains it all. That's hard enough! But there's another, deeper struggle as well—against bias and influence, both in ourselves and in society. This struggle is not just against our ignorance but also against intentional adversaries to the process. Over and over again, blinding has proved itself as key to fighting biases and discovering truth more quickly.

It's ironic that controlled blindness is an engine of insight, and even perhaps a cornerstone of our civilization.

THE LAW OF SMALL NUMBERS

ADAM ALTER
Psychologist; Associate Professor of Marketing, Stern School of Business, NYU; author, *Irresistible*

In 1832, a Prussian military analyst named Carl von Clausewitz explained that "three quarters of the factors on which action in war is based are wrapped in a fog of . . . uncertainty." The best military commanders seemed to see through this "fog of war," predicting, on the basis of limited information, how their opponents would behave. Sometimes, though, even the wisest generals made mistakes, divining a signal through the fog when no such signal existed. Often, their mistake was endorsing the law of small numbers—too readily concluding that the patterns they saw in a small sample of information would hold for a much larger sample.

Both the Allies and the Axis powers fell prey to the law of small numbers during World War II. In June 1944, Germany flew several raids on London. Allied experts plotted the position of each bomb as it fell and noticed one cluster near Regent's Park and another along the banks of the Thames. This clustering concerned them, because it implied that the German military had designed a new bomb that was more accurate than any existing bomb. In fact, the *Luftwaffe* was dropping bombs randomly, aiming generally at the heart of London but not at any particular location over others. What the experts had seen were clusters that occur naturally through random processes—misleading noise masquerading as a useful signal.

That same month, German commanders made a similar mistake. Anticipating the raid later known as D-Day, they assumed the Allies would attack, but they weren't sure precisely when. Combing old military records, a weather expert named Karl Sonntag noticed that the Allies had never launched a major attack when there was even a small chance of bad weather. Late May and much of June were forecast to be cloudy and rainy, which "acted like a tranquilizer all along the chain of German command," according to Irish journalist Cornelius Ryan. "The various headquarters were quite confident that there would be no attack in the immediate future. . . . In each case conditions had varied, but meteorologists had noted that the Allies had never attempted a landing unless the prospects of favorable weather were almost certain." The German command was mistaken, and on Tuesday, June 6th, the Allied forces launched a devastating attack amid strong winds and rain.

The British and German forces erred because they had taken a small sample of data too seriously. The British forces had mistaken the natural clustering that comes from relatively small samples of random data for a useful signal, while the German forces had mistaken an illusory pattern from a limited set of data for evidence of an ongoing, stable military policy. To illustrate their error, imagine a fair coin tossed three times. You'll have a 1-in-4 chance of turning up a string of three heads or tails, which, if you make too much of that small sample, might lead you to conclude that the coin is biased to reveal one particular outcome all or almost all the time. If you continue to toss the fair coin, say, 1,000 times, you're far more likely to turn up a distribution that approaches 500 heads and 500 tails. As the sample grows, your chance of turning up an unbroken string shrinks rapidly (to roughly 1-in-16 after five tosses; 1-in-500 after ten tosses; and 1-in-500,000 after twenty tosses). A string is far better evidence of bias after twenty tosses than it is after three tosses—but if you succumb to the law of small numbers, you might draw sweeping conclusions from even tiny samples of data, just as the British and Germans did about their opponents' tactics in World War II.

Of course, the law of small numbers applies to more than military tactics. It explains the rise of stereotypes (concluding that all people with a particular trait behave the same way); the dangers of relying on a single interview when deciding among job or college applicants (concluding that interview performance is a reliable guide to job or college performance at large); and the tendency to see short-term patterns in financial stock charts when in fact short-term stock movements almost never follow predictable patterns. The solution is to pay attention not just to the pattern of data but also to how much data you have. Small samples aren't just limited in value; they can be counterproductive, because the stories they tell are often misleading.

COMMITMENT DEVICES

MICHAEL I. NORTON
Harold M. Brierley Professor of Business Administration, Harvard Business School; co-author (with Elizabeth Dunn), *Happy Money: The Science of Smarter Spending*

Arguments over which species makes the best pet are unproductive: Clearly, those with views different from your own are deeply misguided. Turtles, say, are easy: They sit around, chewing slowly. Dogs, say, are difficult: They run around, chewing rapidly. But there's a hidden benefit to some choices, unbeknownst to their owners. It's the fact that turtles are passive and dogs active that's the key. A dog, it turns out, needs to go for walks, and so dog owners get a little exercise every day. And what is the dog's absolute favorite activity in the world? Meeting (and sniffing) other dogs, who happen to be attached to humans via their leashes—and so dog owners get a little socializing every day as well. Research shows that getting a little exercise and chatting with strangers contributes to our well-being. Now, we could just decide, "I'm going to go for a walk and chat with new people today." And repeat that to ourselves as we press "Play" on another episode of *Breaking Bad*. But because dogs importune us with, well, puppy-dog eyes, they prompt us in a way we're unable to prompt ourselves.

Dogs beat turtles because they serve as commitment devices— decisions we make today that bind us to be the kind of person we want to be tomorrow. (The most famous example is Odysseus tying himself to the mast to resist the lure of the sirens; he wanted to hear today but not be shipwrecked tomorrow.)

Researchers have documented a wide array of effective commitment devices. In one study, would-be exercisers were granted free access to audio versions of trashy novels—the kind they might usually feel guilty about reading. The commitment device? They were allowed to listen only while exercising at the gym, which increased their subsequent physical activity. In another, shoppers who qualified for a 25-percent discount on their groceries were given the chance to make their discount contingent on committing to increase their purchase of healthy food by 5 percent; not only did many commit to "gamble" their discount but also the gamble paid off in healthier buying. Commitments can even seem irrational. People will agree to sign up for savings accounts that don't allow money to be withdrawn, for any reason, for long periods of time. They will even sign up for accounts that not only offer zero interest but also charge huge penalties for withdrawals. Committing to such accounts makes little sense economically but perfect sense psychologically: People are seeking commitment devices to force themselves to save.

The decision about which pet to choose seems trivial compared with health and financial decisions, but it suggests the broad applicability of commitment devices in everyday life. Thinking of life as a series of commitment devices—of not just wanting to be your ideal self tomorrow but designing your environment to commit yourself to it—is a critical insight from social science. In a sense, most relationships can be seen as commitment devices. Siblings, for example, commit us to experiencing decades-long relationships (whether we like it or not). Want to better understand different political viewpoints? You could decide to read Ayn Rand or Peter Singer—or you can drag yourself to Thanksgiving with the extended family. Want to spend more time helping others? You could sign up to volunteer and then never show—or you can have a baby, whose importuning skills trump those even of puppies. And, finally, want to avoid pointless arguments? This one isn't a commitment device, just advice: Never discuss pet preferences.

ILLUSORY CONJUNCTION

DIANA DEUTSCH

Professor of Psychology, UC San Diego; author, *The Psychology of Music*

The concept of an *illusory conjunction* is not sufficiently explored in studies of perception and memory and rarely discussed in philosophy. Yet this concept is of considerable importance to our understanding of perceptual and cognitive function. For example, when we hear a musical note, we attribute a pitch, a loudness, a timbre, and we hear the note as coming from a particular spatial location; so each perceived note can be described as a bundle of attribute values. This bundle is thought to reflect the characteristics and location of the sound emitted. But when many note sequences arise simultaneously from different regions of space, these bundles of attribute values sometimes fragment and recombine incorrectly, so that illusory conjunctions result. This causes several illusions of sound perception, such as the octave illusion and the scale illusion, in which the melodies we "hear" are quite different from those presented. The effect can even be found in live musical performances—for example, in the final movement of Tchaikovsky's Sixth Symphony.

Illusory conjunctions can also occur in vision. Under certain circumstances, when people are shown several colored letters and asked to report what they saw, they sometimes combine the colors and shapes of the letters incorrectly; for example, having been shown a blue cross and a red circle, viewers sometimes report having seen a red cross and a blue circle.

Hallucinations—both auditory and visual—frequently involve illusory conjunctions. In musical hallucinations, many aspects of a piece of music may be heard accurately in detail, while some aspect is altered or appears corrupted. A familiar piece of music may be "heard" as played by a different or even unknown musical instrument, as transposed to a different pitch range, or as played much faster or slower than usual. In vision, hallucinated faces may be "seen" to have inappropriate components; in one report, a woman's face appeared with a long, white Santa Claus beard attached.

Presumably, when we see and hear in the normal way, we pro-

cess the information in modules or circuits that are each specific
to some attribute, and we combine the outputs of these circuits
so as to obtain the final integrated percept. Usually this process
leads to veridical perception, but under certain circumstances—
such as in some orchestral music, or during hallucinations—this
process breaks down, and our percepts are influenced by illusory
conjunctions. An understanding of how this happens could shed
valuable light on perceptual and cognitive processing in general.

BISOCIATION

JAMES GEARY
Deputy Curator, Nieman Foundation for Journalism, Harvard
University; author, *I Is an Other*

Charles Lamb once remarked that when the time came for him
to leave this Earth, his fondest wish would be to draw his last
breath through a pipe and exhale it in a pun. And he was indeed
a prodigious punster. Once, when a friend, about to introduce
the notoriously shy English essayist to a group of strangers, asked
him, "Promise, Lamb, not to be so sheepish," he replied, "I wool."

Lamb and his close friend Samuel Taylor Coleridge shared
a passion for punning, not just as a fireside diversion but as a
model for the way the imaginative mind works. "All men who
possess at once active fancy, imagination, and a philosophical
spirit, are prone to punning," Coleridge declared.

Coleridge considered punning an essentially poetic act, ex-
hibiting sensitivity to the subtlest, most distant relationships, as
well as an acrobatic exercise of intelligence, connecting things
formerly believed to be unconnected. "A ridiculous likeness

leads to the detection of a true analogy" is the way he explained it. The novelist and cultural critic Arthur Koestler picked up Coleridge's idea and used it as the basis for his theory of creativity in the sciences, the humanities, and the arts.

Koestler regarded the pun, which he described as "two strands of thought tied together by an acoustic knot," as among the most powerful proofs of "bisociation," the process of discovering similarity in the dissimilar that he suspected was the foundation for all creativity. A pun "compels us to perceive the situation in two self-consistent but incompatible frames of reference at the same time," Koestler argued. "While this unusual condition lasts, the event is not, as is normally the case, associated with a single frame of reference, but *bisociated* with two."

Newton was bisociating when, as he sat in contemplative mood in his garden, he watched an apple fall to the ground and understood it as both the unremarkable fate of a piece of ripe fruit and a startling demonstration of the law of gravity. Cézanne was bisociating when he rendered his astonishing apples as both actual produce arranged so meticulously before him and as impossibly off-kilter objects that existed only in his brushstrokes and pigments. St. Jerome was bisociating when, translating the Old Latin Bible into the simpler Latin Vulgate in the 4th century, he noticed that the adjectival form of "evil," *malus*, also happens to be the word for "apple," *malum*, and picked that word as the name of the previously unidentified fruit that Adam and Eve ate.

There's no sharp boundary splitting the bisociation experienced by the scientist from that experienced by the artist, the sage, or the jester. The creative act moves seamlessly from the "Aha!" of scientific discovery to the "Ah . . ." of aesthetic insight to the "Ha-ha" of the pun and the punchline. Koestler even found a place for comedy on the bisociative spectrum of ingenuity: "Comic discovery is paradox stated—scientific discovery is paradox resolved." Bisociation is central to creative thought, Koestler believed, because "the conscious and unconscious processes underlying creativity are essentially combinatorial activities—the bringing together of previously separate areas of knowledge and experience."

Bisociation is a form of improvised, recombinant intelligence that integrates knowledge and experience, fuses divided worlds, and links the like with the unlike—a model and a metaphor for the process of discovery itself. The pun is at once the most profound and the most pedestrian example of bisociation at work.

CONCEPTUAL COMBINATION

LISA FELDMAN BARRETT

University Distinguished Professor of Psychology, Northeastern University; Research Neuroscientist, Massachusetts General Hospital; Lecturer in psychiatry, Harvard Medical School

Right now, as your eyes glide across this text, you're effortlessly understanding letters and words. How does your brain accomplish this remarkable feat—converting blobs of ink or patterns of tiny pixels into full-fledged ideas? Your brain uses concepts you've accumulated throughout your life. Each letter of the alphabet, each word, and each sequence of words standing for an idea is represented in your brain by concepts. Even more remarkably, you can often comprehend things you've never seen before, like a brand-new word in the middle of a sentence. You can see an unfamiliar breed of dog and still instantly know it's a dog. How does your brain achieve these everyday marvels? The answer is concepts in combination.

Most scientists will tell you that your brain contains a storehouse of concepts to categorize the objects and events around you. In this view, concepts are like dictionary definitions stored in your brain, such as "A pet is an animal companion that lives with people." Each concept is said to have an unchanging core shared by all individuals. Decades of research, however, show that this isn't the case. A concept is a dynamic pattern of neural activity. Your brain doesn't store and retrieve concepts; it makes concepts on the fly, as needed, in its network of billions of communicating neurons. Each time you construct the "same" concept, such as "dog," the neural pattern is different. A concept is a population of variable instances, not a single static instance, and your mind is a computational moment within a constantly predicting brain.

Whenever your brain encounters sensory inputs, whether familiar or novel, it tries to produce an answer to the question "What is this *like*?" In doing so, your brain constructs a concept out of bits and pieces of past experience. This process is called *conceptual combination*. Without it, you'd be experientially blind to anything you hadn't encountered before.

Conceptual combination occurs every time your brain makes a concept for use, but it's easiest to imagine when the combination is explicit, such as "purple elephant with wings." As another example, consider the science-fiction movie *The Matrix*, when the shocking secret is revealed that the Matrix is powered by electrical hookups to live human bodies. In order to experience the horror, you need conceptual combination to construct the new concept "person as battery."

The more familiar a concept—that is, the more frequently you've constructed it—the more efficiently your brain can make it by conceptual combination. Your brain requires less energy to construct the concept "dog" than the combination "hairy, friendly, loyal animal with two eyes, four legs, and a slobbering tongue, who makes a barking sound, eats processed food from a bowl, and rescues children from danger in Disney movies." That sort of combination is what your brain would have to do if it created the concept "dog" for the first time. The word "dog" then helps your brain create the concept efficiently in the future. That's what happened in recent years with the concept "hangry," which began as a combination of "hungry" and "angry" and "irritable" but is now more efficiently constructed in many American brains.

You experience the effort of conceptual combination when you venture to a new culture full of unfamiliar concepts. Some concepts are universally known—a face is a face in any culture—but plenty are culture-specific, such as social concepts serving as the glue for civilization. For example, in the United States we have the concept of a thumbs-up gesture indicating "All's well." To some cultures, the same gesture is an insult. These kinds of conceptual differences are a major reason why culture-switching is stressful and communication across cultures can be perilous.

Conceptual combination can also be fun. Anytime you laugh at a stand-up comic who juxtaposes two unrelated ideas, you're combining concepts. Innovation, the Holy Grail of business success, is effectively conceptual combination for profit.

Some brains cannot manage conceptual combination. Temple Grandin, one of the most eloquent of autistic writers, describes her difficulties with conceptual combination in *How Does Visual Thinking Work in the Mind of a Person with Autism*: "When I was a child, I categorized dogs from cats by sorting the animals by size. All the dogs in our neighborhood were large until our neighbors got a Dachshund. I remember looking at the small dog and trying to figure out why she was not a cat." Naoki Higashida, a teenager with autism, answers the question "What is this like?"

by deliberately searching his memory rather than automatically constructing the best fitting instance, as most people's brains do. "First, I scan my memory to find an experience closest to what's happening now," he writes in *The Reason I Jump.* "When I've found a good close match, my next step is to try to recall what I said at that time. If I'm lucky, I hit upon a usable experience and all is well." If Naoki is unlucky, he becomes flustered, unable to communicate.

Scientists consider conceptual combination one of the most powerful abilities of the human brain. It's not just for making novel concepts on the fly. It's the normal process by which your brain constructs concepts. Conceptual combination is the basis for most perception and action.

BOOLEAN LOGIC

SIMON BARON-COHEN

Professor of Developmental Psychopathology and Director, Autism Research Centre, University of Cambridge; author, *Zero Degrees of Empathy*

George Boole, the son of a shoemaker, left school at sixteen and ended up a professor of mathematics at Queens College in Cork, Ireland. As the only breadwinner in his family, he became a teacher at age sixteen, opened his own school in Lincoln, England, by age nineteen, and fifteen years later was a mathematician and philosopher of logic. Why did Google celebrate his 200th birthday on November 2, 2015? And why should we remember his contributions?

In 1854, he wrote a book called *An Investigation of the Laws of Thought* which distilled the essence of logical thought down to terms like AND (e.g., x AND y), OR (e.g., x OR y), and NOT (e.g., NOT x), or combinations of these. This took Aristotelian logic (the syllogism, expressed in words) and insisted these be formulated as equations, which was a revolutionary step. Equally important, Boolean logic is today seen as the foundation of the Information Age, or what we also call the Computer Age. This is because each "value" in these logic statements or equations comes down to being either true or false, with zero ambiguity. The logic is binary. No wonder they could be applied, more than a century later, in the design of electronic circuits in computers.

For me, the importance of Boolean logic goes even beyond its contribution to the logic of computers. Each of his terms is an operation, like the familiar mathematical ones of addition and multiplication. These terms describe what happens if you take something as input and perform an operation on it. You end up with an output. This is at the core of what I call systemizing: Take input, perform an operation, and observe the output. Boole, without realizing it, was describing how the human mind systemizes. In this way, he anticipated how to describe a uniquely human aspect of cognition, one enabling humans to do engineering (design a system) and to innovate (change a system).

Consider a much discussed example: Vinton Cerf, co-inventor of the Internet, poured peppercorns into a funnel and found if he dropped handfuls of peppercorns into it the funnel got blocked. Nothing came out. But if he poured the peppercorns in one a time, they didn't get stuck and they flowed out smoothly. In system 1, the input is a handful of peppercorns, the operation is pouring them into a funnel, and the output is disappointingly nothing! In system 2, the input is one-peppercorn-at-a-time, the operation remains the same, but now the output is a pleasing flow of peppercorns. There are lessons here for how we as humans design not just pepper grinders but also traffic systems (which either cause or avoid congestion), or how we design the post office to cope with a volume of letters. Indeed, Boolean logic allows us to describe not just how an engineer designs systems but also how we humans systemize.

Here's one more legacy of Boolean logic. We know that people with autism have logical minds and a strong drive to systemize. If we can generalize: They have a preference for information that can be systemized (such as factual information, or repeating, lawful patterns of information that don't change unlawfully). They don't cope well with information that's hard to systemize because it contains ambiguity or changes unexpectedly (such as social interaction, where what people do or say is rarely the same except in highly ritualized contexts). Our modern understanding of autism as a hypersystemizing mind owes a huge debt to Boolean logic, which enables us to characterize the beauty and extraordinary power of binary thinking, and also where such thinking is best used.

NEURODIVERSITY

JOICHI ITO

Director, MIT Media Lab; Professor of the Practice in Media Arts and Sciences, MIT

Humans have diverse neurological conditions. While some of these conditions, such as autism, are considered disabilities, many claim that they're the result of normal variations in the human genome. The neurodiversity movement is an international civil rights movement arguing that autism shouldn't be "cured" and that it's an authentic form of human diversity that should be protected.

In the early 1900s, eugenics and the sterilization of people considered genetically inferior were scientifically sanctioned ideas, with advocates like Theodore Roosevelt, Margaret Sanger, Winston Churchill, and U.S. Supreme Court Justice Oliver Wendell Holmes, Jr. The horror of the Holocaust, inspired by the eugenics movement, demonstrated the devastation those programs could exact when put into practice.

Temple Grandin, an outspoken spokesperson for autism and neurodiversity, argues that Albert Einstein, Wolfgang Mozart, and Nikola Tesla would have been diagnosed as on the autistic spectrum if they were alive today. She also believes that autism has long contributed to human development and that "without autism traits we might still be living in caves." Today, non-neurotypical children often suffer through remedial programs in the traditional educational system, only to be revealed as geniuses later. Many of them end up at MIT and other research institutes.

With the invention of CRISPR, the possibility of editing the human genome at scale has suddenly become feasible. The initial applications being developed involve the "fixing" of genetic mutations that cause debilitating diseases, but they also open a path to potential elimination not only of autism but of much of the diversity that makes human society flourish. Our understanding of the human genome is rudimentary enough that it will be some time before we can enact complex changes involving things like intelligence or personality, but it's a slippery slope. I saw a business plan a few years ago claiming that autism was just "errors" in the genome which could be identified and "corrected," like "de-noising" a grainy photograph or audio recording.

Clearly, some children born with autism have debilitating issues that require intervention. But our attempts to "cure" autism, either through remediation or eventually through genetic engineering, could result in the eradication of a neurological diversity that drives scholarship, innovation, arts, and many of the essential elements of a healthy society.

We know that diversity is essential for healthy ecosystems. We see how agricultural monocultures have created fragile and unsustainable systems. My concern is that even if we believe that neurological diversity is essential for society, we'll still develop tools for designing away any risky traits that deviate from the norm, and so, given a choice, people will tend to opt for a neurotypical child.

As we march down the path of genetic engineering to eliminate disabilities and disease, it's important to realize that this path, while scientifically sophisticated, has sometimes ended in unintended and possibly irreversible consequences and side effects.

CASE-BASED REASONING

ROGER SCHANK

CEO, Socratic Arts, Inc.; John Evans Professor Emeritus of Computer Science, Education, and Psychology, Northwestern University; author, *Make School Meaningful—and Fun!*

We do case-based reasoning all the time, without thinking that that's what we're doing. Case-based reasoning is essential to personal growth and learning. Although we hear people proclaim that mathematics teaches one to think, or knowing logic will help one reason more carefully, humans do a different kind of reasoning quite naturally.

When we go to a restaurant, we think about what we ordered the last time we were there and whether we want to order the same thing again. When we go out on a date, we think about how that person reminds us of someone we went out with before, and we think about how that turned out. When we read a book, we're reminded of other books with similar themes or situations, and we tend to predict outcomes on that basis. When we hear someone tell us a story about their life, we're immediately reminded of something similar that happened to us.

Reminding, based on the examination of an internal library of cases, is what enables learning and is the basis of intelligence. In order to be reminded of relevant prior cases, we create those cases subconsciously by thinking about them and telling someone about them. Then, again subconsciously, we label the previously experienced cases in some way. A classic example is "The Steak and the Haircut," a story about a colleague of mine who responded to my complaint about my wife's inability to cook steak as rare as I wanted by saying that twenty years earlier, in London, he couldn't get his hair cut as short as he wanted. While this may sound like a brain-damaged response, the two stories are identical at the right level of abstraction. They're both about asking someone to do something who, although capable of doing it, refused because they thought the request too extreme. My friend had been wondering about his haircut experience for twenty years. My story reminded him of it and helped him explain to himself what had happened.

We're all case-based reasoners, but no one ever teaches us how to do this (except possibly in medical school, business school, and law school). No one teaches you how to label cases or retrieve cases from your memory. Yet our entire ability to reason depends on this capability. We need to see something as an instance of something we've seen before, in order to make a judgment about it and learn from it.

We do case-based reasoning naturally and without conscious

thought, so we tend to ignore its importance in thinking and learning. Whenever you participate in a conversation with someone about a subject of mutual interest, you're having a kind of case-based-reasoning party—exchanging stories and being reminded of new stories to tell. Both participants come out slightly changed from the experience. That experience itself is, of course, a new case to be remembered and reasoned from in the future.

MEDIA RICHNESS

NINA JABLONSKI

Evan Pugh University Professor of Anthropology, Pennsylvania State University

The term *media richness* was coined by information theorists Richard Daft and Robert Lengel in 1986. Media richness describes the density of learning that can be conveyed through a particular communications medium. Face-to-face communication is the richest medium, according to media richness theory (MRT), because it allows for the simultaneous interpersonal exchange of cues from linguistic content, gesture, posture, tone of voice, facial expression, and direction of gaze. MRT was developed before the rise of electronic communications media to help managers in business contexts decide which medium was most effective for communicating a message. Rich media, such as conversations and phone calls, were deemed best for non-routine messages, while "lean" media (e.g., unaddressed memoranda) were considered acceptable for routine messages. In the last two decades, media richness has been extended to describe the strengths and weaknesses of new media—email, Web sites, video conferencing, voicemail, instant messaging. Media richness deserves to be more widely known, because people make choices throughout the day about communications media, often without considering the consequences of the choice of medium and the fit between the content of a message and the medium through which it's communicated.

Humans evolved in media-rich contexts. In stable, tightly knit social groups, face-to-face communication was the only mode of communication for hundreds of thousands of years. The concept of media choice did not exist, because (apart from smoke signals) it was face-to-face or nothing, until about 5,000 years ago. Articulate speech and language complemented the rich repertoire of vocalizations, facial expressions, glances, stares, gestures, and postures upon which our ancestors relied, creating a rich and potentially highly nuanced communications repertoire. Within small groups, people attended closely to what was said, who said it, and how it was said. Pleasantries were exchanged, advice was given, loving whispers traded, and admonitions delivered with a full sensory armada of verbal content: tone of voice, measured eye contact, gesture, and posture. People were bathed in conversation, reassured by touch, verbally upbraided for unreasonableness, publicly shamed by calculated stares, and physically reprimanded for antisocial behavior. Although interpersonal communication has probably never been without the potential for guile and social manipulation, deception was hard to pull off, because information flowed through visual, auditory, and even tactile and olfactory channels. Communication had immediate effects and consequences. Mass communication reached only as far as the human voice could carry.

The consequences of media richness and the concept of media choice became relevant for people only with the introduction of writing in early agricultural societies. Initially developed to facilitate clerical and payroll functions, writing was soon marshaled in support of military, political, and religious causes, and much later for the exchange of personal information and the composition of poetry and philosophical treatises. Communication through writing was augmented early in the 20th century by modes of remote voice communication (telephone, microphone, and radio) and later by combined visual and auditory modes (movies, television, Web sites), which provided unprecedented scope for unidirectional communication. Historians and scholars of communication theory talk about trade-offs between the richest face-to-face channels and the leaner modalities of email, voicemail, and text messaging, which provide fewer cues, slower feedback, and limited scope for redress. Media richness has been criticized in recent years because it fails to predict why people choose lean over rich media, especially in situations where a richer medium clearly would be more effective. Many will have experienced the shock of an email reporting the death of a loved one, or been anguished because a misplaced comma or an inappropriate emoticon in a text message botches an expression of friendship or love. This is exactly why media richness is important—and interesting from an evolutionary perspective.

In a world of unconstrained media choice, people often choose leaner and unidirectional modalities because they want to make a point, or at least think they want to make a point. A need for incessant and immediate connection (or just a need to save money) can provoke the blurting of something through a cheap, low-grade channel rather than waiting for the chance to use a richer one. Leaner media also carry lower risk of rejection or immediate retribution. Regardless of the reason, we now live in a world where people are opting for leaner modes of communication because they've been socialized inadequately in richer ones and are functionally ignorant of the concept of media richness. The chances of misunderstanding have never been greater; the opportunities for providing comfort and solace, or for exacting meaningful and appropriate retaliation, have never been more limited. We still yearn to see one another, but contacts often consist more of broadcast static faces and less of breathing exchanges of shared wonder, love, tribulation, and loss.

Like all primates, humans have nurtured harmonious relationships and maintained social cohesion by being intensely good, high-bandwidth communicators. Media richness is a concept worthy of wider propagation, because it will help ensure the future of individuals and societies in times of increasing individual social isolation, electronic bullying, touch aversion, personal anxiety, and social estrangement.

PEIRCEAN SEMIOTICS

DANIEL L. EVERETT

Linguist; Professor of Global Studies and Sociology, Dean of Arts and Sciences, Bentley University; author, *How Language Began*

The course followed by humans on the path to language was a progression through natural signs to human symbols. Signs and symbols are explained in reference to a theory of semiotics, the study of signs, in the writings of Charles Sanders Peirce (1839–1914). Peirce was perhaps the most brilliant American philosopher who ever lived. Bertrand Russell said of him, "Beyond doubt . . . he was one of the most original minds of the later nineteenth century, and certainly the greatest American thinker ever." He contributed to mathematics, science, the study of language, and philosophy. He is the founder of multiple fields of study, including semiotics as well as pragmatism, the only uniquely American school of philosophy, further developed by William James and others.

Peirce's theory of semiotics outlines a conceptual progression of signs, from indexes to icons to human-created symbols. It is a progression not only of the increasing complexity of types of signs but also of the evolution of *Homo* species' language abilities.

A sign is any pairing of a form (a word, a smell, a sound, a street sign, Morse code) with a meaning (what the sign refers to). According to Peirce, an index, as the most primitive part of the progression, is a form that has a physical link to what it refers to. The footprint of a cat refers us to, makes us think of, a cat. The smell of a grilling steak brings to mind the steak and the grill. Smoke indicates fire.

An icon is something that is physically somehow like what it refers to. A sculpture represents the real-life object it's about. A portrait, likewise, is an icon of whomever is portrayed. An onomatopoeic word like "bam" or "clang" bears an iconic sound resemblance to another sound.

It turns out that Peirce's theory also predicts the order of language evolution we discover in the fossil record. First, we discover indexes being used by creatures far predating the emergence of genus *Homo*. Second, we discover the use of icons by Australopithecines in South Africa some 3 million years ago. And finally, through recent archaeology on their sea voyages, settlement, and burial patterns, we discover that 1.9 million years ago the first *Homo, erectus*, had and used symbols, almost certainly indicating that language—the ability to communicate almost anything we can communicate today, in principle—began well before our species appeared. What is most fascinating is that Peirce's semiotics is a theory of philosophy that inadvertently makes startlingly accurate predictions about the fossil record.

The spread of Peirce's semiotics throughout world philosophy, influencing figures such as Ferdinand de Saussure and Alfred North Whitehead, extends to industry, science, linguistics, anthropology, philosophy, and beyond. Peirce introduced the concept of "infinite semiosis" long before Noam Chomsky raised the issue of recursion as central to language.

Perhaps only Peirce, in the history of inquiry into human language, has come up with a theory that at once predicts the order of language evolution from the earliest hominins to *Homo sapiens*, while enlightening researchers from across the intellectual spectrum about the nature of truth and meaning and the conduct of scientific inquiry.

Peirce himself was a cantankerous curmudgeon. For that reason, he never enjoyed stable employment, living in part off the donations of friends such as William James. But his semiotics has brought intellectual delight and employment to hundreds of academics since the late 19th century, when he first proposed his theory. His work on semiotics is worthy of being much more widely known, as relevant to current debates. It's far more than a quaint relic of 19th-century reflection.

HISTORIOMETRICS

HOWARD GARDNER
Hobbs Professor of Cognition and Education, Harvard Graduate School of Education; author, *Truth, Beauty, and Goodness Reframed*

Over the centuries, reflective individuals have speculated about the causes of differences among individuals (who is talented and why; what makes certain persons influential) and among societies (why certain cultures have thrived and others vanished; which societies are bellicose and why). In the 19th century, scholars began systematic study of such issues. Work in this vein, which has continued and expanded, has been dubbed *historiometrics*—or, alternatively, *cliometrics*. These terms, while literally and etymologically appropriate, are hardly transparent or snappy—one of the reasons historiometric scholars and approaches have yet to receive due credit.

Among continental scholars, the Belgian statistician Adolphe Quetelet is often credited with opening up this line of work. But in the Anglo-American scholarly community, the British polymath Francis Galton is generally seen as the patron saint of historiometric studies. As a member of the distinguished Darwin family, Galton was particularly interested in the nature and incidence of genius. Using statistical methods, he demonstrated how genius thrived in certain families and attributed this distribution to hereditary factors. Yet Galton was also sensitive to the possible confounds between hereditary and environmental contributions, so he pioneered in comparisons of identical twins, fraternal twins, and other members of the same family. On both sides of the English channel, historiometric studies had been launched!

In the 20th century, these lines of work, whether or not officially labeled as historiometry, continued in many locales. Deserving special mention is the American psychologist Dean Keith Simonton, who has devoted several decades to historiometric study—carrying out dozens of intriguing studies as well as explicitly laying out the methods available in the armamentarium of the historiometric scholar. If you're speculating about the kinds of issues just alluded to—for example, during which decade of life do scholars in specific domains conduct their most influential research; during which decade of life do artists create their most enduring works—chances are that Simonton has already done relevant research. And if not, he'll readily suggest how such studies might be done and their results interpreted.

I've often wondered why Simonton's work isn't more widely known and appreciated. I suspect it's because the work spans standard social science (psychology, sociology) and humanistic studies (history, the arts)—and most scholars work comfortably within rather than across these two cultures. It's also notable that Simonton has worked largely alone, with neither a big staff nor a large research budget, and that's an unusual research profile in our time.

Enter Big Data. Neither Quetelet nor Galton had significant computational aids; pencil and paper was the medium of choice. When Simonton began his studies in the 1970s, we were in the era of large mainframes, punch cards, and limited computing power. To be sure, Simonton (and other self-styled cliometricians, like Charles Murray) has kept up with advances in technology. But only in the last decade or so has it become possible—indeed, easy—to pursue historiometric puzzles, drawing on vast amounts of data sitting on one's lap, or, more precisely, on one's laptop.

Occasionally, historiometric findings have found their way into the mainstream media. Last year, a team of researchers led by Roberta Sinatra introduced the Q phenomenon. Curious about the distribution of influential work over the investigative life span of productive scientists, the researchers examined publication records of scientists drawn from seven disciplines. And they discovered—presumably to their and others' surprise—that "the highest-impact work in a scientist's career is randomly distributed within her body of work." Many commentators, prominent among them Simonton, have reacted to this claim, and it's safe to say we've not heard the last of the Q phenomenon.

Has the moment for historiometry finally arrived? Given the fascination of historiometric questions and the relative ease these days of researching them, what was once an exotic exercise of eccentric European savants can now become a regular part of the disciplinary terrain.

I'm not quite persuaded that the moment has arrived. To be sure, the availability of vast sources of data and powerful data-mining techniques have greatly enhanced the "metric" part of historiometrics. Yet the "historio" part is equally important. Historians should be judged by the quality of the questions they raise and the sense they make of what they've uncovered. These are issues of judgment, not mere measurement. (As has long been quipped, "Garbage in, garbage out.") And so, to return to the Q phenomenon, the unexpected findings of the Sinatra team open up a slew of possible explanations and interpretations. But the available data will never tell us which issues to pursue next; you need a solid historical sense as well as a dollop of historical humility. Whether the Sinatra team or some other team or individual will significantly raise the stock of historiometry depends on its historical wisdom as well as its data-analytic prowess.

There may be broader lessons here. Scholarship, scientific or humanistic, always entails a dialectic between issues worth pursuing and the methods available for pursuing them.

Historiometric curiosity dates back to Classical times, but the advent of measurement techniques (statistics, data analytics) has allowed this curiosity to be pursued with increasing power and elegance. It's desirable to maintain a balance between questions/curiosity/judgment on the one hand and analytic measures on the other; when either becomes dominant, the pursuit itself can be compromised.

And for extra credit: When you want to explain to others what you're up to, find a succinct and memorable descriptor!

POPULATION THINKING

DAN SPERBER

Social and cognitive scientist; Professor Emeritus, CNRS, Paris; Director, International Cognition and Culture Institute, Central European University, Budapest; co-author (with Hugo Mercier), *The Enigma of Reason*

What were Darwin's most significant contributions? The evolutionary biologist Ernst Mayr answered: (1) producing copious evidence of evolution, (2) explaining it in terms of natural selection, and (3) thinking about species as populations.

"Population thinking"? Philosophers are still debating what this means. For scientists, however, to think of living things in terms of populations rather than types having each its own essence is a clear and radical departure both from earlier scholarly traditions and from folk biology.

Species evolve; early features may disappear; novel features may appear. From a populationist point of view, a species is a population of organisms that share features not because of a common "nature" but because they're related by descent. A species so understood is a temporally continuous, spatially scattered entity that changes over time.

Population thinking readily extends beyond biology to the study of cultural evolution (as argued by Peter Richerson, Robert Boyd, and Peter Godfrey-Smith). Cultural phenomena can be thought of as populations whose members share features because they influence one another, even though they don't beget one another as organisms do and aren't exact copies of one another. Here are three examples.

What is a word—the word "love," for instance? It's commonly described as an elementary unit of language that combines sound and meaning. Yes, but a word so understood is an abstraction without causal powers. Only concrete uses of the word "love" have causes and effects: An utterance of the word has, among its causes, mental processes in a speaker and, among its effects, mental processes in a listener (not to mention hormonal and other biochemical processes). This speech event is causally linked, on another timescale, to earlier similar events from which the speaker and listener acquired their ability to produce and interpret "love" the way they do. The word endures and changes in a linguistic community through all these episodes of acquisition and use.

So the word "love" can be studied as a population of causally related events taking place inside people and in their shared environment—a population of billions and billions of such events, each occurring in a different context, each conveying a meaning appropriate at that instant, and all nevertheless causally related. Scholarly or lay discussions about the word "love" and its meaning are themselves a population of mental and public metalinguistic events evolving on the margins of the "love" population. All words can similarly be thought of, not (or not just) as abstract units of language but as populations of mental and public events.

What is a dance? Take tango. There are passionate arguments about the true character of tango. From a populationist perspective, tango should be thought of as a population of events of producing tango music, listening and dancing to it, watching others dance, commenting on and discussing the music and the dance—a population that originated in the 1880s in Argentina and spread around the world. The real question is not what is a true tango but how attributions or denials of authenticity evolve on the margins of this population of acoustic, bodily, mental, and social "tango" events. All culturally identified dances can be thought of as populations in the same way.

What is a law? Take the United States Constitution. It's commonly thought of as a text—or more accurately, since it has been repeatedly amended, as a text with several successive versions. Each version has millions of paper and now electronic copies. Each article and amendment has been interpreted on countless occasions in a variety of ways. Many of these interpretations have been quoted again and again and reinterpreted, and their reinterpretations reinterpreted in turn. Articles, amendments, and interpretations have been invoked in a variety of situations. In other words, there's a population whose members are all these objects and events in the environment plus all the relevant mental representations and processes in the brains of the people who have produced, interpreted, invoked, or otherwise considered versions and bits of the Constitution. All the historical effects of the Constitution have been produced by members of this population of material things and not by the Constitution considered in the abstract. The Constitution, then, can usefully be thought of as a population, and so can all laws.

Population thinking is itself a population of mental and public things. Philosophers' discussions of what population thinking really is are members of this population. So is the text you just read, and so is your reading of it.

BOUNDED OPTIMALITY

TOM GRIFFITHS

Professor, Department of Psychology and Cognitive Science Program,
UC Berkeley; Director, Institute of Cognitive and Brain Sciences;
co-author (with Brian Christian), *Algorithms to Live By*

How are we supposed to act? To reason, to make decisions, to
learn? The classic answer to this question, hammered out over
hundreds of years and burnished to a fine luster in the middle
of the last century, is simple: Update your beliefs in accordance
with probability theory and choose the action that maximizes
your expected utility. There's only one problem with this answer:
It doesn't work.

There are two ways in which it doesn't work. First, it doesn't
describe how people actually act. People systematically deviate
from the prescriptions of probability and expected utility. Those
deviations are often taken as evidence of irrationality—of our
human foibles getting in the way of our aspirations to intelligent action. However, human beings remain the best examples
we have of systems capable of anything like intelligent action in
many domains. Another interpretation of these deviations is thus
that we're holding people to the wrong standard.

The second way in which the classic notion of rationality
falls short is that it's unattainable by real agents. Updating beliefs
in accordance with probability theory and choosing the action
that maximizes expected utility can quickly turn into intractable
computational problems. If you want to design an agent capable
of intelligent action in the real world, you need to take into ac-

count not just the quality of the chosen action but also how long it took to choose that action. Deciding to pull a pedestrian out of the path of an oncoming car isn't useful if it takes more than a few seconds to make the decision.

What we need is a better standard of rational action for real agents. Fortunately, artificial-intelligence researchers have developed one: bounded optimality. The bounded-optimal agent navigates the trade-off between efficiency and error, optimizing not the action taken but the algorithm used to choose that action. Taking into account the computational resources available to the agent and the cost of using those resources to think rather than act, bounded optimality is about thinking just the right amount before acting.

Bounded optimality deserves to be more widely known because of its implications for both machines and people. As artificial-intelligence systems play larger roles in our lives, understanding the trade-offs that inform their design is critical to understanding the actions they take; machines are already making decisions that affect the lives of pedestrians. But understanding the same trade-offs is just as important for thinking about the design of the pedestrians. Human cognition is finely tuned to make the most of limited onboard computational resources. With a more nuanced notion of what constitutes rational action, we might better understand human behavior that otherwise seems irrational.

SATISFICING

MICHAEL HOCHBERG
Evolutionist; Distinguished Research Director, CNRS, University of
Montpellier; External Professor, Santa Fe Institute

AI pioneer Herbert Simon contributed importantly to our understanding of a number of problems in a wide array of disciplines, one of which is the notion of achievement. Achievement depends not only on one's ability and the problem at hand (including information available and the environment) but also on one's motivation and targets. Two of Simon's many insights were into how much effort it takes to become an "expert" at an endeavor requiring a special skill (using chess as a model, the answer is approximately 10,000 hours) and how objectives are accomplished—whether individuals maximize, optimize, or just accept (or even seek) an apparently lesser outcome: that is, *satisfice*. Satisficing recognizes constraints on time, ability, and information and the risk and consequences of failure. In Simon's words, from his 1956 paper on the subject:

> Since the organism, like those of the real world, has neither the senses nor the wits to discover an "optimal" path—even assuming the concept of optimal to be clearly defined—we are concerned only with finding a choice mechanism that will lead it to pursue a "satisficing" path, a path that will permit satisfaction at some specified level of all of its needs.

Will a marginal increase in effort result in an acceptable increase in achievement? It's evidently unusual to be so calculating when deciding how to commit to an endeavor, competitive or not. In a noncompetitive task, such as reading a book, we may be time-limited or constrained by background knowledge. Beyond completing the task, there's no objective measure of achievement. At the other extreme—for example, competing in a 100-yard dash—it's not only about victory and performance relative to other contestants but also about outcomes relative to past and future races. Effort is maximized from start to finish.

Obviously, many endeavors are more complex than those, and it's not as easy to question the default objectives of maximization or optimization. Evolutionary thinking is a useful framework, in this regard, for gaining a richer understanding of the processes at work. Consider the utility of running speed for a predator (me) trying to catch a prey. I could muster that extra effort to improve on my performance, but at what cost? For example, if I were to run as fast as I could, then if I failed, not only would I miss my dinner but I might also need to wait to recover the energy to run again. In running at full speed over uneven terrain, I would also risk injury and could become dinner for another predator.

Endeavors have not only risks but also constraints—what evolutionists refer to as trade-offs. Trade-offs, such as between running speed and endurance, may appear simple, but imagine trying to adapt running speed to preserve endurance so as to catch any prey you've spotted, regardless of the prospects of actually catching and subduing it. Your time and effort are probably wasted, resulting in insufficient numbers of prey caught overall, but it's also possible that more prey are caught than needed, meaning less time spent on other important tasks. A satisficer chases only enough of the easier-to-catch prey to satisfy basic needs and so can spend more time on other useful tasks.

Satisficing should be more widely known because it's a different way of looking at nature in general, as well as at certain facets of human endeavor. More, higher, faster is "better" only up to a point, and perhaps only in a small number of contexts. Indeed, it's a common misconception that natural selection optimizes or maximizes whatever it touches; the evolutionary mantra "survival of the fittest" can be misleading. Rather, the evolutionary process *tends* to favor fitter genetic alternatives, and performance ability will vary between individuals. Winners will sometimes be losers and vice versa. Most finish somewhere in between, and for some, this is success.

We humans will increasingly satisfice, because our environments are becoming ever richer, more complex, and more challenging. Some may fear that satisficing will create a world of laziness, apathy, substandard performance, and economic stagnation. On the contrary, if norms in satisficing embody certain standards, this could lead to a ratcheting-up of individual well-being, social stability, and sustainability.

DE-ANONYMIZATION

ROSS ANDERSON
Professor of Security Engineering, University of Cambridge

We keep hearing about Big Data as the latest magic solution for all society's ills. The sensors surrounding us collect ever more data on everything we do; companies use that data to work out what we want and sell it to us. But how do we avoid a future in which the secret police know everything?

We're often told our privacy will be safe because our data will be made anonymous. But Dorothy Denning and other computer scientists discovered around 1980 that anonymization doesn't work very well. Even if you write software that will answer a query only if the answer is based on the data of six or more people, there's a lot of ways to cheat it. Suppose university professors' salaries are confidential but statistical data are published, and suppose that one of the seven computer science professors is a woman. Then I just need to ask, "average salary computer science professors?" and "average salary male computer science professors?" And given access to a database of "anonymous" medical records, I can query the database before and after the person I'm investigating visits their doctor and look at what changed. There are many ways to draw inferences.

For about ten years now, we've had a decent theoretical model of this. Microsoft researcher Cynthia Dwork's work on differential privacy established bounds on how many queries a database can safely answer, even if it's allowed to add some noise and permitted a small probability of failure. In the best general case, the bound is of the order of n^2 where there are n attributes. So if your medical record has about 100 pieces of information about you, then it's impractical to build an anonymized medical record system that will answer more than about 10,000 queries before a smart interrogator will be able to learn something useful about someone. Common large-scale systems, which can handle more than that many queries an hour, simply cannot be made secure—except in a handful of special cases.

One such case may be where your navigation app uses the locations of millions of cell phones to work out how fast the traffic is moving where. But even this is hard to do right. You have to let programmers link up some successive sightings of each phone within each segment of road to get average speeds, but if they can link between segments they might be able to reconstruct all the journeys that any phone user ever made. To use anonymization effectively—in the few cases where it can work—you need smart engineers who understand inference control and are incentivized to do the job properly. Both understanding and incentive are usually lacking.

A series of high-profile data scandals has hammered home the surprising power of de-anonymization. And it's getting more powerful all the time, as we get ever more social data and other contextual data online. Better machine-learning algorithms help, too; they've recently been used, for example, to de-anonymize millions of mobile-phone-call data records by pattern-matching them against the public friendship graph visible in online social media. So where do we stand now?

It's reminiscent of the climate-change debate. Just as Big Oil has lobbied for years to undermine the science of global warming, so also Big Data firms have a powerful incentive to pretend that anonymization works, or at least will work in the future. When people complain of some data grab, we're told that research on differential privacy is throwing up lots of interesting results and our data will be protected better real soon now. (Never mind that differential privacy teaches exactly the reverse; namely, that such protection is usually impossible.) And many people who earn their living from personal data follow suit. This is an old problem: It's hard to get people to understand anything if their job depends on not understanding it.

In any case, the world of advertising pushes toward ever more personalization. Knowing that people on Acacia Avenue are more likely to buy big cars, and that forty-three-year-olds are, too, is of almost no value compared with knowing that the forty-three-year-old who lives on Acacia Avenue is looking to buy a new car right now. Knowing how much he's able to spend opens the door to ever more price discrimination, which, although unfair, is both economically efficient and profitable. We know of no technological silver bullet, no way to engineer an equilibrium between surveillance and privacy; boundaries will have to be set by other means.

FUNCTIONAL EQUATIONS

JASON WILKES

Graduate student in psychology, UC Santa Barbara; author, *Burn Math Class*

Where does mathematics come from? I'm not talking about the philosophical question of whether mathematical truths have an existence independent of human minds. I mean concretely. What on earth is this field? When a mathematician makes cryptic pronouncements like "We define the entropy of a probability distribution to be such-and-such," who or what led them to explore that definition over any other? Are they accepting someone else's definition? Are they making up the definition themselves? Where exactly does this whole dance begin?

Mathematics textbooks contain a mixture of the timeless and the accidental, and it isn't always easy to tell exactly which bits are necessary inevitable truths and which are accidental social constructs that could have easily turned out differently. An anthropologist studying mathematicians might notice that for some unspecified reason our species seems to prefer the concept of "even numbers" over the barely distinguishable concept of "(s?)even numbers," where a "(s?)even number" is defined to be a number that's either (a) even, or (b) seven. Now, when it comes to ad-hoc definitions, this example is admittedly extreme. But in practice not all cases are quite this clear-cut, and our anthropologist still has an unanswered question: What is it that draws the mathematicians to one definition and not the other? How do mathematicians decide which of the infinity of possible mathematical concepts to define and study in the first place?

The secret is a piece of common unspoken folk knowledge among mathematicians, but being an outsider, our anthropologist had no straightforward way of discovering it, since for some reason the mathematicians don't often mention it in their textbooks. The secret is that although it's legal in mathematics to arbitrarily choose any definitions we like, the best definitions aren't just chosen—they're derived.

Definitions are supposed to be the starting point of a mathematical exploration, not the result. But behind the scenes the distinction isn't always so sharp. Mathematicians derive definitions all the time. How do you derive a definition? There's no single answer, but in a surprising number of cases the answer turns out to involve an odd construct known as a functional equation. To see how this happens, let's start from the beginning.

Equations are mathematical sentences describing the behaviors and properties of some (often unknown) quantity. A functional equation is just a mathematical sentence that says something about the behaviors not of an unknown number but of an entire unknown function. This idea seems mundane at first glance. But its significance becomes clearer when we realize that, in a sense, what a mathematical sentence of this form gives us is a quantitative representation of qualitative information. And it's exactly that kind of representation that's needed to create a mathematical concept in the first place.

When Claude Shannon invented information theory, he needed a mathematical definition of uncertainty. What's the right definition? There isn't one. But whatever definition we choose, it should act somewhat like our everyday idea of uncertainty. Shannon decided he wanted his version of uncertainty to have three behaviors. Paraphrasing heavily: (1) Small changes in our state of knowledge cause only small changes in our uncertainty (whatever we may mean by "small"); (2) Dice with more sides are harder to guess (and the dice don't actually have to be dice); and (3) If you stick two unrelated questions together (e.g., "What's your name?" and "Is it raining?"), your uncertainty about the whole thing should just be the first one's uncertainty plus the second one's (i.e., independent uncertainties add). These specs all seem reasonable.

In fact, even though we're allowed to define uncertainty whatever way we want, any definition that didn't have Shannon's three properties would have to be at least a bit weird. So Shannon's version of the idea is an honest reflection of how our everyday concept behaves. However, it turns out that just those three behaviors are enough to force the mathematical definition of uncertainty to look a particular way.

Our vague qualitative concept directly shapes the precise quantitative one. (And it does so because the three English sentences above are really easy to turn into three functional equations. Basically, just abbreviate the English until it looks like mathematical symbols.) This is, in a very real sense, how mathematical concepts are created. It's not the only way. But it's a fairly common one, and it shows up in the foundational definitions of other fields, too.

It's the method Richard Cox employed to prove (assuming degrees of certainty can be represented by real numbers) that the formalism we call Bayesian probability theory isn't just one ad-hoc method among many but is in fact the only method of inference under uncertainty that reduces to standard deductive logic in the special case of complete information, while obeying a few basic qualitative criteria of rationality and internal consistency.

It's the method behind the mysterious assertions you may have heard if you've ever spent any time eavesdropping on economists: A "preference" is a binary relation that satisfies such-and-such. An "economy" is an n-tuple behaving like . . . , et cetera. These statements aren't as crazy as they might seem from the outside. The economists are doing essentially the same thing Shannon was. It may involve functional equations, or it may take some other form, but in every case it involves the same translation from qualitative to quantitative that functional equations so elegantly embody.

The pre-mathematical use of functional equations to derive and motivate our definitions exists on a curious boundary between vague intuition and mathematical precision. It's the DMZ where mathematics meets psychology. And although the term "functional equation" isn't nearly as attention-grabbing as the underlying concepts deserve, they offer valuable and useful insights into where mathematical ideas come from.

DECENTERING

GARY A. KLEIN
Psychologist; Senior Scientist, MacroCognition LLC; author, *Seeing What Others Don't: The Remarkable Ways We Gain Insights*

You may have worked so closely with a partner that you reached a point where each of you could finish the other's sentences. You have a pretty good idea of how your partner would respond to a particular event. What's behind such a skill?

Decentering is the activity of taking the perspective of another entity. When we look into the past, we try to explain why that entity behaved in a way that surprised us. Peering into the future, we decenter in order to anticipate what that entity is likely to do.

Skilled decentering comes into play not just with partners but also with strangers and even with adversaries. It gives us an edge in combat, readying us to ward off an adversary's attack. It helps authors write more clearly, as they anticipate what might confuse a reader or what a reader expects to find out next. Police officers who are good at decentering can de-escalate a tense situation by steering an encounter in less volatile directions. Teachers rely on decentering to keep the attention of students by posing questions or creating suspense, anticipating what will catch their students' attention. Team members who can quickly decenter can predict how one another will react to unexpected changes in conditions, increasing the team's coordination.

Decentering is not about empathy—intuiting how others might be feeling. Rather, it's about intuiting what others are thinking. It's about imagining what's going through another person's mind. It's about getting inside someone else's head.

Despite its importance, particularly for social interaction, decentering has received very little attention. Military researchers have struggled to find ways to study decentering and train commanders to adopt the perspective of an adversary. Social psychologists haven't made much progress in unpacking decentering. Part of the problem might be that researchers examine the average decentering accuracy of observers, whereas they should be investigating those observers whose accuracy is consistently above average. What are their secrets? What's behind their expertise?

It could be even more valuable to study decentering outside the laboratory, in natural settings. I suspect that some people don't even try to decenter, others may try but aren't particularly effective, and still others may be endowed with an uncanny ability to take another person's perspective.

Decentering also comes into play when we interact with inanimate objects, such as intelligent technologies that are intended to help us make decisions. That's why the definition at the beginning of this essay refers to entities rather than people. The human-factors psychologist Earl Wiener once described the three typical questions people ask when interacting with information technology: What is it doing? Why is it doing that? What will it do next? We're trying to take the perspective of the decision aid.

It would be nice if the decision aid could help us decenter the way a teammate can. After all, when we interact with a person, it's natural to ask them to explain their choice. But intelligent systems struggle to explain their reasons. Sometimes they just recite all the factors that went into the choice, which isn't particularly helpful. We want to hear less, not more. We want the minimum necessary information. And the minimum necessary information depends on decentering.

For example, if you're using a GPS system to drive somewhere, and the device tells you to turn left whereas you expected to turn right, you would *not* want the system to explain itself by showing the logic it followed. Rather, the system should understand why you expected a righthand turn at that juncture and should provide the central data element (e.g., an accident up ahead is creating thirty-minute delays on that route) you need to understand the choice.

If you were driving with a human navigator who unexpectedly advised you to turn left, all you'd have to do is say, "Left?"

Most navigators instantly determine from your intonation that you want an explanation, not a louder repetition. Good navigators grasp that you, the driver, didn't expect a lefthand turn and would quickly (remember, there's traffic getting ready to honk at you) explain, "Heavy traffic ahead." Your navigator would convey the minimum necessary information, the gist of the explanation. But the gist depends on your own beliefs and expectations—it depends on the navigator's ability to decenter and get inside your head. Smart technology will never be really smart until it can decenter and anticipate. That's when it will become a truly smart partner.

When we're uncomfortable with the recommendations offered by a decision aid, we don't have any easy ways to enter into a dialogue. In contrast, when we dislike a person's suggestions we can examine their reasons. Being able to take someone else's perspective lets people disagree without escalating into conflicts. It allows people to be more decent in their interactions with each other. If they can decenter, then they can become decenter.

TRANSFER LEARNING

PETER LEE
Computer scientist; Corporate Vice President, Microsoft Research

"You can never understand one language until you understand at least two." This statement by the English author Geoffrey Willans feels intuitive to anyone who has studied a second language. The idea is that learning to speak a foreign language inescapably conveys deeper understanding of one's native tongue. Goethe, in fact, found this such a powerful concept that he felt moved to make a similar but more extreme assertion: "He who knows no foreign languages knows nothing of his own."

As compelling as this idea may be, what's surprising is that its essence—that learning or improvement in one skill or mental function can positively influence another—applies not only to human intelligence but also to machine intelligence. The effect is called *transfer learning*, and besides being an area of fundamental research in machine learning, it has potentially wide-ranging practical applications.

The field of machine learning, which is the scientific study of algorithms whose capabilities improve with use, has been making startling advances. Some of these have led to computing systems that are competent in skills associated with human intelligence, sometimes at levels not just approaching human capabilities but exceeding them—for example, in the ability to understand, process, and even translate languages. In recent years, much of the research in machine learning has focused on the algorithmic concept of deep neural networks (DNNs), which learn by inferring patterns, often of remarkable complexity, from large amounts of data. A DNN-based machine can be fed many thousands of snippets of recorded English utterances, each one paired with its text transcription, and from this discern the patterns of correlation between the speech recordings and the paired transcriptions. These inferred correlation patterns are precise enough that eventually the system "understands" English speech. In fact, today's DNNs are so good that when given enough training examples and a powerful enough computer, they can listen to a person speaking and make fewer transcription errors than a human would.

What may surprise some is that computerized learning machines exhibit transfer learning. Consider an experiment involving two machine-learning systems: machines A and B. Machine

A uses a brand-new DNN, whereas machine B uses a DNN that has been trained to understand English. Suppose we train both A and B on identical sets of recorded Mandarin utterances, along with their transcriptions. What happens? Machine B, the English-trained one, ends up with better Mandarin capabilities than machine A. In effect, the system's prior training in English transferred capabilities to the related task of understanding Mandarin. But there's an even more astonishing outcome: not only does Machine B end up better in Mandarin but also B's ability to understand English is improved! It seems that Willans and Goethe were on to something—learning a second language enables deeper learning about both languages, even for a machine.

The idea of transfer learning is still the subject of basic research, and many fundamental questions remain open. For example, not all "transfers" are useful, because, at a minimum, for transfer to work well the learned tasks apparently need to be related in ways eluding precise definition or scientific analysis. Connections to related concepts in other fields, such as cognitive science and learning theory, remain to be elucidated. It's intellectually dangerous for a computer scientist to anthropomorphize computer systems, but we have to acknowledge that transfer learning creates a powerful, alluring analogy between learning in humans and machines. Surely if general artificial intelligence is ever to become real, transfer learning will probably be a fundamental factor in its creation. For the more philosophically minded, formal models of transfer learning may contribute new insights and taxonomies for knowledge and knowledge transfer.

There's also high potential for applications of transfer learning. So much of the practical value of machine learning—for example, in search and information retrieval—has focused on systems that learn from the huge data sets available on the World

Wide Web. But what can Web-trained systems learn about smaller communities, organizations, or even individuals? Can we expect a future in which intelligent machines can learn useful tasks highly specialized to a specific individual or small organization? Transfer learning offers the possibility that all the intelligence of the Web can form the foundation of machine-learned systems, from which more individualized intelligence is learned through transfer learning. Achieving this would be another step toward the democratization of machine intelligence.

THE SYMBOL-GROUNDING PROBLEM

JOSHUA BONGARD

Veinott Professor of Computer Science, University of Vermont; author, *How the Body Shapes the Way We Think*

I. J. Good, a collaborator of Alan Turing's during World War II, once wrote that a machine more intelligent than humans would be the "last invention" we make. Many believe that our society is about to reach this point, as smart machines increasingly invade our daily lives and imaginations. This invasion is proceeding along two fronts. Such machines are increasingly to blame for economic change, leading some people to question the necessity of work altogether. But, more deeply, we wonder whether such ultra-intelligent machines will be "like us": They may think, but will they feel?

To help anchor such questions, the cognitive scientist Stevan Harnad formulated the symbol-grounding problem. For a machine, it's problematic to learn what the symbols making up human languages mean by simply transforming heard words into spoken words. It's like trying to learn Egyptian hieroglyphs using a dictionary in which each hieroglyph is defined using other hieroglyphs. Consider this exchange I just had with the most high-powered chatbot on the Web:

Me: Hi there.
It: Hello, how are you?
Me: Not too bad. You?
It: Fine, thank you.
Me: I've been thinking lately.
It: Thinking about what?
Me: The symbol-grounding problem.
It: Can you tell me a happy story?

The solution to the symbol-grounding problem is to ground symbols in the rich soil of physical experience. For example, a robot may learn that the symbol "jump" correlates with sudden loss of pressure on the soles of its feet. This robot may begin to empathize with humans by asking one of them to jump, observing her feet leaving the ground, and inferring that she feels what it feels when it jumps. Finally, such a machine may recursively ground increasingly abstract symbols in these already-grounded symbols, until eventually it understands the English idiom "Don't jump to conclusions" the first time it hears it.

The public should care about this seemingly esoteric corner of cognitive science, because machines that don't ground symbols in this way will be dangerous. Consider my exchange with the chatbot. Not only did it not know that its final comment was "wrong" but it failed to predict that I would be frustrated or amused by the comment. Similarly, another machine may fail to predict my terrified response to its actions.

Current machines can now, after receiving a million photographs containing a human and another million that don't, tell you whether or not a new photograph contains a human without having to ground symbols in experience. But consider another data set, composed of 2 million conversations: In the first million, the speakers are discussing how best to help Bob; in the second million, they're conspiring to harm him. Current state-of-the-art machines cannot tell you whether the speakers in a new conversation intend to help or harm Bob.

Most humans can listen to a conversation and predict whether the person being discussed is in danger. It may be that we can do so because we've heard enough such discussions in real life, books, and movies to be able to generalize to the current conversation, not unlike computers that recognize humans in previously unseen photographs. But we can also empathize by connecting words, images, and physical experience: We can put ourselves in the shoes of the people talking about Bob, or in the shoes of Bob himself. If one speaker says, "One good turn deserves another" and follows it with a sarcastic sneer, we can take those verbal symbols ("one," "good," . . .), combine them with the visual cue, and do some mental simulation.

First, we can go back in time to inhabit Bob's body mentally and imagine him/us acting in a way that lessens the speaker's hunger by providing her with food or assuages another of her physical or emotional pains. We can then return to the present as ourselves and imagine saying what she said. We won't follow up the statement with a sneer, as she did. Our prediction has failed.

So, our brain will return to the past, inhabit Bob's body again, but this time mentally simulate hurting the speaker in some way. During the act, we transfer into the speaker's body and suffer her pain. Back in the present, we would imagine ourselves saying the same words. Also, feelings of anticipated revenge would be bubbling up inside us, bringing a sneer to our lips, thus matching the speaker's sneer. So: We predict that the speakers wish to harm Bob.

Growing evidence from neuroscience indicates that heard words light up most parts of the brain, not just some localized language module. Could this indicate a person twisting words, actions, their own former felt experiences and mental body-snatching (placing themselves in another's shoes) into sensory/action/experiential braided cables? Might these cables support a bridge from the actions and feelings of others to our own actions and feelings, and back again?

Robots with the ability to connect with people by simulating what they sense and feel may be useful and even empathetic. But would they be conscious? Consciousness is currently beyond the reach of science, but one can wonder. If I "feel" your pain, the subject and the object are clear: I'm the subject and you're the object. But if I feel the pain of my own stubbed toe, the subject and object aren't as obvious. Or are they? If two humans can connect by empathizing with each other, cannot two parts of my brain empathize with each other when I hurt myself? Perhaps feelings are verbs instead of nouns: They may be specific exchanges between cell clusters. May consciousness then not simply be a fractal arrangement of ever smaller sensory/motor/experiential braids grounding the ones above them? If myths tell us that the Earth is flat and rests on the back of a giant turtle, we might ask what holds up the turtle. The answer, of course, is that it's turtles all the way down. Perhaps consciousness is simply empathy between cell clusters, all the way down.

ABSTRACTION

URSULA MARTIN

Professor of Computer Science, University of Oxford

Open up Ada Lovelace's 1843 paper about Charles Babbage's unbuilt Analytical Engine, and, if you're geek enough and can cope with long 19th-century sentences, it's astonishingly readable today.

The Analytical Engine was entirely mechanical. Setting a heavy metal disc with ten teeth stored a digit, a stack of fifty such discs stored a 50-digit number, and the store, or memory, would have contained 100 such stacks. A basic instruction to add two numbers moved them from the store to the mill (i.e., the CPU), where they'd be added together and moved back to a new place in the store to await further use—all mechanically. The engine was to be programmed with punch cards representing variables and operations, with further elaborate mechanisms to move the cards around and reuse groups of them when loops were needed. Babbage estimated that his gigantic machine would take three minutes to multiply two 20-digit numbers.

The paper is so readable because Lovelace describes the machine not in terms of elaborate ironmongery but using abstractions—store, mill, variables, operations, etc. These abstractions and the relations between them capture the essence of the machine, identifying the major components and the data passing among them. They capture, in the language of the day, one of the core problems in computing then and now—that of exactly what can and cannot be computed with different machines. The

397

paper identifies the elements needed "to reproduce all the operations which intellect performs in order to attain a determinate result, if these operations are themselves capable of being precisely defined" and these—arithmetic, conditional branching, etc.—are exactly the elements Alan Turing needed 100 years later to prove his results about the power of computation.

You can't point to a variable or an addition instruction in Babbage's machine—only to the mechanical activities representing them. What Lovelace could tackle only with informal explanation was made more precise in the 1960s, when computer scientists such as Oxford's Dana Scott and Christopher Strachey used separate abstractions to model both the machine and the program running on it, so that precise mathematical reasoning could predict its behavior. These concepts have been further refined, as computer scientists such as Samson Abramsky, also of Oxford, seek out subtler abstractions using advanced logic and mathematics to capture not only classical computers but quantum computation as well.

Identifying a good abstraction for a practical problem is an art as well as a science, portraying building blocks of a problem and the elements connecting them with just the right amount of detail—not too little and not too much—and abstracting away from the intricacies of the internals of the blocks so the designer need focus only on the elements needed to interact with other components. Jeannette Wing of Carnegie Mellon characterizes these kinds of skills as computational thinking, a concept relevant to many situations, not just programming.

Lovelace herself identified the wider power of abstraction and wrote of her ambition to understand the nervous system by developing "a law, or laws, for the mutual actions of the molecules of the brain." And computer scientists today are indeed extending their techniques to develop suitable abstractions for this purpose.

NETWORKS

SHEIZAF RAFAELI

Director, Center for Internet Research, University of Haifa, Israel

Biological, human, and organizational realities are networked. Complex environments are networks. Computers are networked. Epidemics are networks. Business relations are networked. Thought and reasoning are in neural networks. Emotions are networked. Families are networks. Politics are networked. Culture and social relations are, too. Your network is your net worth.

Yet the general public does not yet "speak networks."

Network concepts are still new to many and not widely enough spread. I have in mind structural traits like peer-to-peer and packet switching; process qualities like assortativity, directionality, and reciprocity; indicators like in- and out-degree, density, centrality, betweenness, multiplexity, and reciprocity; and ideas like bridging versus bonding, Simmelian cliques, network effects, and strength of ties. These amount to a language and analytical approach whose time has come.

Much of scientific thought in the last century, especially in the social and life sciences, was organized around notions of central tendency and variance. These statistical lenses magnified and clarified much of the world beyond earlier pre-positivist and less evidence-based approaches. But these same terms miss and mask the network. It's time to open minds to an understanding of the somewhat more complex truths of networked existence. We need to see more networks in public coverage of science, in media reporting, in writing and rhetoric, even in the teaching of expression and composition. "Network speaking" beckons more post-linear language.

Some of the best minds of the early 21st century are working on developing a language not yet known, integrated, or spoken outside their own small circle, or network. It's this language of metaphors and analytical lenses, which focuses on networks, that I propose be shared more widely, now that we're beginning to see its universal value in describing, predicting, and even prescribing reality.

In an era of fascination with Big Data, it's too easy to be dazzled by the entities ("vertices"), and their counts and measures, at the expense of the links ("edges"). Network ideas bring the connections back to the fore. Whether these are hidden or in plain view, they're the essence. Statistics, and especially variance-based measures such as standard deviation and correlation analyses, are reductionist. Network lenses allow and even encourage a much needed pulling back to see the broader picture. In all fields, we need more topology and metrics that recognize the mesh of connections beyond the traits of the components.

As with literacy, recent generations saw an enormous leap in numeracy. More people know numbers, are comfortable with calculations, and can see the relevance of arithmetic and even higher math to their daily lives. Easy access to calculators followed by widespread access to computation devices have accelerated the public's familiarity and comfort with numbers as a way of capturing, predicting, and dealing with reality.

We don't yet have the network equivalent of the pocket calculator. Let's make that a next improvement in the public awareness of science. Whereas a few decades ago the public had to be taught about mean, median, and mode, standard deviations and variance, percentages and significance, it's now time for network concepts to come to the fore. For us to understand the spread of truth and lies, political stances and viruses, wealth and social compassion, we need to internalize the mechanisms and measures of the networks along which such dynamics take place. The opposite of networking is not working.

MORPHOGENETIC FIELDS

ROBERT PROVINE

Research Professor/Professor Emeritus, University of Maryland, Baltimore County; author, *Curious Behavior: Yawning, Laughing, Hiccupping, and Beyond*

A morphogenetic field is a region of an embryo that forms a discrete structure, such as a limb or heart. Morphogenetic fields became known through the experimental work of Ross G. Harrison, one of the most deserving scientists never to have received a Nobel Prize. The regions are described as fields instead of discrete cells because they can recover from the effects of partial destruction. For example, if half of a salamander's forelimb field is destroyed, it will still develop into a reasonable approximation of a complete limb, not a half-limb. If the limb field is transplanted to a new region, such as the mid-flank of a host embryo, it will develop into an extra limb. These remarkable discoveries were widely reported in the scientific and popular media during the Golden Age of experimental embryology in the first half of the 20th century, but have been partly eclipsed by the emergence of more modern, reductionistic approaches to developmental problems.

The morphogenetic field offers important lessons about the nature of development and genetic determination. A morphogenetic field has the property of self-organization, forming the best possible whole from available cells. The field is a cellular ecosystem that won't work if the fates of component cells are predetermined entities lacking the requisite plasticity. The cellular community of a field is coordinated by a chemical gradient and therefore isn't scalable, which is why all embryos are small and about the same size, whether that of a mouse or a great blue whale. Self-organization of a morphogenetic field enables error correction, a tremendous advantage to a complex, developing system, where a lot can and does go wrong. For example, if a cell in the field is missing, another will be programmed to take its place, or if an errant cell wanders into the field, its developmental program will be overridden by its neighbors, in both cases forming the best possible whole.

The probabilistic, epigenetic processes of morphogenetic fields force a reconsideration of what it means to be "genetically determined" and illustrate why genes are better understood as recipes than as blueprints: Genes provide instructions for assembly, not a detailed plan for the final product. For psychologists and other social scientists whose developmental studies are based more on philosophical than biological foundations, morphogenetic fields provide a good starting point for learning how development works. Embryos provide excellent instruction about development for those knowing where to look and how to see.

HERD IMMUNITY

BUDDHINI SAMARASINGHE
Molecular biologist; science writer, Medical Research Council, U.K.

The eradication of smallpox was one of the most significant achievements of modern medicine. It was possible because of an effective vaccine, coupled with global vaccination programs. Theoretically, we can eradicate other diseases, such as measles or polio, the same way; if enough of the global population is vaccinated, those diseases will cease to exist. We've come tantalizingly close to eradication in some cases. In 2000, the Centers for Disease Control and Prevention declared that measles had been eliminated from the United States. Sixteen years later, the Pan American Health Organization announced that measles had been eradicated from the Americas. Polio is now endemic in only three countries in the world. Infectious diseases that routinely killed young children are now preventable, thanks to childhood vaccination programs. Yet, despite these milestones, there have been several outbreaks of vaccine-preventable diseases in recent times. How can this be?

A significant reason is that many people, despite evidence to the contrary, view vaccine efficacy and safety as a matter of opinion instead of scientific fact. This has serious consequences not just for individuals who choose to avoid vaccines but also for public-health initiatives as a whole.

To be effective, vaccination strategies for contagious diseases rely on a scientific concept known as *herd immunity*. Herd immunity can be considered a protective shield that prevents unvaccinated people from coming into contact with the disease, thus stopping its spread. Herd immunity is particularly important for people who cannot be vaccinated, including infants, pregnant women, or immuno-compromised individuals. The required level of immunization to attain benefit from herd immunity varies for each disease and is calculated based on the infectious agent's reproductive number—how many people, on average, each infected person goes on to infect. For measles, which can cause about eighteen secondary cases for each infected person, the required level of immunization to attain herd immunity is about 95 percent. In other words, at least 95 percent of the entire population must be immune to prevent the spread of measles following an infection. Low vaccination levels are failing to provide protection through herd immunity, stripping one of the greatest tools in public health of its power.

Vaccines work by imitating an infection, thereby helping the body's own defenses to be prepared in case of actual infection. Unfortunately, no vaccine is 100-percent effective, and the immunity given by vaccines can wane over time; these facts are often cited by anti-vaccine activists in an effort to discredit the entire concept of vaccination. But even waning immunity is better than no immunity. For example, the smallpox vaccine was generally thought to be effective for seven to ten years, but a recent analysis shows that even individuals who were vaccinated up to thirty-five years ago would still have substantial resistance to a smallpox infection.

The concept of herd immunity also applies to the annual flu vaccine. Unlike vaccines for measles or polio, the flu vaccine needs to be given every year, because the influenza virus evolves rapidly. And because it isn't as infectious as measles, only half the population needs to be immune to prevent spread of the disease. Herd immunity protects us from the common circulating variations of the flu, while the annual vaccine protects us from new versions that have escaped the existing immune response. Without a vaccine and herd immunity, a far greater number of people would be infected each year with the flu.

Vaccines are one of the greatest successes of public health. They've helped us conquer diseases such as smallpox and polio, granting us longer, healthier, more productive lives. Yet because of decreasing levels of vaccination, the threshold required to provide protection through herd immunity becomes unattainable; as a result, previously eradicated diseases are starting to reappear. Vaccination may be seen as an act of individual responsibility, but it has a tremendous collective impact. On a large scale it not only prevents disease in an individual but also helps protect the vulnerable in a population. To convince the general public of its necessity and encourage more people to get vaccinated, the concept of herd immunity must be more widely understood.

SOMATIC EVOLUTION

ITAI YANAI

Director, Institute for Computational Medicine; Professor, Department of Biochemistry and Molecular Pharmacology, NYU School of Medicine; co-author (with Martin Lercher), *The Society of Genes*

Cancer seems inscrutable. It has been variously described as a disease of the genome, a result of viral infection, a product of misbehaving cells, a change in metabolism, and cell signaling gone wrong. Like the six blind men touching different parts of the elephant, these all indeed describe different aspects of cancer. But the elephant in the room is that cancer is evolution.

Cancer is a form of evolution within our body, the "soma." Cancer is somatic evolution. Take that spot on your arm as proof that some of your cells are different from others—some darker, some lighter. This difference is also heritable when one of your body's cells divides into two daughter cells, encoded as a mutation in the cell's DNA, perhaps caused by sun exposure. Much of somatic evolution is inconsequential. But some of the heritable variation within a human body may be of a kind that makes a more substantial change than color: It produces cells that divide faster, setting in motion a chain of events following from the inescapable logic of Darwin's natural selection. The cells carrying such a mutation will become more popular in the body over time. This must happen, since the criteria of natural selection have been met: heritable change providing an advantage over neighboring cells—in this case, in the form of faster growth.

But no single mutation can produce a cancerous cell—i.e., one able to mount a threat to the body's well-being. As in the evolution of a species, change occurs upon change, allowing a population to adapt sequentially. As the clones of faster-dividing cells amass, there is power in numbers, and it becomes probable for another random mutation to occur among them, which further increases the proliferation. These mutations and their selection allow the cancer to adapt to its environment—to ignore the signaling of its neighbors to stop dividing, to change its metabolism to a quick and dirty form, to secure access to oxygen. Sometimes the process starts with an infection by a virus—consider this just another form of heritable variation, as the viral genome becomes a part of the DNA in the cell it attacks. Evolution is rarely fast, and this is also true of somatic evolution. The development of a cancer typically takes many years, while the mutated cells acquire more and more mutational changes, each increasing their ability to outcompete the body's other cells. When the cancer finally evolves the ability to invade other tissues, it becomes nearly unstoppable.

Evolution is sometimes confused with progress. From the perspective of a cancer patient, somatic evolution certainly isn't. Rather, as cancer develops, changes in the composition of the body's gene pool occur—the very definition of evolution. The notion that cancer is evolution is not an analogy but a matter-of-fact characterization of the process. It is humbling indeed that evolution is not only an ancient process that explains our existence on this planet but also constantly happening within our bodies, our soma.

CRITICALITY

CÉSAR HIDALGO

Associate Professor of Media Arts and Sciences, MIT; head, Macro Connections group, MIT Media Lab; author, *Why Information Grows*

In physics we say a system is in a critical state when it's ripe for a phase transition—like water turning into ice or a cloud pregnant with rain. Both are examples of physical systems in a critical state.

The dynamics of criticality, however, aren't intuitive. Consider the abruptness of freezing water. For an outside observer, there's no difference between cold water and water that's just about to freeze. This is because water just about to freeze is still liquid. Yet, microscopically, cold water and water just about to freeze are not the same.

When close to freezing, water is composed of gazillions of tiny ice crystals—crystals so small that the water remains liquid. But this is water in a critical state, a state in which any additional cooling will result in the crystals touching one another, generating the solid mesh we know as ice. Yet the ice crystals that form during the transition are infinitesimal; they're just the last straw. So freezing cannot be considered the result of these last crystals. They represent only the instability needed to trigger the transition; the real cause of the transition is the criticality of the state.

But why should anyone outside statistical physics care about criticality?

The reason is that history is full of individual narratives that should perhaps be interpreted in terms of critical phenomena.

Did Rosa Parks start the civil rights movement? Or was the movement already running in the minds of those who had been promised equality and were instead handed discrimination? Was the collapse of Lehman Brothers an essential trigger for the Great Recession? Or was the financial system so critical that any disturbance would have done the trick?

As humans, we love individual narratives. We evolved to learn from stories and communicate almost exclusively in terms of them. But as Richard Feynman said repeatedly, the imagination of nature is greater than that of man. Maybe our obsession with individual narratives is nothing but a reflection of our limited imagination. Going forward, we need to remember that systems often make individuals irrelevant. Just as none of your cells can claim to control your body, society also works in systemic ways.

So the next time the house of cards collapses, remember to focus on why we were building a house of cards in the first place, instead of on whether the last card was the queen of diamonds or the two of clubs.

INFORMATION PATHOLOGY

THOMAS A. BASS

Professor of English and Journalism, State University of New York, Albany; author, *The Spy Who Loved Us*

Our modern world of digitized bits moving with ever-increasing density and speed through a skein of channels resembling an electronic nervous system is built on information. The theory of information was born full-blown from the head of Claude Shannon in a seminal paper published in 1948. Shannon provided the means—but not the meaning—for this remarkable feat of engineering. Now, as we're coming to realize with growing urgency, we have to put the meaning back into the message.

Information theory has given us big electronic pipes, data compression, and wonderful applications for distinguishing signal from noise. Internet traffic is ballooning into the realm of zettabytes—250 billion DVDs' worth of data—but the theory underlying those advances provides no way to get from information to knowledge. Awash in propaganda, conspiracy theories, and other signs of information sickness, we're giving way to the urge to exit from modernity itself. What is it about information that's making us sick? Its saturation and virulence? Its vertiginous speed? Its inability to distinguish fact from fiction? Its embrace of novelty, celebrity, distraction?

Information theory, as defined by Shannon in his paper on "A Mathematical Theory of Communication" (republished in book form the following year as "*The* Mathematical Theory of Communication"), deals with getting signals transmitted from information sources to receivers. "The fundamental problem of communication is that of reproducing at one point either exactly or approximately a message selected at another point," wrote Shannon in the paper's second paragraph. "Frequently the messages have *meaning*, that is they refer to or are correlated according to some system with certain physical or conceptual entities. These semantic aspects of communication are irrelevant to the engineering problem."

Shannon's theory has proved remarkably fruitful for signal processing and other aspects of a modern world bathed in *bits* (Shannon was the first to use this word in print), but he had nothing to say about messages that "frequently . . . have *meaning*."

As Marshall McLuhan said of Shannon, "Without an understanding of causality there can be no theory of communication. What passes as information theory today is not communication at all, but merely transportation."

Take, for example, the comedian George Carlin's Hippy Dippy Weatherman, who announces, "Tonight's forecast, dark. Continued dark tonight. Turning to partly light in the morning." Since this message conveys information already known to us in advance, the amount of information it carries, according to Shannon, is zero. But according to McLuhan—and anyone who has watched George Carlin pace the stage as he delivered Al Sleet's weather report—the message contains a raft of information. The audience guffaws at mediated pomposity, the unreliability of prediction, and the dark future, which, if we survive it, might possibly turn partly light by morning.

Information theory hasn't budged since Shannon conceived it in 1948, but the pathologies surrounding information have begun to metastasize. We're overwhelmed by increasing flows of information, while our capacity for understanding this information remains as primitive as ever. Instead of interpreting it, teasing knowledge from data, we shrug our shoulders and say, "I dunno. It's a wash. You have a lot of information on your side. I have a lot on my side." Whether it's verified information or disinformation or lies, who cares? There's a *lot* of it. So let's raise our shoulders in a big cosmic shrug.

Addressing the "common angst" of our age, Jared Bielby, co-chair of the International Center for Information Ethics, describes "the fallout of information pathologies following information saturation, dissolution, and overload." The theory for understanding the causes of information sickness is being put together by people like Luciano Floridi, a professor in the philosophy and ethics of information at Oxford. According to Floridi, we're transforming ourselves into "informational organisms (inforgs), who share with other kinds of agents a global environment, ultimately made of information, the infosphere. . . ." In this global environment of information—which is related, of course, to the other environment made of *stuff*—the task is to find the meaning in the message. Tonight's forecast is dark, to be sure, but we're hoping for signs of light by morning.

IATROTROPIC STIMULUS

GERALD SMALLBERG

Practicing neurologist, New York City; playwright, Off-Off Broadway Productions: *Charter Members, The Gold Ring*

This concept comes from epidemiology, the field of medicine comprising methods used to find the causes of health outcomes and diseases in populations. The meaning and relevance of "iatrotropic stimulus" require some historical background.

When I was a medical student almost fifty years ago, I learned this concept from Dr. Alvan Feinstein, a professor of both medicine and epidemiology at Yale. He taught a course in clinical diagnosis that would prepare us for seeing patients in the hospital. A strict and exacting teacher, he demanded that we learn to take a detailed and carefully crafted patient history, combined with a meticulous physical examination; both were crucial, he believed, to the art and science of medicine. Dr. Feinstein, as a cardiologist, had helped delineate the criteria defining rheumatic heart disease. The critical role that rheumatic fever—caused by streptoccocal infection—plays in its pathogenesis could then be firmly established, so that early treatment of this infection became the standard of care to prevent the debilitating heart condition.

Feinstein's clinical research inspired his lifelong study of the natural history of disease, with the hope that this would lead to better diagnosis and therapy. While I was his student, his research was focused on the epidemiology of lung cancer. He was trying to analyze the optimal use of cancer-screening studies, an issue that still bedevils medicine.

Feinstein believed in ensuring that the medical record contained the best data possible, so that when the data were reviewed retrospectively, unknown variables unappreciated when they were first obtained could be used to better classify patients into appropriate categories for clinical studies. He stressed to us that medical students had the best chance of recording this vital information in their own case histories, as we were the most inclusive and least biased in data collection. Every patient note composed at each rung of the medical ladder, from the intern to the attending physician, was progressively streamlined and abbreviated to reflect the preceding impressions and conclusions. He read our reports fastidiously and underlined in red what he liked or didn't like, in the rigorous manner befitting his role as a teacher, journal editor, and clinical researcher. Along with the chief complaint—that is, the patient's answer to the question "What's bothering you?" or "Why are you here?"—he also wanted to know why the patient had decided to see a doctor at that particular time.

It was the latter response that led Feinstein to coin the term "iatrotropic stimulus," a phrase combining the Greek *iatros* (physician) with *trope* ("to turn"). In other words, what led the patient to seek help *that* day as opposed to another time when they may have been experiencing the same symptoms? Perhaps the chronic cough that had been ignored was now associated with a fleck of blood or had become of greater concern because a friend or an acquaintance had just been diagnosed with cancer. In Feinstein's view, this question would unleash information that not only could provide further epidemiological insights but also would be invaluable in better understanding the fears, concerns, and motivation that drove the patient to seek medical care. Although well grounded in science and having studied mathematics before becoming a physician, he taught us that clinical judgment depended not on a knowledge of causes, mechanisms, or names for diseases, but on "a knowledge of patients."

The iatrotropic stimulus did not find its rightful place in the medical literature. After his course, I never used the term in any of my subsequent reports. Yet its clinical importance forever left its mark on me. It formed the backstory, or *mise-en-scène*, of my interaction, whenever possible, with my patients. As we get more and more data indelibly inscribed in an electronic record derived from encoded questionnaires, algorithm-generated inputs and outputs that yield problem lists and diagnoses, in our striving for more evidence-based decision-making, we can get lost in the fog of information. In the end, it's the relationship of the patient and the treating physician that's most important. Together they must deal with complexity and uncertainty, the perfect Petri dish for incubating the fear and anxiety that, despite all technological progress, will remain the lot of humankind. As we confront our medical (as well as many other) problems, the iatrotropic stimulus—the "Why now?"—is an important concept for us to keep in mind.

MISMATCH CONDITIONS

DANIEL LIEBERMAN

Evolutionary biologist; Edwin M. Lerner II Professor of Biological Sciences, Harvard University; author, *The Story of the Human Body*

Assuming that you fear getting sick and dying, you really ought to think more about *mismatch conditions*. They're a fundamental evolutionary process.

Mismatch conditions are problems, including illnesses, caused by organisms being imperfectly or inadequately adapted to new environmental conditions. As extreme examples, a chimpanzee adapted to the rainforests of Africa would be hopelessly mismatched in Siberia or the Sahara, and a hyena would be mismatched to a diet of grass or shrubs. Such radical mismatches almost always cause death and sometimes extinction.

Mismatches, however, are typically more subtle and most commonly occur when climate change, dispersal, or migration alters a species' environment, including its diet, predators, and more. Natural selection occurs when heritable variations to these sorts of mismatches affect offspring survival and reproduction. For instance, when tropically adapted humans who evolved in Africa dispersed to such temperate habitats as Europe about 40,000 years ago, selection acted rapidly in these populations to favor shifts in body shape, skin pigmentation, and immune systems, which lessened any resulting mismatches.

Although mismatches have been going on since life first began, the rate and intensity of mismatches that humans now face has been magnified, thanks to cultural evolution, arguably now a more rapid and powerful force than natural selection. Just think how radically our bodies' environments have been transformed because of the agricultural, industrial, and post-industrial revolutions, in terms of diet, physical activity, medicine, sanitation, even shoes. While most of these shifts have been beneficial in terms of survival and reproduction, everything comes with costs, including several waves of mismatch diseases.

The first great wave of mismatches was triggered by the origins of farming. As people transitioned from hunting and gathering to farming, they settled down in large, permanent communities with high population densities, not to mention lots of sewage, farm animals, and various other sources of filth and contagion. Farmers also became dependent on a few cereal crops that yielded more calories but less nutrition than what hunter-gatherers could obtain. The resulting mismatches included all sorts of nasty infectious diseases, more malnutrition, and a greater chance of famine.

A second great wave of mismatch, which is ongoing, arose from the industrial and post-industrial revolutions. The standard description of this shift, generally known as the epidemiological transition, is that advances in medicine, sanitation, transportation, and government vastly reduced the incidence of communicable diseases and starvation, thus increasing longevity and resulting in a concomitant increase in chronic, noninfectious diseases. According to this logic, as people became less likely to die young from pneumonia, tuberculosis, or the plague, they became more likely to die in old age from heart disease and cancer—now the cause of two out of three deaths in the developed world. The epidemiological transition is also thought to be responsible for other diseases of aging, such as osteoporosis, osteoarthritis, and type 2 diabetes.

The problem with this explanation is that aging is not a

cause of mismatch, and we too often confuse diseases that occur more commonly with age with diseases that are actually *caused* by aging. To be sure, some diseases, like cancers, are caused by mutations accruing over time, but the most common age of death among hunter-gatherers who survive childhood is between sixty-eight and seventy-eight, and studies of aging among hunter-gatherers and subsistence farmers routinely find little or no evidence of so-called diseases of aging, such as hypertension, coronary heart disease, osteoporosis, diabetes, and more. Instead, these diseases are mostly caused by recent environmental changes: physical inactivity, highly processed diets, smoking. In other words, they're primarily novel mismatch diseases caused by industrial and post-industrial conditions.

In short, there are three reasons you should pay attention to the concept of mismatch. First, mismatches are a powerful evolutionary force that always has and always will drive much selection. Second, you're most likely to get sick and then die from a mismatch condition. And, most important, mismatches are by nature partly or largely preventable, if you can alter the environments that promote them.

ACTIONABLE PREDICTIONS

ROGER HIGHFIELD

Director, External Affairs, Science Museum Group, U.K.; co-author (with Martin Nowak), *Supercooperators: Altruism, Evolution, and Why We Need Each Other to Succeed*

You might be forgiven for thinking that this is so blindingly obvious that it's hardly worth stating, let alone arguing, that it should become a popular meme. After all, "pre-" means "before," so surely you should be able to take action in the wake of a prediction to change your future—like buying an umbrella when a deluge is forecast, for example.

Weather forecasting is indeed a good example of an actionable prediction, a beautiful marriage of real-time data from satellites and other sensors with modeling. But when you shift your gaze away from the physical sciences toward medicine, those predictions are harder to discern.

We're a long way from doctors being able to make routine and reliable actionable predictions about individual patients—which treatments will help them the most, which drugs will cause them the fewest side effects, and so on. The answers to many simple questions remain frustratingly elusive. Should I take an antibiotic for that sore throat? Will immunotherapy work for me? Which of that vast list of possible drug side effects should I take seriously? What's the best diet for me? If only we could predict the answers for a particular patient as reliably as meteorologists predict tomorrow's weather.

Today there's much talk of Big Data providing all the answers. In biology, for example, data from the Human Genome Project once kindled widespread hope that if we sequenced a patient's DNA we'd get a vivid glimpse of their destiny. Despite the proliferation of genomes, epigenomes, proteomes, and transcriptomes, that crystal ball looks cloudier than at first thought, and the original dream of personalized medicine in genomics has been downgraded to precision medicine, where we assume that a given person will respond in a similar way to a previously studied group of genetically similar people.

Blind gathering of Big Data in biology continues apace, however, emphasizing transformational technologies such as machine learning—artificial neural networks, for instance—as a way to find meaningful patterns in all the data. But no matter their "depth" and sophistication, neural nets merely fit curves to the available data. They may be capable of interpolation, but extrapolation beyond their training domain can be fraught.

The quantity of data isn't the whole story, either. We're gathering a lot, but are we gathering the right data and of sufficient quality? Can we discern a significant signal in a thicket of false correlations? Given that bodies are dynamic and ever-changing, can data snapshots really capture the full complexities of life?

To make true actionable predictions in medicine, we also need a step change in mathematical modeling in biology, which is relatively primitive compared to physics, and for understandable reasons: Cells are hugely complicated, let alone organs and bodies.

We need to promote interest in complex systems, so that we can truly predict the future of individual patients rather than infer what might be in store for them from earlier population studies of who responded to a new treatment, who did not, and who suffered serious side effects. We need deeper insights, not least to end making diagnoses postmortem and prevent tens of thousands of people perishing at the hands of doctors every year through iatrogenic effects.

Ultimately we need better modeling based on mechanistic understanding in medicine, so that one day your doctor can carry out timely experiments on a digital *Doppelgänger* before she experiments on you. Modern medicine needs more actionable predictions.

THE TEXAS SHARPSHOOTER

CHARLES SEIFE

Professor of Journalism, New York University; former writer, *Science*; author, *Virtual Unreality*

A city-slicker statistician was driving through the backwoods of rural Texas, so the story goes, when she slammed on the brakes. There, right by the side of the road, was a barn that bore witness to a nigh-impossible feat of marksmanship. The barn was covered with hundreds of neat little white bullseyes, each of which was perforated with a single bullet hole in the dead center.

The incredulous statistician got out of her car and examined the barn, muttering about Gaussian distributions and probability density functions. She didn't notice that Old Joe had sidled up to her, ancient Winchester slung over his shoulder.

Joe cleared his throat. With a start, the amazed statistician looked at him. "Four hundred and twelve targets, every single one of which is hit in the center with less than 2-percent deviation every time. The odds against that are astronomical! You must be the most accurate rifleman in history. How do you do it?"

Without a word, Joe walked ten paces away from the barn, spun round, raised his rifle, and fired. A slug thunked dully into the wood siding. Joe casually pulled a piece of chalk out of his overalls as he walked back toward the barn, and, after finding the hole his bullet had just made, drew a neat little bullseye around it.

There are far too many scientists who have adopted the Texas Sharpshooter's methods, and we're beginning to feel the effects. For example, there's a drug on the market—just approved—to treat Duchenne muscular dystrophy, an incurable disease. Too bad it doesn't seem to work.

The drug, eteplirsen, received a lot of fanfare when researchers announced that it had hit two clinical bullseyes: It increased the amount of a certain protein in patients' muscle fibers, and patients did better on a certain measure known as the six-minute-walk test (6MWT). The drug was effective! Or so it seemed, if you didn't know that the 6MWT bullseye was painted on the wall well after the study was underway. Another bullseye (the number of certain white blood cells in muscle tissue), drawn before the study started, was hastily erased. That's almost certainly the sign of a missed target.

Looking at all the data and all the scientists' prognostications makes it clear that the drug didn't behave as researchers had hoped. Eteplirsen's effectiveness is highly questionable, to put it mildly. Yet it was trumpeted as a big breakthrough and approved by the FDA. Patients can now buy a year's supply for about $300,000.

Texas-style sharpshooting—moving the goalposts and cherry-picking data so that results seem significant and important when they're not—is extremely common; check out any clinical-trials registry and you'll see just how frequently endpoints are tinkered with. It goes almost without saying that a good number of these changes effectively turn the sow's ear of a negative or ambiguous result into the silk purse of a scientific finding worthy of publication. No wonder, then, that many branches of science are mired in replicability crises; there's no replicating a finding that is the result of a bullseye changing positions instead of reflecting nature's laws.

The Texas Sharpshooter problem should be more widely known—and not just by scientists—so that we can move toward a world with more transparency about changing protocols and unfixed endpoints. Maybe, just maybe, that will make us a little less impressed by the never-ending procession of supposed scientific marksmen—researchers whose results are little more permanent than chalk marks on the side of a barn.

DIGITAL REPRESENTATION

JON KLEINBERG
Tisch University Professor of Computer Science, Cornell University

Three people stand in front of a portrait in a museum, each making a copy of it: an art student producing a replica in paint, a professional photographer taking a picture of it with an old film camera, and a tourist snapping a photo with a phone. Which one is not like the others?

The art student is devoting much more time to the task, but there's a sense in which the tourist with the phone is the odd one out. Paint on canvas, like an exposed piece of film, is a purely physical representation—a chemical bloom on a receptive medium. There's no representation distinct from this physical embodiment. In contrast, the cell phone camera's representation of the picture is fundamentally numerical. To a first approximation, the phone's camera divides its field of view into a grid of tiny cells, and stores a set of numbers to record the intensity of the colors in each of the cells it sees. These numbers are the representation; they're what gets transmitted (in a compressed form) when the picture is sent to friends or posted online.

The phone has produced a digital representation—a recording of an object using a finite set of symbols, endowed with meaning by a process for encoding and decoding the symbols. The technological world has embraced digital representations for almost every imaginable purpose—to record images, sounds, the measurements of sensors, the internal states of mechanical devices—and it has done so because digital representations offer two enormous advantages over physical ones. First, digital representations are transferable: After the initial loss of fidelity in converting a physical scene to a list of numbers, this numerical version can be stored and transmitted with no further loss, forever. A physical image on canvas or film, in contrast, degrades at least a little essentially every time it's reproduced or even handled, creating an inexorable erosion of information. Second, digital representations are manipulable: You can brighten, sharpen, or add visual effects to an image represented by numbers simply by using arithmetic on the numbers.

Digital representations have been catalyzed by computers, but they're fundamentally about the symbols, not the technology that records them, and they were with us long before any of our current electronic devices. Musical notation, for example— the decision made centuries ago to encode compositions using a discrete set of notes—is a brilliant choice of digital representation, encoded manually with pen and paper. And it conferred the benefits we still expect today from going digital. Musical notation is transferable: A piece by Mozart can be conveyed from one generation to the next with limited subjective disagreement over which pitches were intended. And musical notation is manipulable: We can transpose a piece of music, or analyze its harmonies using the principles of music theory, by working symbolically on the notes, without ever picking up an instrument to perform it. To be sure, the full experience of a piece of music isn't rendered digitally on the page; we don't know exactly what a Mozart sonata sounded like when originally performed by its composer. But the core is preserved in a way that would have been essentially impossible without the representation by an alphabet of discrete symbols.

Other activities, like sports, can also be divided on a digital-or-not axis. Baseball is particularly easy to follow on the radio, because the action has a digital representation—a coded set of symbols conveying the situation on the field. If you follow baseball and you hear that the score is tied 3–3 in the bottom of the ninth inning, with one out, a 3-and-2 count, and a runner on second, you can feel the tension in the representation itself. It's a representation that's transferable—it can communicate a finely resolved picture of what happened in a game to people far away in space or time. And it's manipulable—we can evaluate the advisability of various coaching decisions from the pure description alone. By comparison, sports like hockey and soccer lack a similarly expressive digital representation; you can happily listen to them on the radio, but you can't reconstruct the action on the field with anything approaching the same fidelity.

And digital representation goes beyond any human construction; complex digital representations predate us by at least a billion years. With the discovery that a cell's protein content is encoded using three-letter words written in an alphabet of four genetic bases, the field of biology stumbled upon an ancient digital representation of remarkable sophistication and power. And we can check the design criteria: It's transferable, since you need only have an accurate symbol-by-symbol copying mechanism in order to pass your protein content to your offspring; and it's manipulable, since evolution can operate directly on the symbols in the genome rather than on the molecules they encode.

We've reached a point now in the world where the thoughtful design of digital representations is becoming increasingly critical. They're the substrates on which large software systems and Internet platforms operate, and the outcomes we get will depend on the care we take in the construction. The algorithms powering these systems don't just encode pictures, videos, and text; they encode each of us as well. When one of these algorithms recommends a product, delivers a message, or makes a judgment, it's interacting not with you but with a digital representation of you. And so it becomes a central challenge for all of us to think deeply about what such a representation reflects, and what it leaves out. Because it's what the algorithm sees, or thinks it sees: a transferable, manipulable copy of you, roaming across an ever-widening landscape of digital representations.

EMBODIED THINKING

BARBARA TVERSKY
Professor Emerita of Psychology, Stanford University

Many of these *Edge* Question essayists have chosen the cosmic—appropriately, in these heady times of gravity waves and Einstein anniversaries. The secrets of the universe. But how did Einstein arrive at his cosmic revelations? Through his body, imagining being hurled into space at cosmic speed. Not through the equations that proved his theories or the words that explain them.

Imagining bodies moving in space. This is the very foundation of science, from the cosmic—bright stars and black holes and cold planets—to the tiny and tinier reverberating particles inside particles inside particles. It's the foundation of the arts: figures swirling or erect on a canvas, dancers leaping or motionless on a stage, musical notes ascending and descending, staccato or adagio. It's the foundation of sports and wars and games.

And the foundation of us. We are bodies moving in space. You approach a circle of friends, the circle widens to embrace you. I smile or wince and you feel my joy or my pain, perhaps smiling or wincing with me. Our most noble aspirations and emotions, and our most base, crave embodiment, actions of bodies in space, close or distant. Love, from which spring poetry and sacrifice, yearns to be close and to intertwine, lovers, mothers suckling infants, roughhousing, handshakes, and hugs.

That foundation—bodies moving in space, in the mind, or on the Earth—seeks symbolic expression in the world: rings and trophies, maps and sketches and words on pages, architectural models and musical scores, chess boards and game plans, objects that can be touched and treasured, scrutinized and transformed, stirring new thinking and new thoughts.

When I was seven, we moved from the city to the country. There were stars then, half a hollow indigo sphere of sparkling stars encompassing me, everyone—the entire universe right before my eyes. Exhilarating. The speck that was me could be firmly located in that cosmos. In the return address on letters to my grandfather, I wrote my name, my house number, my street, my town, my state, my country, my continent, Planet Earth, the Milky Way, the Universe. A visible palpable route linking the body, my own body, to the cosmic.

THE TROLLEY PROBLEM

DANIEL ROCKMORE

Professor of Mathematics and Computer Science; Director, Neukom Institute for Computational Science, Dartmouth College

The history of science is littered with *Gedankenexperiments* ("thought experiments"), a term dreamed up by Albert Einstein for an imagined scenario that sharply articulates the crux of some intellectual puzzle. Among the most famous are Einstein's tale of chasing a light beam, which led him to the special theory of relativity, and Erwin Schrödinger's illustration of the measurement problem in quantum mechanics by putting a cat in a box so fiendishly designed that the animal is both alive and dead until someone opens the box.

"The trolley problem" is another thought experiment, one that arose in moral philosophy. There are many versions, but here's one: A trolley is rolling down the tracks and reaches a branchpoint. To the left, one person is trapped on the tracks, and to the right, five people. The trolley can't brake in time. You can throw a switch that diverts the trolley from the track with the five trapped people to the track with the one. Do you? What if we know more about the people on the tracks? Maybe the one is a child and the five are elderly? Maybe the one is a parent and the others are single? How do all these different scenarios change things? What matters? What are you valuing and why?

It's an interesting thought experiment, but these days it's more than that. As we offload more and more of our decisions to machines and the software that manages them, developers and engineers will increasingly be confronted with having to encode important and potentially life-and-death decision-making into machines. Decision-making always comes with a value system, a "utility function," whereby we do one thing or another because one choice reflects a greater value for the outcome than the other. Sometimes the value might seem obvious or trivial: This blender is recommended to you over that one, given your past buying history. This pair of shoes is a more likely purchase than another (or perhaps not *the* most likely, because they're kind of expensive, but worth a shot; this gets us to probabilistic calculations and expected returns). This song versus that song, etc.

But sometimes more is at stake: this news or that news? This piece of information or that piece of information on a given subject? The values embedded in the program may start shaping your own values and, with that, society's. Those are some pretty high stakes. The trolley problem shows us that the value systems that pervade programming can be a matter of life and death. Soon we'll have driverless trolleys, driverless cars, driverless trucks. They'll come with a moral compass. Shit happens, and choices need to be made: the teenager on the bike in the breakdown lane or the Fortune 500 CEO and his assistant in the stopped car ahead? What does your vehicle's algorithm do and why?

The same conundrum will apply to our robot companions. They will have values and will necessarily be moral machines—ethical automata whose morals and ethics are engineered by us. The trolley problem is a *Gedankenexperiment* for our age, shining a bright light on the complexities of engineering our new world of humans and machines.

MENTAL EMULATION

STEPHEN M. KOSSLYN

Cognitive neuroscientist, psychologist; Founding Dean, Minerva Schools at the Keck Graduate Institute; co-author (with G. Wayne Miller), *Top Brain, Bottom Brain*

When he was sixteen years old, Albert Einstein imagined he was chasing after a beam of light and observed what he "saw." And this vision launched him on the path to developing his special theory of relativity. Einstein often engaged in such thinking; he stated, "The psychical entities which seem to serve as elements in thought are certain signs and more or less clear images which can be 'voluntarily' reproduced and combined. . . . [T]his combinatory play seems to be the essential feature in productive thought before there is any connection with logical construction in words or other kinds of signs which can be communicated to others."

Einstein was relying on *mental emulation*, a kind of thought that many of us use and could all probably use more frequently and more productively if we became aware of it. Mental emulations range from the sublime to the ordinary, such as imagining impossible events (including chasing a light-beam) or the best route for climbing a hill, or visualizing the best way to pack your refrigerator before starting to take the cans, boxes, and jugs out of your shopping bag.

A mental emulation is a way to simulate what you'd expect to happen in a specific situation. Just as Einstein reported, most mental emulations appear to involve mental imagery. By definition, mental imagery is an internal representation that mimics perception. You can visualize "running" an event in your head and "seeing" how it unfolds. Mental emulations are partly a way of letting you access knowledge you have only implicitly, to enable such knowledge to affect your mental images.

My colleague Sam Moulton and I characterize mental emulation as follows: Each step of what you imagine represents the corresponding step in the real world, and the transitions between steps represent the corresponding transitions between steps in the event. Moreover, the processes that transition from state to state in the representation mimic the actual processes that transition from state to state in the world. If you imagine kicking a ball, you'll activate motor circuits in the brain, and neural states will specify the transitional states of how your leg is positioned as you kick. And as you imagine tracking the flying ball, your mental images will transition through the intermediate states in ways analogous to what would happen in the real world.

Mental emulation is a fundamental form of thought, and should be recognized as such.

PREDICTION ERROR MINIMIZATION

ANDY CLARK

Philosopher, cognitive scientist; Chair in Logic and Metaphysics, University of Edinburgh; author, *Surfing Uncertainty: Prediction, Action, and the Embodied Mind*

This sounds dry and technical. But it may well be the key thing brains do that enables us to experience a world populated by things and events that matter to us. If so, it's a major part of the solution to the mind-body problem. It's also a concept that can change how we feel about our own daily experience. Brains like ours, if recent waves of scientific work using this concept are on track, are fundamentally trying to minimize errors concerning their predictions of the incoming sensory stream.

Consider something as commonplace as it is potentially extremely puzzling—the ability of humans and many other animals to find specific absences salient. A repeated series of notes followed by an omitted note results in a distinctive experience, one that presents a world in which that very note is strikingly absent. How can a specific absence make such a strong impression on the mind?

The best explanation is that the incoming sensory stream is processed relative to a set of predictions about what should be happening at our sensory peripheries right now. These mostly unconscious expectations prepare us to deal rapidly and efficiently with the stream of signals coming from the world. If the sensory signal is as expected, we can launch responses we've already started to prepare. If it's not as expected, then a distinctive signal results—a so-called prediction-error signal. These signals, calculated in every area and at every level of neuronal processing, highlight what we got wrong and invite the brain to try again. Brains like this are forever trying to guess the shape and evolution of the current sensory signal, using what we know about the world.

Human experience here involves a delicate combination of what the brain expects and what the current waves of sensory evidence suggest. Thus, in the case of the unexpectedly omitted sound, there's evidence that the brain briefly starts to respond as if the missing sound were present, before the absence of the expected sensory "evidence" generates a large prediction error. We thus faintly hallucinate the onset of the expected sound before becoming aware of its absence. And what we thus become aware of isn't just any old absence but the absence of that specific sound. This explains why our experiential world often seems full of real absences.

When things go wrong, attention to prediction and prediction error can be illuminating, too. Schizophrenic individuals have been shown to rely less heavily on their sensory predictions than neurotypical folk. Schizophrenic subjects outperformed neurotypical ones in tests that involved tracking a dot that unexpectedly changed direction but were worse at dealing with predictable effects. Autistic individuals are also impaired in the use of "top-down" predictions, so that the sensory flow seems continually surprising and hard to manage. Placebo and nocebo effects are also grist for the mill. For predictions can look inward, too, allowing how we expect to feel to make a strong and perfectly real contribution to how we actually do feel.

By seeing experience as a construct that merges prediction and sensory evidence, we begin to see how minds like ours reveal a world of human-relevant stuff. For the patterns of sensory stimulation we most strongly predict are the patterns that matter most to us both as humans (with distinctive needs and capacities) and as individuals (with different histories and interests). The world revealed to the predictive brain is a world permeated with human mattering.

IMPOSSIBLE

CHIARA MARLETTO

Quantum physicist; Junior Research Fellow, Wolfson College;
Postdoctoral Research Associate, Materials Department, University
of Oxford

The concept of "impossible" underlies all fundamental theories of physics, yet its exact meaning is little known. The impossibility of cloning, or copying, certain sets of states is at the heart of quantum theory and of quantum information. The impossibility of exceeding the speed of light is a fundamental feature of the geometry of spacetime in relativity. The impossibility of constructing perpetual-motion machines is the core idea of thermodynamics.

But what do we mean, exactly, by "impossible"?

The concept in physics is profound and has beautiful implications; it sharpens the everyday meaning of the word, giving to that airy nothing a firm, deep connotation rooted in the laws of physics. That something is impossible means that the laws of physics set a fundamental draconian limit to how accurately it can be brought about.

For instance, one can construct real machines approximating a perpetual-motion machine to *some* degree, but there's a fundamental limit to how well that can be done. Since energy is conserved overall, the energy supplied by any such machine must come from somewhere else in the universe, and since there are no infinite sources of energy, perpetual motion is impossible. All finite sources eventually run out.

The impossibility of something is different from its not happening at all under the particular laws of motion and initial conditions of our universe. For example, it may be that under the actual laws and initial conditions an ice statue of the pirate Barbarossa will never arise in the whole history of our universe, and yet that statue need not be impossible. Unlike the case for perpetual-motion machines, it might still be possible to create arbitrarily accurate approximate copies of such a statue under different initial conditions. The impossibilities I mentioned above are instead categorical: A perpetual-motion machine cannot be realized to arbitrarily high accuracy under any laws of motion and initial conditions.

The exact physical meaning of the word "impossible" is also illuminating because it provides a deeper understanding of what is possible.

That something is possible means that the laws of physics set no limit to how well it can be approximated. For example, thermodynamics says that a heat engine is possible. We can come up with better and better ways of approaching the ideal behavior of the ideal engine, with higher and higher efficiencies, with no limitation on how well that can be done. Each realization will have a different design; they'll use different, ever-improving technologies; and there's no limit to how much any given heat engine can be improved upon.

So, once we know what's impossible under the laws of physics, we're left with plenty of room for our ideas to try and create approximations to things that are possible. This opening up of possibilities is the wonderful unexpected implication of contemplating the fundamental physical meaning of "impossible." May it be as widely known as is physically possible.

OPTIMIZATION

SABINE HOSSENFELDER
Theoretical physicist; Research Fellow, Frankfurt Institute for Advanced Studies, Germany

Lawns in public places all suffer from the same problem: People don't like detours. Throughout the world we search for the fastest route, the closest parking spot, the shortest way to the restroom—we optimize. Incremental modification, followed by evaluation and readjustment, guides us to solutions that maximize a desired criterion. These little series of trial and error are so ingrained we rarely think about them. But optimization, once expressed in scientific terms, is one of the most versatile scientific concepts we know.

Optimization under variation isn't only a human strategy but a ubiquitous selection principle you can observe everywhere in nature. A soap bubble minimizes surface area. Lightning uses the path of least resistance. Light travels the path that takes the least amount of time. And with only two slight twists, optimization can be applied even more widely.

The first twist is that many natural systems don't actually perform the modifications—they work by what physicists call "virtual variations." This means that from all the possible ways a system could behave, the one we observe is quantifiably optimal: It minimizes a function called the "action." Using the mathematical procedure known as "principle of least action," we can then obtain equations that let us calculate how the system will behave.

Accommodating quantum mechanics requires a second twist. A quantum system doesn't do only one thing at a time; it does everything that's possible, all at the same time. But properly weighted and summed up in the "path integral," this collection of all possible behaviors again describes observations. And usually the optimal behavior is still the most likely one, which is why we don't notice quantum effects in everyday life.

Optimization isn't a new concept. It's the scientific variant of Leibniz's hypothesis that we live in the "best of all possible worlds." But while the idea dates back to the 18th century, it's still the most universal law of nature we know. Modern cosmology and particle physics both work by specifying exactly how our world is "the best." (Though I have to warn you that "specifying exactly" requires a whole lot of mathematics.)

And if physics isn't your thing, optimization also underlies natural selection and free-market economies. Our social, political, and economic systems are examples of complex adaptive systems; they're collections of agents who make incremental modifications and react to feedback. By doing so, the agents optimize adaptation to certain criteria. Unlike in physics, we can't calculate what these systems will do, and that's not what we want—we just use them as tools to work to our ends. Exactly what each system optimizes is encoded in the feedback cycle. And there's the rub.

It's easy to take the optimization done by adaptive systems for granted. We're so used to this happening that it seems almost unavoidable. But how well such systems work depends crucially on the setup of the feedback cycle. Modifications should be neither too random nor too directed, and the feedback must— implicitly or explicitly—evaluate the optimization criteria.

When we use adaptive systems to facilitate our living together, we therefore have to make sure they optimize what they're supposed to. An economic system pervaded by monopolies, for example, doesn't optimize supply to customers' demands. And a political system that gives agents insufficient information about their current situation and doesn't let them extrapolate probable consequences of their actions doesn't optimize the realization of its agents' values.

Science, too, is an adaptive system. It should optimize knowledge discovery. But science, too, doesn't miraculously self-optimize what we hope it does; the implementation of the feedback cycle requires careful design and monitoring. It's a lesson even scientists haven't taken sufficiently to heart: If you get something from nothing, it's most likely not what you wanted.

When we use optimization to organize our societies, we have to decide what we mean by "optimal." There's no Invisible Hand to take this responsibility off us. And that ought to be more widely known.

THE CANCER SEED AND SOIL HYPOTHESIS

AZRA RAZA

Chan Soon-Shiong Professor of Medicine, Columbia University Medical Center

One in two men and one in three women in the U.S. will get cancer. Five decades after declaring war on the disease, we're still muddling our way rather blindly from the slash-poison-burn (surgery-chemo-radiation) strategies to such newer approaches as targeted therapies, nanotechnology, and immunotherapies, which benefit only a handful of patients. Among other reasons for the slow progress is the study of cancer cells in isolation, which de-links the seed from its soil.

The English surgeon Stephen Paget was the first to propose, in 1889, that "seeds" (cancer cells) preferentially grew in the hospitable "soil" (microenvironment) of select organs. The crosstalk between seed and soil hypothesized by Paget indeed proved to be the case whenever the question was examined (such as in the elegant studies of I. R. Hart and I. J. Fiddler in the 1980s). Yet consistent research combining studies of seed and soil weren't pursued, largely because in the excitement generated by the molecular revolution and the discovery of oncogenes, the idea of creating animal models of human cancers seemed far more appealing. This led to the field of cancer research being hijacked by investigators studying animal models, xenografts, and tissue-culture cell lines in patently artificial conditions. The result of this de-contextualized approach, akin to looking for car keys under the lamppost, is nothing short of a tragedy for cancer patients, whose pain and suffering some of us witness and try to alleviate on a daily basis.

Many of my fellow researchers are probably rushing to attack me for making misleading statements and ignoring the great advances they've accomplished in oncology using the very systems I'm criticizing. I'm still receiving hate mail for answering the 2014 *Edge* Question (about what idea is ready for retirement) by saying that mouse models as surrogates for developing cancer therapeutics need to go. I'm sorry to remind them that we've failed to improve the outcome for the vast majority of our cancer patients. The point is that if strategies we've been using aren't working, it's time to let them go. Or at least to stop pretending that these mutated, contrived systems have anything to do with malignant diseases in humans. Both the funding agencies and leaders of the oncology field need to admit that the paradigms of the last several decades are failing.

The concept I want to promote is that of Paget's seed-and-soil approach to cancer and a serious examination of cancer cells as they exist in their natural habitats. Basic researchers ask what they should replace their synthetic models with; my answer is that first and foremost they should work directly with clinical oncologists. If methods to recapitulate human cancers in vitro don't exist, then we must prepare to study them directly, in vivo. We have a number of effective drugs, but these usually help only subsets of patients. It would be a tremendous step forward if we could match the right drug to the right patient.

For example, in the study of leukemia, we could start by treating patients with a study drug while simultaneously studying freshly obtained pre- and post-therapy blood/bone-marrow samples with pan-omics (genomics, proteomics, metabolomics, transcriptomics) technologies. A proportion of patients will respond and a proportion will fail. Compare the pan-omics results of the two groups and then design studies to enrich for subjects predicted to respond. It's likely that a few more patients will respond in round two. Repeat all the studies in successive clinical trials until identifiable reasons for response and non-response are determined.

If this exercise is undertaken for each drug that has shown efficacy, within the foreseeable future we will not only have accurate ways of identifying which patients should be treated but also be able to protect the patients unlikely to respond from receiving ineffective and toxic therapies. Besides, the pan-omics results are likely to identify new targets for more precise drug development. In this strategy, each successive trial design would be informed by the previous one on the basis of data obtained from cancer cells as they exist in their natural soil.

Readers are probably wondering why such obvious studies based on patient samples aren't being done already. Sadly, there's little incentive for basic researchers to change, not only because of the precious nature of human tissue and the difficulties of working with harassed, overworked clinical oncologists (mice are easier to control) but also because of resources. I'm aghast at funding agencies like the NIH which continue to prefer funding grants that use an animal model or a cell line. After all, who makes the decisions at these agencies? As Gugu Mona, the South African writer and poet, has noted: "The right vision to a wrong person is like the right seed to wrong soil."

SIMPLISTIC DISEASE PROGRESSION

ERIC TOPOL

Professor of Genomics, Scripps Translational Science Institute; author,
The Patient Will See You Now

The two leading causes of death are heart attacks and cancer. For much too long, we've had the wrong concept about the natural history of these conditions. The two have something in common: People generally don't die of cholesterol buildup in their arteries (atherosclerosis) unless they have a heart attack or stroke; similarly, cancer rarely causes death unless it metastasizes.

For decades it was believed that cholesterol buildup inside an artery supplying the heart muscle followed a slow, progressive course: As the plaque grew bigger and bigger, it would eventually clog the artery and interrupt blood supply—a heart attack. That turned out not to be true; the minor cholesterol narrowings are by far the most common precursors of a heart attack. The attack results from a blood clot, as the body tries to seal a sudden crack in the artery wall; the crack itself is an outgrowth of inflammation, which doesn't cause any symptoms until a blood clot forms and heart muscle is starved for oxygen. That's why people who have heart attacks often have no warning symptoms or easily pass an exercise stress test but keel over days later.

There's another aspect of heart attacks that should be more widely known. The media often mislabel a sudden death caused by a heart-rhythm problem as a "heart attack." That's wrong. A heart attack is defined by loss of blood supply to the heart. If that leads to a chaotic heart rhythm, it can result in death. But most people with heart attacks have chest pain and other symptoms. Someone can have an electrical heart-rhythm event that has nothing to do with atherosclerosis but can cause death. That's not a heart attack.

Regarding cancer, the longstanding dogma was that it slowly progressed over years until it began to spread throughout the body. But it has recently been shown that metastasis can occur with early lesions, defying the linear model of cancer's natural history. We knew that mammography picked up early breast cancer, for example, but that hasn't had a meaningful effect on saving lives. So much for the simplistic notion of how cancer develops—a notion that may be one of the reasons it has been so difficult to treat.

With the global health burden so largely due to these two killer diseases, it's vital that we raise awareness of their courses and reboot our preventive strategies.

EFFECT MODIFICATION

BEATRICE GOLOMB
Professor of Medicine, UC San Diego

Enshrined in the way much medical research is done is the tacit assumption that an exposure has an effect on an outcome. To quote the Web site of the Boston University School of Public Health: "Effect modification occurs when the magnitude of the effect of the primary exposure on an outcome (i.e., the association) differs depending on the level of a third variable. In this situation, computing an overall estimate of association is misleading." The reach of effect modification is wildly underappreciated. Implications challenge the core of how medicine—and, indeed, science—is practiced.

The results of randomized, double-blind, placebo-controlled trials (RCTs) are the foundation on which modern medicine rests. RCTs are the gold standard of study designs. These define treatment approaches, which are encouraged by clinical practice guidelines, with teeth in their implementation imposed through "performance pay" to doctors. But there's a problem. Often, a single estimate of effect is generated for a given outcome and presumed to apply generally (and perhaps, if favorable, to impel treatment for those outside the study). But results of studies—including but not exclusively RCTs—may not apply to those outside the study. They may not apply to some enrolled in the study. They may not, in fact, apply to anyone in the study.

The chief recognized problem inherent in RCTs is generalizability (sometimes called external validity). Results need not apply to types of people outside the study (e.g., studies of men may not apply to women) because of potential for effect modification. Less appreciated is that effect modification also means that results of a trial need not apply to all those *within* the study. To ascribe what's true for a whole to its parts is to succumb to the fallacy of division.

It's seldom appreciated that effect modification can engender differences not only in the magnitude but in the sign of effect. When subsets experience opposite effects, a neutral finding for the overall study can apply to *none* of its participants. Such bidirectional effects—let's call them Janus effects—aren't rare. Penicillin can save lives but can cost life in the highly allergic. A surgery may save lives but take lives of poor surgical candidates. Fluoroquinolone antibiotics can raise and lower blood sugar.

Benzodiazepines anxiolytics can "paradoxically" increase anxiety. Bisphosphonates to prevent fractures cause "pathological fractures." Statins prevented new diabetes (WOSCOPS trial) and promoted it (JUPITER trial); reduced cancer deaths (JUPITER trial) but significantly increased new cancer (PROSPER trial, the sole trial at ages over seventy).

How is this possible? One factor is that agents that can yield anti-oxidant effects (like statins) are almost always pro-oxidant in some patients and settings, including at high doses, where co-antioxidants are depleted. Conversely, agents that can have pro-oxidant effects can have anti-oxidant effects in sufficiently low doses for some people via oxidative preconditioning, by which a bit of oxidative stress ramps up endogenous anti-oxidant defenses. For agents meant to alter an aspect of physiology, counterregulatory mechanisms—imposed by evolution—may partly offset the intended effects, and in some people overshoot. So drugs and salt restriction meant to lower blood pressure paradoxically raise blood pressure in some.

Also, many exposures activate multiple mechanisms, which may act in opposition on an outcome. Imbibing alcohol can prevent stroke, via anti-oxidant polyphenols and thinning the blood, and can promote stroke via mitochondrial dysfunction, arrhythmia, and hypertension (or by thinning the blood too much). Swilling coffee is linked to reduced heart attacks in genetically fast caffeine-metabolizers (probably through anti-oxidant effects) but to increased heart attacks in slow caffeine-metabolizers (probably via caffeine-induced adrenergic-type ones).

Implications are rife. When studies of the same intervention produce different (or even opposite) results, this needn't mean there were study flaws, as is often presumed. Contradictory results may all be true. For Janus effects, whether exposure effects on an outcome are favorable, adverse, or neutral may depend on the composition of the study group. Evidence supports such bidirectional effects for statins with outcomes including diabetes, cancer, and aggression. Selection of a study group that yields sizable benefit (or, for environmental exposures, no harm) may drive a product to be recommended, or exposure mandated, for vast swaths of the populace—with potential for harm to many.

Individual experiences that controvert RCT results shouldn't be scorned, even if a source of effect modification is not (yet) known. Those who observed their blood sugar rise on statins were dismissed—given neutral average statin effects on glucose in RCTs. Then they were disparaged more contemptuously after the WOSCOPS trial reported that statins reduced diabetes risk. Later, many other trials and meta-analyses showed that statins can increase diabetes incidence. That was true, of course, before those later trials were published. (Now that it's accepted that statins can increase glucose, recognition that statins can reduce glucose has faded.)

So, conventional thinking about studies' implications must be jettisoned. It would be convenient if the observed association in a good-quality study could be thought the final word. But effect modification may be more the rule than the exception—at least in complex domains like biology and medicine where "computing the overall effect" can be misleading. An effect cannot be presumed to reliably hew to what any study shows, in magnitude or even direction. The pesky play of effect modification must be borne in mind.

THE POWER LAW

LUCA DE BIASE

Journalist; Editor-in-Chief, *Nòva24, Il Sole 24 Ore*

The way predictions are made is changing. Data scientists are competing with traditional statisticians, and Big Data analysis is competing with the study of statistic samples. This change mirrors a wider paradigm shift in the conception of society and what rules its structural dynamics. In order to understand this change, you need to know the *power law.*

If facts happen randomly in a two-axis world, it's quite possible that they will distribute as in a Gaussian bell curve, with most happenings concentrated around the average. But if facts are interlinked and co-evolve so that a change in one quantity results in a proportional change in the other quantity, it's more probable that they will distribute as in a power-law graph—in a ski-jump shape, in which the average isn't important and polarization is unavoidable.

The distributions of a large set of phenomena observed in physics, biology, and astronomy follow a power law, and this kind of curve was much discussed when it was applied to the understanding of the Internet. Study of the number of links to specific Web pages soon demonstrated that some pages were attracting more links, and that with the growth of the network it was more and more likely that new pages would link to those pages. In such a network, some nodes became hubs and other pages were only destinations. The number of links to pages followed the power law, and one could predict that the dynamics of the network would bring about a polarization of resources, as in the Barabási–Albert model, an algorithm invented by Albert-László Barabási and Réka Albert. This kind of understanding has consequences.

As the Internet became more and more important for society, the network theory became part of the very notion of social dynamics. In a network society, the power law is becoming the fundamental pattern.

In social sciences, prediction has often been more a kind of shaping the future than a description of what will actually happen. That sort of shaping by predicting has often relied on the assumptions used in the predicting process: Predicting something that will happen in a society relies on an *idea* of society. When scholars shared the assumptions defined in the notion of "mass society" (mass production, mass consumption, mass media, in which almost everybody behaves the same, both at work and when consuming), in their view fundamental characters were the same and diversity was randomly distributed. Gauss ruled. In a mass society, most people were average, different people were rare and extreme, thus society was described by a bell curve—a Gaussian curve, the "normal" curve. Polls based on statistic samples could predict behaviors.

But in a network society the fundamental assumptions are quite different. In a network society, all characters are linked and co-evolve, because a change in a character will probably affect other characters. In such a society, the average doesn't predict much, and scholars need a different fundamental pattern.

The power law is such a fundamental pattern. In this kind of society, resources aren't random; they co-evolve and they polarize. In finance, as in knowledge, resources are attracted by abundance. The rich get unavoidably richer.

Understanding this pattern is the only way for a network

society to oppose inequality without looking for solutions that were good in a mass society. Bernardo Huberman, a network theorist, has observed that the winner takes all in a category—that is, in a meaningful context. For example, the best search engine wins in the search-engine category but not necessarily in the whole of the Web, thus not necessarily in the social-network category. In such a network, innovation is the most important dynamic to oppose inequality, and real competition is possible only if new categories can emerge. If finance is only one big market, then the winner takes all. If rules ensure that different banks can play only in different categories of financial services, then there's less concentration of resources and less global risk.

In a mass society, everything tends to go toward the average: The middle class wins, in a normal distribution of resources. In a network society, resources are attracted by the hub, and differences inevitably grow. The mass society is an idea of the past, but the network society is a challenge for the future.

The power law can help our understanding of, and maybe correct the dynamics of, networks, given a growing awareness of its fundamental patterns. Predictions are narratives, and good narratives need some empirical observations. Moore's law is useful to those who share the technocentered narrative of the exponential growth of computer abilities. The power law is useful to those who want to critically study the evolution of a network.

TYPE I AND TYPE II ERRORS

PHIL ROSENZWEIG

Professor of Business Administration, IMD, Lausanne, Switzerland;
author, *Left Brain, Right Stuff*

A few years ago, a reporter at a leading financial daily called with an intriguing request:"We're doing a story about decision-making and asking researchers whether they follow their own advice." I must have chuckled, because she continued: "I just talked with one man who said, 'For heaven's sake, you don't think I actually use this stuff, do you? My decisions are as bad as everyone else's!'"

I suspect my colleague was being ironic, or perhaps tweaking the reporter. I gave a few examples I've found useful—sunk-cost fallacy, regression toward the mean, and more—but then focused on one concept that everyone should understand: Type I and Type II errors, or, respectively, false positives and false negatives.

The basic idea is straightforward: We should want to go ahead with a course of action when it will succeed and refrain when it will fail; accept a hypothesis when it's true and reject it when it's false; convict a defendant who's guilty and free one who's innocent. Naturally, we spend time and effort to improve the likelihood of making right decisions. But since we cannot know for sure that we'll be correct, we also have to consider the possibility of error. Type I and Type II thinking forces us to identify the ways we can err and ask which error we prefer.

In scientific research, we want to accept new results only when they're demonstrably correct. We want to avoid accepting a false result, or minimize the chance of a Type I error, even if it means committing a Type II error and failing to accept results that turn out to be true. That's why we insist that claims be supported by evidence deemed statistically significant—often set (by convention) as the probability of an observation being due to random effects as less than one in twenty ($p < .05$) or (more stringent but also by convention) less than one in 100 ($p < .01$). (How we know the probability distribution in the first place leads us into the debate between frequentists and Bayesians, an exceedingly interesting question but beyond the scope of this note.)

Similarly, in criminal trials, we want to convict a defendant only when we're certain of guilt. Here again, a Type II error is far preferable to a Type I error, a view expressed in Blackstone's law, which says that it's better to let ten guilty men go free than to convict an innocent man, since the severity of a false positive is not only great but perhaps irreversible. The preference for Type II error is reflected in such cornerstones of Anglo-Saxon law as the presumption of innocence and the burden of proof resting with the prosecution.

In other settings, the greater danger is to commit a Type II error. Consider competition in business, where rival firms seek payoffs (revenues and profits) that accrue disproportionately to a few or perhaps are winner-take-all. Although bold action may not lead to success, inaction almost inevitably leads to failure, because it's highly likely that some rival will act and succeed. Hence the dictum made famous by Intel's Andy Grove: *Only the paranoid survive.* Grove did *not* say that taking risky action predictably leads to success; rather, he observed that in situations of intense competition those who survive will have taken actions involving high risk. They will have understood that when it comes to technological breakthroughs or launching new products, it's better to act and fail (Type I error) than fail to act (Type II error).

To summarize, a first point is that we should consider not only the outcome we desire but also the errors we wish to avoid. A corollary is that different kinds of decisions favor one or the other type of error. A next point is that for some kinds of decisions, our preference may shift over time. A young adult considering marriage may prefer a Type II error, finding it more prudent to wait rather than marry too soon. As the years pass, he or she may be more willing to marry even if the circumstances don't seem perfect (a Type I error) rather than never marry (a Type II error).

A final point: Discussion of Type I and Type II errors can reveal different preferences among parties involved in a decision. To illustrate: Expeditions to the summit of Mt. Everest are very risky in the best of times. Of course, climbers want to push on if they can reach the summit safely and will want to turn back if pushing on would result in death—that's the easy part. More important is discussing *in advance* preferences for error: Would they rather turn back when they could have made it (Type II) or keep going and die in the effort (Type I)? If the choice seems obvious to you, it may be because you haven't invested the time and money in attempting to fulfill a dream—nor have you ever been in a state of exhaustion and oxygen deprivation when the moment of decision arrives. As past Everest expeditions have shown, sometimes tragically, members of a team—leaders, guides, and climbers—should do more than focus on the outcomes they most desire. They should also discuss the consequences of error.

THE IDEAL FREE DISTRIBUTION

LAURA BETZIG
Anthropologist and historian

The gist of the "ideal free distribution" is that individuals, in the best of all possible worlds, should distribute themselves in the best of all possible ways. They should sort themselves out across space and time so as to avoid predators, find prey, get mates, and leave as many descendants as they can behind. Where information is imperfect, the best spots will be missed, and where mobility is blocked, distributions will be despotic. But where information is unlimited and mobility is unrestrained, distributions will be ideal and free.

The idea is intuitively obvious, and it has predictive power. It works for aphids. It works for sticklebacks. And it works for us.

Over most of the long stretch of the human past, our distributions were more or less ideal free distributions. We usually moved around with our prey, following the plants we collected and the animals we tracked. Some foragers, even now, are more footloose than others: On the Kalahari, hunters with access to n!oresi (waterholes) make more claims to being a big man, or n!a, and have bigger families. But across continents—from Africa to the Americas to Australia—the most reproductively successful forager fathers raise children numbering in the low double digits and so do the most reproductively successful forager mothers. Genetic evidence backs that up. In contemporary populations, including the Khoisan, elevated levels of diversity on the X chromosome and low diversity on the Y suggest consistent sex differences in reproductive variance. A larger fraction of the female population has reproduced; a larger fraction of the male population has not. But overall, those differences are small.

After plants and animals were domesticated across the Fertile Crescent, our distributions became more despotic. Agriculture spread up and down the Nile Valley, and within a millennium Menes, who founded the First Dynasty, had created an empire that endured for some 3,000 years. When the Eighteenth Dynasty pharaoh Amenhotep III married a daughter of the king of Mitanni, she brought along a harem of 317 women, with their hand-bracelets, foot-bracelets, earrings, and toggle-pins, says a letter dug up at Amarna. And from ostraca and scarabs in and beyond the Nineteenth Dynasty pharaoh Rameses II's tomb, archaeologists have uncovered the names of ninety-six of his sons. From one end of the map to the other—from Egypt to Mesopotamia to the Ganges to the Yellow River to the Valley of Mexico to the Andes—overlords rose up wherever the subjected were trapped, and they left behind hundreds of daughters and sons. Again, the genetic evidence matches. Geographically diverse samples of Y chromosome sequences suggest a couple of reproductive bottlenecks over the course of human evolution. One from around 40,000 to 60,000 years ago coincides with our moves out of Africa into Eurasia, when the effective breeding population among women became more than twice the effective breeding population among men. A second bottleneck, from around 4,000 to 8,000 years ago, overlaps with the Neolithic, when the effective breeding population among women became seventeen times the effective breeding population among men.

Then, in 1492, Columbus found a New World. And over the last half-millennium, the flow of bodies—and the flow of information—grew in an unprecedented way. Despotisms collapsed. And distributions from sea to sea became relatively ideal and free.

Our pursuit of the ideal free distribution is as old as we are. We've always pushed back against ignorance and forward against borders. From our migrations around Africa; to our migrations out of Africa and into Eurasia; to our migrations out of Europe, Africa, and Asia into the Americas; to our first intrepid Earth orbits aboard *Friendship* 7 and beyond—we've risked our lots for a better life. I hope we never stop.

CHRONOBIOLOGY

DOUGLAS RUSHKOFF
Media analyst; documentary writer; author, *Throwing Rocks at the Google Bus*

The time-is-money ethos of the Industrial Age and wage labor, combined with the generic quality of computerized time-keeping and digital calendars, has all but disconnected us from the temporal rhythms on which biological life has oriented itself for millennia. Like all organisms, the human body has evolved to depend on the cyclical ebbs and flows of light, weather, and even the gravitational pull of the moon, in order to function effectively.

But our culture and its technologies are increasingly leading us to behave as if we could defy these cycles or simply ignore them completely. We fly ten time zones in as many hours, drink coffee and take drugs to wake ourselves up, pop sedatives to sleep, and then take SSRIs to counter the resulting depression. We schedule our work and productivity oblivious to the lunar cycle's influence on our moods and alertness and those of our students, customers, and workforces as well.

Chronobiology is the science of the biological clocks, circadian rhythms, and internal cycles that regulate our organs, hormones, and neurotransmitters. And although most of us know that it's probably healthier to be active during the day and sleep at night, we still tend to act as if one moment were as good as any other, for anything. It's not.

For instance, new research suggests that our dominant neurotransmitters change with each of the four weeks of a lunar cycle. The first week of a new moon brings a surge of acetylcholine, the next brings serotonin, then comes dopamine, and finally norepinephrine. During a dopamine week, people would tend to be more social and relaxed. Norepinephrine would make people more analytic. A serotonin week might be good for work, and an acetylcholine week should be full of pep.

Ancient cultures learned of these sorts of cycles through experience—trial and error. They used specific cyclical schedules for everything from planting and harvesting to rituals and conflict. But early science and its emphasis on repeatability treated all time as the same and saw chronobiology as closer to astrology than to physiology. The notion that wood taken from trees dries faster if it's cut at a particular time in the lunar cycle when its sap is at "low tide" seemed more like witchcraft than botany.

But like trees, we humans are subject to the cycles of our biological clocks, most of which use external environmental cues to set themselves. Divorced from these natural cues, we experience the dis-ease of organ systems that have no way to sync up—and an increased dependence on artificial signals for when to do what. We become more at the mercy of artificial cues, from news alerts to the cool light of our computer screens, for a sense of temporality.

If we were to become more aware of chronobiology, we wouldn't necessarily have to obey all our evolutionary biases. Unlike our ancestors, we do have light to read by at night, heat and air-conditioning to insulate us from the cycle of the seasons, 24/7 businesses that cater to people on irregular schedules. But we'd have the opportunity to reacquaint ourselves with the natural rhythms of our world and the grounding, orienting sensibilities that come with operating in harmony with them.

A rediscovery and wider acknowledgment of chronobiology would also go a long way toward restoring the solidarity and interpersonal connection so many of us are lacking without it. As we all became more aware and respectful of our shared chronobiology, we'll be more likely to sync up, or even "phase lock," with one another. This would allow us to recover some of the peer-to-peer solidarity and social cohesiveness we've lost to a culture that treats the time like a set of flashing numbers instead of the rhythm of life.

DELIBERATE IGNORANCE

GERD GIGERENZER
Psychologist; Director, Center for Adaptive Behavior and Cognition, Max Planck Institute for Human Development, Berlin; author, *Risk Savvy*

Ignorance is generally pictured as an unwanted state of mind, and the notion of deliberate ignorance may raise eyebrows. Yet people often choose to be ignorant, demonstrating a form of negative curiosity at odds with concepts such as ambiguity aversion, a need for certainty, and the Bayesian principle of total evidence. Such behavior also contrasts with the standard belief that more knowledge and data are always preferable, expressed in various forms, from Aristotle ("All men by nature desire to know") to the view of humans as informavores to the mission of national surveillance programs.

Deliberate ignorance can be defined as the willful decision not to know the answer to a question of personal interest, even if the answer is free—that is, with no search costs. The concept differs from the study of *agnotology,* or the sociology of ignorance, which investigates the systematic production of ignorance by deflecting, covering up, and obscuring knowledge—like the tobacco industry's efforts to keep people unaware of the evidence that smoking causes cancer. Deliberate ignorance isn't inflicted by third parties but self-chosen.

Why would people not want to know? The few existing studies and even fewer explanations suggest at least four motives.

The first is to avoid potential bad news, particularly if no cure or other prevention is available. According to Greek mythology, Apollo granted Cassandra the power of foreseeing the future but added a curse that her prophecies would never be believed. Cassandra foresaw the fall of Troy, the death of her father, and her own murder; anticipating the approach of future horrors became a source of endless pain. Technological progress steadily shifts the line between what we can and cannot know, as Cassandra did. When having his full genome sequenced, James Watson, the co-discoverer of the structure of DNA, stipulated that his *APOE* ε4 allele, which indicates risk of Alzheimer's disease, be both kept from him and deleted from his published genome sequence. Researchers claim to have discovered biomarkers that predict when a person will likely die and from what cause. Others claim to be able to predict whether a marriage will end in divorce. But do you want to know the date of your death? Or whether you should consult a divorce lawyer? The few available studies indicate that 85 to 90 percent of the general public don't want to know the details surrounding their death or their marital stability. Yet unlike the curse condemning Cassandra to foresee the future, technological progress means that we'll increasingly often have to decide how much foresight we want.

The second motive is to maintain surprise and suspense. Depending on the country, some 30 to 40 percent of parents don't want to know the sex of their unborn child even after ultrasound or amniocentesis. For these parents, knowing the answer would destroy their pleasurable feeling of surprise, a feeling that seems to outweigh the benefit of knowing and planning ahead.

A third motive is to profit strategically from remaining ignorant, as proposed by economist Thomas Schelling in the 1950s. The game of chicken is an example: pedestrians staring at their smartphones and ignoring the possibility of a collision, thereby forcing others to do the work of paying attention. Similarly, it has been argued that since the crisis of 2008, bankers and policymakers strategically display blindness in order to ignore the risks in which they continue to engage and to stall effective reform.

Finally, deliberate ignorance is a tool for achieving fairness and impartiality. In keeping with Lady Justice, often depicted as wearing a blindfold, many U.S. courts don't admit evidence about a defendant's criminal record. The idea is that the jury should remain ignorant of the defendant's previous crimes in order to reach an impartial verdict. The political philosopher John Rawls' "veil of ignorance" is another form of ignorance in the service of fairness.

Despite these insights, however, the phenomenon of deliberate ignorance has been largely treated as an oddity. Science and science fiction celebrate the value of prediction and total knowledge through Big Data analytics, precision medicine, and largely unquestioned surveillance programs. As with Cassandra, foreknowledge may not suit every person's emotional makeup. How we decide between wanting and not wanting to know is a topic calling for more scientific attention and greater curiosity.

THE NEED FOR CLOSURE

DYLAN EVANS
Founder and CEO, Projection Point; author, *The Utopia Experiment*

The poet John Keats coined the term *negative capability* to refer to the ability to remain content with half-knowledge "without any irritable reaching after fact and reason." The opposite of negative capability is known by psychologists as the *need for closure* (NFC). NFC refers to an aversion to ambiguity and uncertainty and the desire for a firm answer to a question. When NFC becomes overwhelming, *any* answer, even a wrong one, is preferable to remaining in a state of confusion and doubt.

If we could represent the knowledge in a given brain as dry land and ignorance as water, then even Einstein's brain would contain just a few tiny islands scattered in a vast ocean of ignorance. Yet most of us find it hard to admit how little we really know. How often, in the course of our everyday conversations, do we make assertions for which we have no evidence, or cite statistics that are nothing but guesses? Behind all this apparently innocuous confabulation lies NFC.

There's nothing wrong with wanting to know the answer to a question or feeling disturbed by the extent of our ignorance. It isn't the reaching after fact and reason that Keats condemns but the *irritable* reaching after fact and reason. However great our desire for an answer may be, we must make sure that our desire for truth is even greater, with the result that we prefer to remain in a state of uncertainty rather than filling in the gaps in our knowledge with something we've made up.

Greater awareness of the dangers of NFC would lead to more people saying, "I don't know" much more often. In fact, everyday conversation would overflow with admissions of ignorance. This would represent a huge leap forward toward the goal of greater rationality in everyday life.

POLYTHETIC ENTITATION

TIMOTHY TAYLOR
Professor of the Prehistory of Humanity, University of Vienna; author, *The Artificial Ape: How Technology Changed the Course of Human Evolution*

When is a wine glass not a wine glass? This question fascinated the archaeological theorist David Clarke in the late 1960s, but his elegant solution, critical for a correct conceptualization of artifacts and their evolution over time, is shamefully ignored. Understanding polythetic entitation opens the door to a richer view of the built world, from doors to computers, cars to chairs, torches to toothbrushes. It's a basic analytical tool for understanding absolutely anything people make. It indicates limits to the idea of the meme, and that it may be reasonable to consider the intentional patterning of matter by *Homo sapiens* as a new, separate kind of ordering in the universe.

Celebrating the end of our archaeological excavation season, someone tops up my glass with wine . . . except it isn't a glass. That is, it's not made of glass; it's a clear plastic disposable object with a stem. Nevertheless, when I put it down on a table to get some food and then cannot find it, I say, "Who took my glass?" In this context, I'm effectively blind to its plastic-ness. But with the situation transposed to an expensive restaurant and a romantic evening with my wife, I should certainly expect my glass to be made of glass.

Clarke argued that the world of wine glasses was different from the world of biology, where a simple binary key could lead to the identification of a living creature. (Does it have a backbone? If so, it's a vertebrate. Is it warmblooded? If so, it's a mammal or bird. Does it produce milk? . . . and so on.) A wine glass is a polythetic entity, which means that none of its attributes is simultaneously sufficient and necessary for group membership. Most wine glasses are made of clear glass, with a stem and no handle, but there are flower vases with all these, so they aren't definitionally sufficient attributes. And a wine glass may have *none* of these attributes; they're not absolutely necessary. It's necessary that the wine glass be able to hold liquid and be of a shape and size suitable for drinking from, but this is also true of a teacup. If someone offered me a glass of wine and then filled me a fine ceramic goblet, I wouldn't complain.

We continually make fine-tuned decisions about which types of artifacts we need for particular events, mostly unconscious of how category definitions shift according to social and cultural contexts. A Styrofoam cup could hold wine, mulled wine, or hot soup, but a heat-proof, handled punch glass would not normally be used for soup, although it would function just as well. Cultural expectations allow a Styrofoam cup to be a wine glass at a student party but not on the lawn at Buckingham Palace; a wine glass in a Viennese café is often a stemless beaker, which would be unusual in London, where it would be read as a water or juice glass. And a British mulled-wine glass in a metal holder, transposed to Russia, would not formally or materially differ from a tea glass to use with a samovar.

Our cultural insider, or *emic*, view of objects is both sophisticated and nuanced, but typically maps poorly onto the objectively measurable, multidimensional, and clinal formal and material variance—the scientific analyst's *etic* of polythetic entitation. Binary keys are no use here.

Asking at the outset whether an object is made of glass takes us down a different avenue from first asking if it has a stem, or if it's designed to hold liquid. The first lumps the majority of wine glasses with window panes; the second groups most of them with vases and table lamps; and the third puts them all into a supercategory that includes breast implants and Lake Mead, the Hoover Dam reservoir. None of the distinctions provides a useful classificatory starting point. So grouping artifacts according to a kind of biological taxonomy will not do.

As a prehistoric archaeologist, David Clarke knew this, and he also knew that he was continually bundling classes of artifacts into groups and subgroups without knowing whether his classification would have been recognized emically—that is, in terms understandable to the people who created and used the artifacts. Although the answer is that probably they did have different functions, how might one work back from the purely formal etic variance—the measurable features or attributes of an artifact—to securely assign it to its proper category? Are Bronze Age beakers, with all their subtypes, really beakers, and were they all used for the same purposes? Were they "memes" of one another, like the sort of coded repeatable information that enables the endless reproduction of individual computer-generated images? Or were some beakers non-beakers, with a distinct socially acceptable deployment?

Clarke's view clashes with the commonsense feeling we have about wine glasses having an essential wine-glass-ness. Granted, there can be a Platonically ideal wine glass if we so wish, but it's specific to times and places as well as contextual expectation. Currently, the heartland territory of the wine glass is dominated by transparent stemmed drinking vessels made of real glass, but such memic simplicity blurs toward the multidimensional edges of the set as attribute configurations trend toward the sweet-points that mark the core—but never at-once-both-sufficient-and-necessary—attributes we expect in our classic ideas of cultural and technological objects. Clarke noted that the space-time systematics of polythetic attribute sets were extraordinarily complex, patterned at a level beyond our immediate grasp.

So the memic turns out simply to be the emic, a shorthand description only, and not part of a valid analytic once our cultural insider knowledge is removed. For the prehistoric archaeologist, the absence of such knowledge is axiomatic. Determining which attributes had original cultural salience, why and how, is endlessly challenging. Those who attempt it without polythetic entitation are flailing in the dark.

QUINES

SAMUEL ARBESMAN

Complexity scientist; Scientist-in-residence, Lux Capital; author, *Overcomplicated*

Within the infinite space of computer programs is a special subset of code: programs that, when executed, output themselves. These short self-replicating programs are often referred to as *quines*, after the philosopher Willard Van Orman Quine, based on the term from Douglas Hofstadter's *Gödel, Escher, Bach*.

When you first hear about quines, they may seem magical. And doubly so if you've ever done any coding, because without knowing the trick for how to create them, they might seem devilishly difficult to construct. They're often elegant little things, and there are now examples of quines in a huge number of computer languages. They can range from short and sweet to ones far more inscrutable to the untrained eye.

But why are they important? Quines are a distillation of numerous ideas from computer science, linguistics, and much more. On a simple level, quines can be thought of as a sort of fixed point, an idea from mathematics: the value of a mathematical function that yields itself unchanged. (Think along the lines of how the square root of 1 is still 1.)

But we can take this further. Quines demonstrate that language, when refracted through the prism of computation, can be both operator and operand—the text of a quine is run, and through a process of feedback upon itself, yields the original code. Text can be both words with meaning and "words with meaning." Think of the sentence "This sentence has five words." We're delighted by this because the words are both describing (acting as an operator) and being described (acting as an operand). But this playfulness is also useful. This relationship between text and function is a building block of Kurt Gödel's work on incompleteness in mathematics, which is in turn related to Alan Turing's approach to the halting problem. These foundational ideas demonstrate certain limitations in both mathematics and computer science: There are certain statements we cannot prove to be either true or false within a given system, and there's no algorithm that can determine whether any given computer program will ever stop running.

More broadly, quines also demonstrate that reproduction is not the distinct domain of the biological. Just as a cell exploits the laws of physics and chemistry in order to persist and duplicate, a quine coopts the rules of a language of programming to persist when executed. While it doesn't quite duplicate itself, there are similar principles at work. You can even see hints of this biological nature in a "radiation hardened" quine—a type of quine in which any character can be deleted and it still will replicate! A radiation-hardened quine looks like gibberish, which is no doubt what the DNA sequence of a gene looks like to many of us. Redundancy and robustness, hallmarks of biology, yield similar structures in the organic and the computational alike.

John von Neumann, one of the pioneers of computing, gave a great deal of thought to self-replication in machines, binding together biology and technology from the dawn of computation. We see this still in the humble quine, tiny snippets of computer code that by their very existence stitch domain after domain together.

VERBAL OVERSHADOWING

N. J. ENFIELD

Professor and Chair, Department of Linguistics, University of Sydney; co-author (with Paul Kockelman), *Distributed Agency*

Suppose that two people witness a crime: One describes in words what she saw, while the other doesn't. When tested later on their memories of the event, the person who verbally described the incident will be worse at remembering or recognizing what actually happened. This is verbal overshadowing.

Putting an experience into words can result in failures of memory about that experience, whether it be the memory of a person's face, the color of an object, or the speed that a car was going. This effect was discovered by the cognitive psychologist Elizabeth Loftus and her students in experiments exploring witness testimony and the malleability of human memory. The stakes are high. As Loftus has shown, imperfect memories can—and often do—put innocent people behind bars. Our memories define much of what we take to be real. Anything that interferes with memory interferes, effectively, with the truth.

The idea that describing something in words can have a detrimental effect on our memory of it makes sense, given that we use words to categorize. To put things in the same category is by definition to set aside any information that would distinguish the things from one another. If I say, "She has three dogs," my act of labeling the three animals with the single word "dog" treats them as the same, regardless of likely differences in color, size, or breed. There are obvious advantages to words' power to group things in this way. Differences between things are often irrelevant, and glossing over those differences reduces effort in both speaking and understanding language. But the findings of Loftus and colleagues suggest that this discarding of distinguishing information by labeling something in a certain way has the effect of overwriting the mental representation of the experience. When words render experience, specific information isn't just left out, it's deleted.

Through verbal overshadowing, your words can change your beliefs, so word choice isn't merely a matter of style or emphasis. And note that the effect isn't just an effect of language but always of a specific language, be it Arabic, German, Japanese, Zulu, or any of the other 6,000 languages spoken in the world today. The facts of linguistic diversity suggest a striking implication of verbal overshadowing: that not just different words but also different languages are distinct filters for reality.

LIMINALITY

CHRISTINE FINN

Archaeologist; journalist; author, *Past Poetic*

I am writing this on the winter solstice. At 5:44 a.m. Eastern Standard Time, I stood on a porch in Vermont and watched the sky. Nothing to discern that was tangible. But somewhere at ancient sites, light passed over etched rock. For millennia, ancient people such as those once here on the banks of the Connecticut River marked a transition. The science is there for the shortest day. But what emerges for me is a search for absolute acuteness—that nano point where some thing changes and every thing changes. So, I'm excited by liminality.

At a time that celebrates fuzziness and mergings and convergence, I'm also intrigued by that absolute movement from one stage to another—one which finesses so acutely that it has a point.

My concept to celebrate comes out of social science but is mediated by ethnography and anthropology—archaeology, too, in the tangible remainders of transition. My mother lode is *The Rites of Passage,* by the prehistorian and ethnographer Arnold van Gennep, published in 1909. When it was translated from the French in the 1960s, it unleashed a torrent of new thinking. The British cultural anthropologist Victor Turner seized on the idea of "liminality." Whereas in van Gennup's better-known thesis we're all engaged in process—separation, transition, reincorporation—it was the transitional that held Turner. The potency of the edge of things, the not-quite-ness, appears to be dwelling in the poetry of ambiguity, but it chimes with so much of science that dwells in the periphery and the stunning space of almost-ness.

As Turner suggested:

> Prophets and artists tend to be liminal and marginal people, "edgemen," who strive with a passionate sincerity to rid themselves of the clichés associated with status incumbency and role-playing and to enter into vital relations with other men in fact or imagination. In their productions we may catch glimpses of that unused evolutionary potential in mankind which has not yet been externalized and fixed in structure.

POSSIBILITY SPACE

IAN BOGOST

Ivan Allen College Distinguished Chair in Media Studies, Professor of Interactive Computing, Georgia Institute of Technology; Founding Partner, Persuasive Games LLC; author, *Play Anything*

Some problems are easy, but most problems are hard. They exceed humans' ability to grasp and reason about possible answers. That's not just true of complex scientific and political problems like making complex economic decisions or building models to address climate change. It's also true of ordinary life. "Let's go out for dinner tonight." "OK, but where?" Without some structure, questions like that quickly descend into existential crisis: "Who am I, even?"

One way mathematicians think about complex problems is in terms of the possibility space (aka solution space, probability space) of their possible solutions. In mathematics, possibility spaces are used as a register, or ledger, of all the possible answers to a problem. For example, the possibility space for a coin toss is heads or tails. Of two coins: heads-heads, heads-tails, tails-heads, and tails-tails. In other cases, the possibility space might be very large or even infinite. The forms of possible life in the universe, for example, or the possible future branches of evolution. Or the possible outcomes of games of Go. Or all the things you might do with an evening.

In such cases, not only is it difficult or impossible to diagram the possibility space completely but also there's usually no sense in trying. An economist might build a model of possible evenings out from, say, the net marginal benefit of a movie or a bike ride or a beef Wellington in relation to the cost of those benefits, but such a practice assumes a rationalism that doesn't exist in ordinary life.

In game design, creators often think of their work as creating possibility spaces for their players. In the ancient Chinese folk game Go, a set of stones, stone-placement rules, and a board provide a large possibility space for overall play. But each individual move is much more limited, reliant on the set of choices each player has made previously. Otherwise it would be impossible for players ever to make a move. You play not within the total mathematical possibility space of all Go games but within the much narrower set of possible legal moves on a given board at a given time.

Some designers exalt the mathematical largesse of games like Go and chess, hoping to produce the largest possibility space with the fewest number of components. But more often what makes a game aesthetically unique isn't how mathematically large or deep it is but how interesting and unique are its components and their possible arrangements. Tetris has just seven different pieces, all of which operate the same way. The delight of Tetris comes from learning to identify and manipulate each piece in various situations.

The exercise of deliberately limiting a possibility space has utility well beyond science, mathematics, and game design. Every situation can be addressed more productively by acknowledging or imposing limitations so as to produce a thinkable, actionable domain of possible action. That doesn't mean you have to make a utility diagram every time you load the dishwasher or debate an evening out with friends. But the first step in any problem is recognizing the wealth of existing limitations waiting to be acknowledged and activated.

Faced with large or infinite possibility spaces, scientists try to impose limits in order to create measurable, recordable work. An astrobiologist might build a possibility space of possible alien life by limiting inquiry to stars or planets of a certain size and composition, for example. When you debate a venue for an evening meal, you do likewise—even though you probably don't think about it that way: What kind of food do you feel like? How much do you want to spend? How far are you willing to travel? Setting even one or two such limits constitutes a path toward progress, while avoiding the descent into an existential spiral, searching ever inward for what you really want or who you really are. In ordinary life, as in science, the answers are already there in the world—more so than they're invented inside your head.

ALTERNATIVE POSSIBILITIES

TANIA LOMBROZO

Associate Professor of Psychology, UC Berkeley

At the heart of scientific thinking is the systematic evaluation of alternative possibilities. The idea is so foundational that it's woven into the practice of science itself. Consider a few examples:

Statistical hypothesis testing is all about ruling out alternatives. With a null hypothesis test, you evaluate the possibility that a result was due to chance alone. Randomized controlled trials, the gold standard for drawing causal conclusions, are powerful precisely because they rule out alternatives; they diminish the plausibility of alternative explanations for a correlation between treatment and effect. And in science classes and laboratories across the globe, students are trained to generate alternative explanations for every observation—an exercise that peer reviewers take on as a professional obligation.

The systematic evaluation of alternative possibilities has deep roots in the origins of science. In the 17th century, Francis Bacon wrote about the special role of *instantia crucis,* "crucial instances," in guiding the intellect toward the true causes of nature by supporting one possibility over stated alternatives. Soon after, Robert Boyle introduced the *experimentum crucis*, or "crucial experiment," a term subsequently used by Robert Hooke and Isaac Newton. A crucial experiment is a decisive test between rival hypotheses—a way to differentiate possibilities. (More than two centuries later, Pierre Duhem would reject the crucial experiment, but not because it involves evaluating alternative possibilities. He rejected crucial experiments because the alternative possibilities they differentiate are too few; there are always more hypotheses available for amendment, addition, or rejection.)

The systematic evaluation of alternative possibilities is a hallmark of scientific thinking, but it isn't restricted to science. To arrive at the truth (in science or beyond), we generate multiple hypotheses and methodically evaluate how they fare against reason and empirical observation. We can't learn without entertaining the possibility that our current beliefs are wrong or incomplete, and we can't seek diagnostic evidence unless we specify the alternatives. Evaluating alternative possibilities is a basic feature of human thinking, a feature that science has successfully refined.

Within psychology, prompting people to consider alternative possibilities is recognized as a strategy for de-biasing judgments. When asked to consider alternatives (and, in particular, to consider the opposite of a possibility being evaluated), people question assumptions and recalibrate beliefs. They realize that an initial thought was misguided, a first impression uncharitable, a plan unrealistic. That such a prompt is effective suggests that in its absence people don't reliably consider the alternative possibilities they should. Yet the basis for doing so was in their heads all along—an untapped potential.

Evaluating alternative possibilities should be better known because it's a tool for better thinking, a tool that doesn't require fancy training or fancy equipment (beyond the fancy equipment we already have in our heads). What it does require is willingness to confront uncertainty and boldly explore the space of discarded or unformulated alternatives. That's a kind of bravery scientists should admire.

INDEXICAL INFORMATION

ANTHONY AGUIRRE

Theoretical physicist, UC Santa Cruz; Associate Scientific Director, Foundational Questions Institute

"Information" is a term with many meanings, but in physics or information theory we can quantify information as the *specificity* of a subset of many possibilities. Consider an eight-digit binary string like 00101010. There are 256 such strings, but if we specify that the first digit is "0" we define a subset of 128 strings with that property; this reduction corresponds (via a base-2 logarithm in the formula) to one bit of information. The string 00101010, one out of 256, has eight bits of information. This information, which is created by pointing to particular instances out of many possibilities, can be called *indexical information*. (A closely related concept is *statistical information*, in which each possibility is assigned a probability.)

This notion of information has interesting implications. You might imagine that combining many strings like 00101010 would necessarily represent more information, but it doesn't! Throw two such strings in a bag, and your bag now contains *fewer* than eight bits of information: two of the 256 possibilities are there, which is less specific than either one.

This paradoxical situation is the basis for Jorge Luis Borges' story *The Library of Babel*, about a library of all possible 410-page books—that is, all ordering of letters and punctuation that add up to 410 pages. Almost all these books are nonsense, and of those that do make sense almost all would be false! This library is information-free and useless. Yet suppose one had an index that identified all the books composed of actual words, and among those the books in sensible English, and of those the books of competent philosophy. Suddenly there's information—and the *index* has created it! This fable also makes clear that more information isn't necessarily better. Pointing out any single book creates the maximal amount of indexical information, but a good index would create a smaller quantity of more useful information.

This creation of indexical information by pointing us to what is important to us underlies many creative endeavors. You could write a computer program to spit out all possible sequences of musical notes, or all possible mathematical theorems provable from some set of axioms. But it wouldn't be writing music or doing mathematics—those endeavors select the tiny subset of possibilities that are interesting and beautiful.

What are most people most interested in? Themselves! Of all the people ever to live, you are *you*, and this creates indexical information particular to you. For example, indexical information results from "our" position in spacetime. Why are there no dinosaurs about? Why is the ocean made of water rather than methane? Those circumstances exist in other times and places—but you are *here* and *now*. Without here and now, there is, of course, no fact of the matter about whether dinosaurs exist, but we are accustomed to considering *here*, and especially *now*, to be objective facts of the world. Modern physics indicates this is unlikely.

Indeed, within some modern physical theories—such as the "many worlds" view of quantum mechanics or that of an infinite universe—this indexical information takes on fundamental importance. In both scenarios, there are many, or infinitely many, indistinguishable copies of "you" in existence. When "you" gather data about the external world, "you" aren't observing a unique external world—copies of "you" are embedded in *all possible* external worlds. Rather, you are narrowing down the set of worlds that "you" might be in, and the accompanying information you've gained is indexical.

In a sufficiently large universe, we can even ask whether there's anything *but* indexical information. Information about *no*-dinosaurs, *yes*-water, and *what*-weather is missing from *the* world, which may contain little or no information, but it's present in *our* world, which contains an immense amount. This information is forged by our point of view and created by us, as our individual view takes a simple objective world with little or no content and draws from it a rich, interesting information structure we call reality.

EMOTION CONTAGION

JUNE GRUBER
Assistant Professor of Psychology, University of Colorado, Boulder

Emotions are contagious. They are rapidly, frequently, and even at times automatically transmitted from one person to the next. Whether it be mind-boggling awe watching the supermoon display its lunar prowess or pangs of anger observing palpable racial injustice, one feature remains salient: We can and often do "catch" the emotions of others. The notion that emotions are contagious dates back to the 1750s, when Adam Smith documented the seamless way people tend to mimic the emotional expressions, postures, and even vocalizations of people they interact with. In the late 1800s, Charles Darwin further emphasized that this highly contagious nature of emotions was fundamental to the survival of humans and nonhumans alike in transmitting vital information among group members. These prescient observations underscore the fact that emotion contagion is pervasive and universal and, hence, why it ought to be more widely known.

Recent scientific models of emotion contagion explicate the ways we're affected by, and affect, the emotions of others. Emotion contagion is robustly supported in laboratory studies eliciting transient positive and negative emotion states among individual participants; efforts outside the laboratory focus on longer-lasting mood states, such as self-reported happiness propagating through large social networks. Emotion contagion matters: It's in the service of critical processes such as empathy, social connection, and relationship maintenance between close partners.

Faulty emotion-contagion processes have been linked to affective disturbances. With the rapid proliferation of online social networks as a main forum for emotion expression, we know, too, that emotion contagion can occur without direct interaction between people or when nonverbal emotional cues in the face and body are altogether absent. This type of contagion spreads across a variety of other psychological phenomena that indirectly or directly involve behaviors such as kindness, health-related eating habits, and even the darker side of human behaviors, including violence and racism. Emotion contagion matters, for better and for worse.

Although well documented, the scientific scope of emotion contagion warrants widening. One uncharted domain my colleagues Nicholas Christakis, Jamil Zaki, Michael Norton, Ehsan Hoque, and Anny Dow have been mapping is that of positive emotion contagion. Surprisingly, we know little about the temporal dynamics of our positive experiences, including those that would seem to connect us with others (and should thus propagate rapidly) such as joy or compassion, versus those that may socially isolate us, such as hubristic pride. Given the vital role positive emotions play in our well-being and physical health, it's crucial to better understand the features of how we transmit these pleasant states within and across social groups. Like waves, emotions cascade across time and geographical space. Yet their ability to cascade across minds is unique and deserves wider recognition.

NEGATIVE EVIDENCE

BART KOSKO

Information scientist, Professor of Electrical Engineering and Law, University of Southern California; author, *Fuzzy Engineering*

Negative evidence is a concept that deserves greater currency in the intellectual trades and popular culture. Negative evidence helps prove that something did not occur. University registrars routinely use negative evidence when they run a transcript check to prove that someone never got a university degree.

Negative evidence is the epistemic dual to positive evidence. It is just that evidence that tends to prove a negative. So it collides headfirst with the popular claim that you cannot prove a negative. A more sophisticated version of the same claim is that absence of evidence is not evidence of absence.

Both claims are false in general.

It may well be hard to prove a negative to a high degree. That does not diminish the probative value of doing so. It took the invasion and occupation of Iraq to prove that the country did not have weapons of mass destruction. The weapons may still turn up someday. But the probability of finding any has long since passed from unlikely to highly unlikely. The search has simply been too thorough.

Absence of evidence can likewise give some evidence of absence. A chest CT scan can give good image-based negative evidence of the absence of lung cancer. The scan does not prove absence to a logical certainty. No factual test can do that. But the scan may well prove absence to a medical certainty. That depends on the accuracy of the scan and how well it searches the total volume of lung tissue.

The CT-scan example shows the asymmetry between positive and negative evidence. The accurate scan of a single speck of malignant lung tissue is positive evidence of lung cancer. It may even be conclusive evidence. But it takes a much larger scan of tissue to count as good negative evidence for the absence of cancer. The same holds for finding just one weapon of mass destruction compared with searching for years and finding none.

Simple probability models tend to ignore this asymmetry. They have the same form whether the evidence supports the hypothesis or its negation.

The most common example is Bayes' theorem. It computes the conditional probability of a converse in terms of a hypothetical condition and observed evidence. The probability that you have a blood clot given a high score on a D-dimer blood test differs from the probability that you would observe such a test score if you in fact had a blood clot. Studies have shown that physicians sometimes confuse these converses. Social psychologists have even called this the fallacy of the inverse or converse. Bayes' theorem uses the second conditional probability (of observing the evidence, given the hypothetical condition) to help compute the first probability (of having the hypothetical condition, given the observed evidence). The simple ratio form of the Bayes calculation ensures that a symmetric ratio gives the probability that the hypothesis is false, given the same observed evidence. The numerical values may differ, but the ratio form does not.

The law is more discerning with negative evidence.

Courts often allow negative evidence if the proponent lays a proper foundation for it. The opponent must also not have shown that the lack of evidence involved something improper or untrustworthy. A convenience store's videotapes of a robbery can give compelling negative evidence that a suspect did not take an active part in the robbery. But the video recordings would have to cover a sufficient portion of the store and its parking lot. The cameras must also have run continuously throughout the robbery.

Federal Rule of Evidence 803 uses a telling phrase to describe when a proponent can use a public record as negative evidence that a prior conviction or other event did not occur. The rule demands a "diligent search" of the public records.

Diligent search is the key to negative evidence.

It is easy to conduct a diligent search to prove the negative that there is not a five-carat diamond in your pocket. It takes far more effort to conduct a diligent search for the diamond in a room or building or in an entire city.

The strength of negative evidence tends to fall off quickly with the size of the search area. That is why we cannot yet technically rule out the existence in the deep seas of kraken-like creatures that can drag ships down to their doom. The ocean contains more than 300 million cubic miles of water. High-resolution sonar has mapped only a fraction of it. That mapping itself has only been a snapshot and not an ongoing movie. It is far from a diligent search of the whole volume.

Search size also justifies patience in the search for extraterrestrial intelligence. Satellites have only recently mapped the surface of Mars. Radio telescopes have searched only a minuscule fraction of the expanding universe for some form of structured energy in interstellar radio signals.

Good negative evidence that we are alone could take thousands or millions of years of diligent search. Positive evidence to the contrary can come at any second.

EMPTINESS

JAEWEON CHO
Professor of Environmental Engineering, UNIST; Director, Science Walden Center

It is said in the *Doctrine of the Mean*, written by the grandson of Confucius, that the greatest knowledge, including scientific concepts and human realizations, comes only from the empty mind. The title of the book, *Doctrine of the Mean*, doesn't mean the middle way between two extremes but the emptiness of the mind in everyday life. We call it moderation. This teaching connects in essence to Nietzsche's *Thus Spake Zarathustra*, as Heidegger explains in his book *What Is Called Thinking?* All three stress that we cannot think without emptying our minds in our everyday lives. If one can think, one can do everything: Confucius's grandson called such a person a scholar or scientist, while Nietzsche called him Zarathustra.

After crossing a river in a raft, we must abandon the raft to climb the mountain. Existing knowledge helps, but it may also hinder us from obtaining new representations of the new corresponding concepts. When we look at a red flower, we may think of the refraction of the color rather than about the role of the color in enabling the flower to survive in nature. When we study microorganisms, we may rely on genetic information rather than paying attention to phenomenological fact. Whenever we observe an object, or experience an event, we can see it in a new light only when our mind is empty.

How to empty a mind filled with knowledge? The discourses of Confucius offer two ways. First, we can empty our mind by empathizing. When we meet someone who suffers, we can let our mind enter the mind of the sufferer (this is called empathy) instead of reminding ourselves of the social welfare system and so forth. When we see a river polluted with algae, we may think only of eutrophication, of nitrogen and phosphorus; it's not easy to go beyond the science in this case. Only when we immerse ourselves in the polluted river may we obtain new knowledge of it.

Second, we can empty our minds by thinking about social justice whenever we have a chance to profit from one or another opportunity. This certainly applies to the issue of climate change. Confucius asks us whether we want to solve the emergent problems of climate change or ignore them in favor of our own interests. He assures us that empathy and acts of justice can lead to knowledge—although he never defines any knowledge but only provides examples from which we can derive it ourselves.

EFFECT SIZE

BRUCE HOOD

Professor of Developmental Psychology in Society, University of Bristol; author, *The Self Illusion*

The media are constantly looking for significant new discoveries to feed to the general public, who want to know what controls our lives and how to better them. Cut down on salt, eat more vegetables, avoid social networking sites, and so on—factors reported to have significant effects on our health and welfare.

Scientists also seek significance—a technical term with a meaning different from its common usage. In science, when a discovery is "highly significant," it's more likely to reflect a real state of nature than a random fluctuation. But when society hears that something is "significant," this is interpreted as an important finding that has major impact. The problem is that patterns can be highly significant but not very meaningful to the individual. This is where effect size comes into play—a concept that ought to be more widely known.

Calculating effect size (e.g., "Cohen's d") involves mathematics beyond the scope of this piece; suffice it to say that population distributions figure in estimating how strong an effect is. Thus effect size, rather than significance, is a more meaningful measure of how important a pattern is in relation to all the other patterns influencing our lives.

Effect size is best calculated from a number of independent studies, to avoid the problems inherent in limited observations. The more scientists studying a phenomenon the better, as there's less opportunity for errors or mendacious manipulation. A meta-analysis is a study of all the studies measuring a phenomenon, and it's the best way to calculate effect sizes. Because there are so many factors that can influence your observations—population differences, sampling errors, methodological differences, and so on—it makes sense to gather as much evidence as you can, in order to estimate effect size.

Consider the reputed difference between the mathematical ability of boys and girls—a highly contentious debate. Meta-analyses of over 3 million children have found significant differences between boys and girls in elementary school, but the effect sizes are so small (*d* less than 0.05) as to be meaningless. The male advantage that does emerge later in schooling is due to factors other than gender which increasingly play a role in mathematical ability.

Humans are complex biological systems affected by a plethora of mechanisms—from genetic inheritance to environmental changes—that interact in ways varying from one individual to another and are too complex to map or predict. In an effort to isolate mechanisms, scientists often strip away extraneous variables to reveal the core factors under investigation, but in doing so they create a false impression of the true influence of a mechanism in relation to all the others that play a role. What they find may be significant, but effect size tells you whether or not it's meaningful.

SURREAL NUMBERS

SIOBHAN ROBERTS

Science journalist; contributor, "Elements," NewYorker.com, *The Guardian*; author, *Genius at Play: The Curious Mind of John Horton Conway*

Merriam-Webster's 2016 word of the year is *surreal*:

> It's a relatively new word in English, and derives from *surrealism*, the artistic movement of the early 1900s that attempted to depict the unconscious mind in dreamlike ways as "above" or "beyond" reality. *Surreal* itself dates to the 1930s, and was first defined in a Merriam-Webster dictionary in 1967. *Surreal* is often looked up spontaneously in moments of both tragedy and surprise. . . .

One of the lesser-known applications of the word belongs to the Princeton mathematician John Horton Conway, who discovered surreal numbers circa 1969. To this day, he wishes more people knew about the surreals, in hopes that the right person will put them to greater use.

Conway happened upon surreal numbers—an elegant generalization and vast expansion of the real numbers—while analyzing games (primarily the game of Go, a popular pastime in math departments). The numbers fell out of the games, as he put it; they were a means of classifying the moves made by each player and determining who seemed to be winning and by how much. And as Conway described, the surreals are "best thought of as the most natural collection of numbers that includes both the usual real numbers . . . and the infinite ordinal numbers discovered by Georg Cantor." Originally, Conway called his new number scheme simply capital "N" Numbers, since he felt they were so natural and such a natural replacement for all previously known numbers.

The surreals are a souped-up continuum of numbers including all the merely real—integers, fractions, and irrationals such as π—and going above and beyond and below and within, gathering in all the infinites. Cantor's transfinite omega is a surreal number, too, as is the square root of omega, omega squared, omega squared plus one, and so on. They're not called surreal for nothing.

Over the years, the scheme won distinguished converts. Most notably, in 1973 the Stanford computer scientist Donald Knuth spent a week sequestered in an Oslo hotel room in order to write a novella introducing the concept to the wider world—*Surreal Numbers*, a love story, in the form of a dialogue between Alice and Bill (now in its 21st printing). It was Knuth, in fact, who gave these numbers their name, which Conway adopted, publishing his own expository account, *On Numbers and Games,* in 1976. Knuth views the surreal numbers as simpler than the reals. He considers the scenario roughly analogous to Euclidean and non-Euclidean geometry; he wonders what the repercussions would be if the surreals had been invented or discovered first.

The Princeton mathematician and physicist Martin Kruskal spent some thirty years investigating the promising utility of surreal numbers. He thought they might help in quantum field theory—such as when asymptotic functions veer off the graph. "The usual numbers are very familiar," he once said, "but at root they have a very complicated structure. Surreals are in every logical, mathematical, and aesthetic sense better."

In 1996, eighteen-year-old Jacob Lurie (now a professor a Harvard) won the top prize in the Westinghouse Science Talent Search, for his project on the computability of surreal numbers. The *New York Times* reported the news and ran a Q&A with Lurie:

Q: "How long have you been working on this?"

A: "It's not clear when I started or when I finished, but at least for now, I'm finished. All the questions that have yet to be answered are too hard."

And that's pretty much where things stand today. Conway still gets asked about the surreals fairly regularly—most recently by some postdocs at a holiday party. He repeated for them what he's said for ages: Of all his work, he's proudest of the surreals, but they're also a great disappointment, since they remain so isolated from other areas of mathematics and science. Per Merriam-Webster, Conway's response is a combination of awe and dismay. The seemingly infinite potential for the surreals continues to beckon but for now remains just beyond grasp.

STANDARD DEVIATION

S. ABBAS RAZA
Founding editor, 3QuarksDaily.com

Suppose we choose an American woman at random. Given just a couple of numbers that we know characterize American women's heights, we can be 95-percent sure (meaning, we'll be wrong only once out of twenty times) that her height will be between 4 feet 10 inches and 5 feet 10 inches. It's the statistical concept of standard deviation that allows us to say this.

To show how, let's take such a data set—the heights of 1,000 randomly chosen American women—and plot the data as points on a graph, where the (horizontal) x-axis shows heights from 0 to 100 inches and the (vertical) y-axis shows the number of women of that height (we'll use only whole numbers of inches). If we then connect all these points in a smooth line, we'll get a bell-shaped curve. Some data sets (including this one) with this characteristic shape are said to have a normal (aka Gaussian) distribution, and this is how a great many kinds of data are distributed.

For data that are strictly normally distributed, the highest point on our graph (in this example, the height in inches that occurs most frequently, which is called the mode) would be at the average (mean) height for American women, which happens to be 64 inches. But with real-world data, the mean and the mode can differ slightly. As we move along the x-axis rightward from that peak to larger heights, the curve will slope downward, become steeper, and then gradually become less steep and peter out to zero as it hits the x-axis just after where the tallest woman (or women) in our sample happens to be. The same thing happens on the other side as we go to smaller heights: This is how we get the familiar symmetrical bell shape.

Suppose the height of Japanese women is also 64 inches on average but varies less than that of American women because Japan is less ethnically and racially diverse than the U.S. In this case, the bell-shaped curve will be thinner and higher and will fall to zero more steeply on either side. *Standard deviation* is a measure of how spread out the bell curve of a normal distribution is. The more the data are spread out, the greater the standard deviation.

The standard deviation is half the distance from one side of the bell curve to the other (so it has the same units as the x-axis), where the curve is about 60 percent of its maximum height. It can be shown that about 68 percent of our data points will fall within +1 or -1 standard deviation around the mean value. So in our example of the heights of American women, if the standard deviation is 3 inches, then 68 percent of American women will have a height between 61 and 67 inches. Similarly, 95 percent of our data points will fall within 2 standard deviations around the mean, so in our case 95 percent of women will have heights between 58 and 70 inches. It can also be calculated that 99.7 percent of our data points will fall within 3 standard deviations around the mean—and so on, for even greater degrees of certainty.

The reason standard deviation is so important in science is that random errors in measurement usually follow a normal distribution. And *every* measurement has some random error associated with it. For example, even for something as simple as weighing a small object on a scale: If we weigh it 100 times, we may get many slightly different values. Suppose the mean of all our measurements comes out to 1,352 grams with a standard deviation of 5 grams. Then we can be 95-percent certain that the object's actual weight is between 1,342 and 1,362 grams (mean weight plus or minus 2 standard deviations).

You may have heard reports before the discovery of the Higgs boson at CERN in 2012 that there was a "3-sigma" result showing a new particle. (The lower-case Greek letter *sigma* is the conventional notation for standard deviation, hence it's often just called "sigma.") The 3-sigma meant we could be 99.7-percent certain the signal was real and not a random error. Eventually a 5-sigma result was announced for the Higgs particle on July 4th, 2012, at CERN, and that corresponds to a 1-in-3.5 million chance that what they detected was due to random error.

Measurement error (or uncertainty in observations) is a fundamental part of science now, but only in the 19th century did scientists start incorporating this idea routinely in their measurements. The ancient Greeks, for example, while sophisticated in some parts of their mathematical and conceptual apparatus, almost always reported observations with much greater precision than was warranted, and this practice often got amplified into major errors. An example is Aristarchus' impressive method of measuring the distance between the Earth and the sun by measuring the angle between the line of sight to the moon when it was exactly half full and the line of sight to the sun. (The line between the Earth and the moon would then be at 90° to the line between the moon and the sun, and along with the line from the Earth to the sun this would form a right triangle.) He measured this angle as 87°, which told him that the distance from the Earth to the sun is twenty times greater than the distance from the Earth to the moon. The problem is that the impeccable geometric reasoning he used is extremely sensitive to small errors in this measurement. The actual angle (as measured today with much greater precision) is 89.853°, which gives a distance between the Earth and the sun as 390 times greater than the distance between the Earth and the moon. If he had made many such measurements and had the concept of standard deviation, Aristarchus would have known that the possible error in his distance calculation was huge, even for a decent reliability of 2 standard deviations, or 95-percent certainty in measuring that angle.

THE BREEDER'S EQUATION

GREGORY COCHRAN
Research associate, Anthropology Department, University of Utah; co-author (with Henry Harpending), *The 10,000 Year Explosion*

$R = h^2S$.

R is the response to selection, S is the selection differential, and h^2 is the narrow-sense heritability. This is the workhorse equation for quantitative genetics. The selective differential S is the difference between the population average and the average of the parental population (some subset of the total population). Almost everything is moderately to highly heritable, from height and weight to psychological traits.

Consider IQ. Imagine a set of parents with IQs of 120, drawn from a population with an average IQ of 100. Suppose the narrow-sense heritability of IQ (in that population, in that environment) is 0.5. The average IQ of their children will be 110. That's what is usually called regression to the mean.

Do the same thing with a population whose average IQ is 85. We again choose parents with IQs of 120, and the narrow-sense heritability is still 0.5. The average IQ of their children will be 102.5—they regress to a lower mean.

You can think of it this way: In the first case, the parents have twenty extra IQ points. On average, half of those points are due to additive genetic factors and the other half are the product of good environmental luck. By the way, when we say "environmental," we mean "something other than additive genetics." It doesn't look as if the usual suspects—the way you raise your kids, or the school they attend—contribute much to this environmental variance, at least for adult IQ. We know what it's not, but not much about what it is, although it must include factors like test error and being hit on the head with a ball-peen hammer.

The kids get the good additive genes but have average environmental luck, so their average IQ is 110. The luck (10 points' worth) goes away.

The 120-IQ parents drawn from the 85-IQ population have 35 extra IQ points, half from good additive genes and half from good environmental luck. But in the next generation, the luck goes away, so they drop 17.5 points.

The luck only goes away once. If you took those kids from the first group, with average IQs of 110, and dropped them on a friendly uninhabited island, they'd eventually get around to mating—and the next generation would also have an IQ of 110. With tougher selection—say, by kidnapping a year's worth of National Merit finalists—you could create a new ethny with far higher average intelligence than any existing. Eugenics is not only possible, it's trivial.

So what can you explain with the Breeder's Equation? Natural selection, for one thing. It probably explains the Ashkenazi Jews. It looks as if there was (once) an unusual reproductive advantage for people who were good at certain kinds of white-collar jobs, along with a high degree of reproductive isolation.

It also explains why the professors' kids are a disproportionate fraction of the National Merit finalists in a college town—their folks, particularly their fathers, are smarter than average—and so are they. Reminds me of the fact that Los Alamos High School has the highest test scores in New Mexico. Our local high school tried copying their schedule, in search of the secret. Didn't work. I know of an approach that would, but it takes about fifteen years.

But those kids, although smarter than average, usually aren't as smart as their fathers, partly because their mothers typically aren't theoretical physicists, partly because of regression toward the mean. The luck goes away.

That's the reason the next generation has trouble running the corporation Daddy founded: regression to the mean, not just in IQ. Dynasties have a similar decay problem. The Otto-man Turks avoided it for a number of generations by a form of delayed embryo screening (the law of fratricide).

And of course the Breeder's Equation explains how average IQ potential is declining today—because of low fertility among highly educated women.

Let me make this clear: The Breeder's Equation is immensely useful in understanding evolution, history, contemporary society, and your own family. And hardly anyone has heard of it. "Breeder's Equation" has not been used by the *New York Times* in the last 160 years.

FIXPOINT

DAVID DALRYMPLE
Computer scientist, neuroscientist; Research Affiliate, MIT Media Lab;
Director, Project Nemaload

Given the operation of squaring a number, there are two numbers that are special because the operation doesn't change them: 0 and 1. 0 squared is 0, and 1 squared is 1, but square any other number and the output will differ from the input. These special numbers are called fixpoints (of squaring). In general, a fixpoint is a value (or state of a system) that's left unchanged by a particular operation. This concept, easily definable by the brief equation $x = f(x)$, is at the essence of several other ideas of great practical significance, from Nash equilibria (used in economics and social sciences to model failures of cooperation) to stability (used in control theory to model systems ranging from aircraft to chemical plants) to PageRank (the foremost Web search algorithm). It even appears in work at the foundations of logic, to give definition to truth itself!

A prevalent example in everyday life is the occurrence of fixpoints in any strategic context ("game") with multiple players. Suppose each player can independently revise his strategy so as to best respond to the others' strategies. If we consider this revision process as an operation, then its fixpoints are those combinations of strategies for which the revision operation yields exactly the same combination of strategies—that is, no players are incentivized to change their behavior in any way. This is the concept of a Nash equilibrium, a kind of fixpoint that directly applies to

490

many real-world situations—ranging in importance from room-mates leaving dirty dishes in the sink to nuclear arsenals. Moving to a different Nash equilibrium (such as disarmament) requires changing the revision operator (e.g., with an agreement that binds multiple players to change their strategies at the same time).

Fixpoints are also an extraordinarily useful framework to consider ideas that seem to be defined circularly. As a concrete example, the golden ratio (*phi*) can be defined as *phi = 1+1/phi,* which actually means that *phi* is the unique fixpoint of the operation that takes the reciprocal and then adds one. In general, self-referential definitions can be classified according to the number of fixpoints of the corresponding operation: More than one, and it's imprecise but potentially serviceable; one, and it's a bona-fide definition; lacking any, and it's a paradox (like "This sentence is false."). In fact, operations lacking fixpoints underlie a host of deeply related paradoxes in mathematics and led to Gödel's discovery that the early 20th-century logicians' quest to completely formalize mathematics is impossible.

Mathematics is not the only reasoning system we might use. In daily life, we use ideologies and belief systems that aren't complete—in the sense that they can't express arbitrarily large natural numbers (nor do we ask them to). We can consider persistent ideologies and belief systems as fixpoints of the revision operation of "changing one's mind"—subject to the types of questions, evidence, and methods of reasoning that the ideology judges acceptable. If there's a fixpoint here, we're unable to change our minds in some cases, even if presented with overwhelming evidence. As with Nash equilibria, the key to escaping such a belief system is to modify the revision operation: Considering just one new question (such as "What questions am I allowed to ask?") can sometimes be enough to abolish ideological fixpoints forever.

One remarkable feature of the scientific belief system is its non-fixedness. New beliefs are constantly integrated, and old beliefs are not uncommonly discarded. Ideally, science would be complete in the limit of infinite time and experiments without losing its openness, analogous to how programs that produce never-ending lists (such as the digits of pi or the prime numbers) are formally given meaning by infinitely large fixpoints that may only be successively approximated. While an absence of fixpoints in the logical foundations of arithmetic dooms it to "incompleteness," fixpoint theorems have recently been used to show that if we relax our notion of completeness to the almost equally satisfying concept of "coherence," there is a revision operation that's guaranteed to have a coherent fixpoint and can even be approximated by computable algorithms!

Perhaps science, too, aspires to an unreachable, infinite fixpoint in which all knowable facts are known and all provable consequences are proved, such that there would be no more room to change one's mind—and we hope that with each passing year our current state of knowledge more closely approximates that ultimate fixpoint.

NON-ERGODIC

STUART A. KAUFFMAN

Theoretical biologist; complex-systems researcher; author, *Humanity in a Creative Universe*

"Non-ergodic" is a fundamental but too little known scientific concept. Non-ergodicity stands in contrast to "ergodicity." "Ergodic" means that the system in question visits all its possible states. In statistical mechanics, this is based on the famous "ergodic hypothesis," which, mathematically, gives up integration of Newton's equations of motion for the system. Ergodic systems have no deep sense of "history." Non-ergodic systems don't visit all their possible states. In physics, perhaps the most familiar case of a non-ergodic system is a spin glass that breaks ergodicity and visits only a tiny subset of its possible states—hence, exhibits history in a deep sense.

Even more profound, the evolution of life in our biosphere is deeply non-ergodic and historical. The universe won't create all possible life-forms. This, together with heritable variation, is the substantial basis for Darwinism, without yet specifying the means of heritable variation, whose basis Darwin did not know. Non-ergodicity gives us history.

CONFUSION

MAXIMILIAN SCHICH

Associate Professor, Arts and Technology, University of Texas at Dallas; Founding member, The Edith O'Donnell Institute for Art History

Commonly, confusion denotes bewildering uncertainty, often associated with delirium or even dementia. From the confusion of languages in the Biblical Genesis to Genesis the band, broader audiences mostly encounter negative aspects of confusion. But confusion can be negative *and* positive, sometimes both at the same time. Moreover, it's a subject of scientific interest, a phenomenon that can't be ignored, that requires scientific understanding, and needs to be designed and moderated.

A convenient tool to measure confusion in a system is the so-called confusion matrix. This is used in linguistics and computer science—in particular, machine learning. In principle, the confusion matrix is a table, where all criteria in the row dimension are compared with all criteria in the column dimension. A simple example is to compare all letters of the alphabet spoken by an English native with the letters actually perceived by a German speaker. An English "e" will often be confused with the German "i," resulting in a higher value in the matrix where the "e" row crosses the "i" column. Ideally, of course, letters are confused only with themselves, resulting in high values exclusively along the matrix diagonal. Actual confusion, in other words, is characterized by patterns of higher values, off the matrix diagonal.

Unfortunately, the use of the confusion matrix is still mostly governed by what Richard Dawkins calls "the tyranny of the discontinuous mind." Processing the confusion matrix, scholars generally derive secondary measures to quantify Type I and Type II errors (false positives and false negatives) and a number of similarly aggregate measures. In short, the confusion matrix is used to make classification by humans and artificial intelligence less confusing. A typical (and useful) example is to compare a machine classification of images with the known ground truth. No doubt, quantifying the confusion of ducks with alligators, just like pedestrians with street signs, is a crucial application that can save lives. Likewise, it's often useful to optimize classification systems in order to minimize the confusion of human curators. A good example would be the effort of the semantic Web community to simplify global classification systems, such as the Umbel ontology or the category system in Wikipedia, to allow for easy data collection and classification with minimal ambiguity. Nevertheless, the almost exclusive focus on optimization by minimizing confusion is unfortunate, as perfect discreteness of categories is undesirable in many real systems, from the function of genes and proteins to individual roles in society. Too little confusion between categories or groups and the system is, in essence, dead. Too much confusion and the system is overwhelmed by chaos. In a social network, total lack of confusion annihilates any base for communication between groups, while complete confusion would be equivalent to a meaningless cacophony of everything meaning everything.

Network science is increasingly curious regarding this situation, dealing with confusion using the concept of overlap in community finding. Multifunctional molecules—genes and proteins, for example—act as drugs and drug targets, where confusion needs to be moderated in order to hit the target while minimizing unwanted side effects. Similar situations arise in social life. Only recently has it become possible in network science to deal with such phenomena in an efficient way. Network science initially focused mostly on identifying discrete communities, since finding them is much simpler in terms of computation. In such a perfect world, one where all communities are discrete, there's no confusion—or, more precisely, confusion is ignored. In such a world, the confusion or co-occurrence matrix can be sorted so that all communities form squares or rectangles along the matrix diagonal. In a more complicated case, neighboring communities overlap, forming subcommunities between two almost discrete communities—say, people belonging to the same company while also belonging to the same family. It's easy to imagine more complicated cases. At the other end of the spectrum, we find all-out, complex overlap, which is hard to imagine or visualize in terms of sorting the matrix. It may well be true, however, that complex overlap is crucial to the survival of the system in question.

There's a known case in which confusion by design is desirable: a highly cited concept in materials science introduced in 1993 in an article in *Nature*. Lindsay Greer's so-called principle of confusion applies to the formation of metallic glass. In short, the principle states that using a greater variety of metal atoms to form a glass is more convenient due to the resulting impurities, which give the material less chance to crystallize. This allows for larger glass objects with interesting material properties, such as being stronger than steel. The convenience of more confusion is counterintuitive; it's increasingly harder to determine the material properties of a glass the greater the variety of metals involved. It wouldn't be surprising to see something like Greer's principle of confusion applied to other systems as well.

While such questions await solution, as a take-home we should expect critical amounts of confusion in many real-life systems, with the optimum in between, but not identical with, perfect discreteness and perfect homogeneity. Further identifying, understanding, and successfully moderating patterns of confusion in real systems is an ongoing challenge. Solving it is likely essential in many fields, from materials and medicine to social justice and the ethics of artificial intelligence. Science will help us clarify—if possible, embrace; if necessary, avoid—confusion. Of course, we should be cautious: The control of confusion can be used for peace or war, much like the rods in a nuclear reactor, with the difference that switching off confusion in a social system may be just as deadly as switching it on.

COALITIONAL INSTINCTS

JOHN TOOBY
Founder of evolutionary psychology; Codirector, Center for
Evolutionary Psychology, Professor of Anthropology, UC Santa Barbara

Every human, not excepting scientists, bears the whole stamp of the human condition. This includes evolved neural programs specialized for navigating the world of coalitions—teams, not groups. (Although the concept of coalitional instincts has emerged over recent decades, there's no mutually-agreed-upon term for this concept yet.) These programs enable us and induce us to form, maintain, join, support, recognize, defend, defect from, factionalize, exploit, resist, subordinate, distrust, dislike, oppose, and attack coalitions. Coalitions are sets of individuals interpreted by their members and/or by others as sharing a common abstract identity (including propensities to act as a unit, to defend joint interests, and to have shared mental states and other properties of a single human agent, such as status and prerogatives).

Why do we see the world this way? Most species don't and can't. Even those with linear hierarchies don't. Among elephant seals, for example, an alpha can reproductively exclude other males, even though beta and gamma are physically capable of beating alpha if only they could cognitively coordinate. The fitness payoff is enormous for solving the thorny array of cognitive and motivational computational problems inherent in acting in groups: Two can beat one, three can beat two, and so on, propelling an arms race of numbers, effective mobilization, coordination, and cohesion.

Ancestrally, evolving the neural code to crack these problems supercharged the ability to successfully compete for access to reproductively limiting resources. Fatefully, we're descended solely from those better equipped with coalitional instincts. In this new world, power shifted from solitary alphas to the effectively coordinated down-alphabet, giving rise to a new, larger landscape of political threat and opportunity: rival groups or factions expanding at your expense or shrinking because of your dominance.

And so a daunting new augmented reality was neurally kindled, overlying the older individual one. It's important to realize that this reality is constructed by and runs on our coalitional programs and has no independent existence. You're a member of a coalition only if someone (such as you) interprets you as being one, and you aren't if no one does. We project coalitions onto everything, even where they have no place, such as in science. We're identity-crazed.

The primary function that drove the evolution of coalitions is the amplification of the power of its members in conflicts with non-members. This function explains a number of otherwise puzzling phenomena. For example, ancestrally, if you had no coalition you were nakedly at the mercy of everyone else, so the instinct to belong to a coalition has urgency, preexisting and superseding any policy-driven basis for membership. This is why group beliefs are free to be so weird. Since coalitional programs evolved to promote the self-interest of the coalition's membership (in dominance, status, legitimacy, resources, moral force, etc.), even coalitions whose organizing ideology originates (ostensibly) to promote human welfare often slide into the most extreme forms of oppression, in complete contradiction to the putative values of the group. Indeed, morally wrong-footing rivals is one point of ideology, and once everyone agrees on something (slavery is wrong), it ceases to be a significant moral issue because it no longer shows local rivals in a bad light. Many argue that there are more slaves in the world today than in the 19th century. Yet because one's political rivals cannot be delegitimized by being on the wrong side of slavery, few care to be active abolitionists anymore, compared to being, say, speech police.

Moreover, to earn membership in a group, you must send signals clearly indicating that you differentially support it, compared to rival groups. Hence, optimal weighting of beliefs and communications in the individual mind will make it feel good to think and express content conforming to and flattering to one's group's shared beliefs and to attack and misrepresent rival groups. The more biased away from neutral truth, the better the communication functions to affirm coalitional identity, generating polarization in excess of actual policy disagreements. Communications of practical and functional truths are generally useless as differential signals, because any honest person might say them regardless of coalitional loyalty. In contrast, unusual, exaggerated beliefs—such as supernatural beliefs (e.g., god is three persons but also one person), alarmism, conspiracies, or hyperbolic comparisons—are unlikely to be said except as expressive of identity, because there's no external reality to motivate nonmembers to speak absurdities.

This raises a problem for scientists: Coalition-mindedness makes everyone, including scientists, far stupider in coalitional collectivities than as individuals. Paradoxically, a political party united by supernatural beliefs can revise its beliefs about economics or climate without revisers being bad coalition members. But people whose coalitional membership is constituted by their shared adherence to "rational," scientific propositions have a problem when—as is generally the case—new information arises which requires belief revision. To question or disagree with coalitional precepts, even for rational reasons, makes one a bad and immoral coalition member—at risk of losing job offers, one's friends, and one's cherished group identity. This freezes belief revision.

Forming coalitions around scientific or factual questions is disastrous, because it pits our urge for scientific truth-seeking against the nearly insuperable human desire to be a good coalition member. Once scientific propositions are moralized, the scientific process is wounded, often fatally. No one is behaving either ethically or scientifically who doesn't make the best case possible for rival theories with which one disagrees.

THE SCIENTIFIC METHOD

NIGEL GOLDENFELD

Swanlund Endowed Chair, Center for Advanced Study Professor in Physics, University of Illinois at Urbana-Champaign

There's a saying that there are no cultural relativists at 30,000 feet. The laws of aerodynamics work regardless of political or social prejudices, and they're indisputably true. Yes, you can discuss to what extent they're an approximation, what their limits of validity are, whether they take into account such niceties as quantum entanglement or unified field theory (of course they don't). But the most basic scientific concept disturbingly missing from today's social and political discourse is the concept that some questions have correct and clear answers. Such questions can be called "scientific" and their answers represent truth. Scientific questions aren't easy to ask. Their answers can be verified by experiment or observation, and they can be used to improve your life, create jobs and technologies, save the planet. You don't need pollsters or randomized trials to determine whether a parachute works. You need an understanding of the facts of aerodynamics and the methodology to do experiments.

Science's main goal is to find the answers to questions. And the rate of advance of science is determined by how well we can ask sharp, scientific questions, not by the rate at which we answer them. In the field of science I identify with—condensed-matter physics—important new discoveries and new questions arise on the scale of about once every five years. They're mostly answered and worked through on a timescale much less than that. Science is also driven by luck and serendipitous discovery. That can also be amplified by asking good questions. For example, the evolutionary biologist Carl Woese discovered a third domain of life by asking how to map out the history of life using molecular sequences of RNA rather than fossils and superficial appearances of organisms. The widely publicized ennui of fundamental physics is a result of the failure to find a sharp scientific question.

It ought to be more widely known that the truth is indeed out there, but only if one knows how to ask sharp and good questions. This is the unifying aspect of the scientific method and perhaps its most enduring contribution.

HUMILITY

BARNABY MARSH

Evolutionary dynamics scholar; Program for Evolutionary Dynamics, Harvard University; Visitor, Institute for Advanced Study, Princeton

You might not think of humility as a scientific concept, but the special brand of humility enshrined in scientific culture is deserving of special recognition for its unique heuristic transformative power.

I reflect upon Sir Richard Southwood's invitation to an incoming class of Oxford undergraduate biologists: "Remember, perhaps fifty percent of the facts that you learn may be not quite right, or even wrong! It is your job to find out where new ideas are needed." In the scientist's toolkit of concepts for solving problems, scientific humility is among the most useful tools for finding the better pathway. It clarifies, it inspires, and it should be more widely known, practiced, and defended.

Respect for scientific humility gives us license to question in ways all too rare in other professional fields. Allow yourself to ponder: When were you last surprised? When were you last wrong? As scientists, we're explorers and need to wonder and play. We need ample freedom to tinker and fail. In contrast to other fields and even the general culture, our field doesn't progress by the brash power of authority, by skillful interpretation, or by rhetorical style. It doesn't advance by the mountain of evidence we amass in our favor but, rather, by how well our ideas stand up to rigorous probing. The humble scientist suspends judgment, remembering that many breakthroughs start with "What if . . . ?" Am I absolutely sure? How do I know that? Is there a better way? The results are compelling. Even the most complex systems become more orderly as different pieces of knowledge fall into place.

As we advance in our scientific careers, it's all too easy to feel overconfident in what we know and how much we know. The same pressures facing us in our everyday lives wait to ensnare us in our professional scientific lives. The human mind looks for certainty and finds comfort in parsimony. We see what we want to see, and we believe what makes intuitive sense. We avoid the complex and the difficult and the unknown. Just look across the sciences, from biochemistry to ecology, where multiple degrees of freedom make many problems seemingly intractable. But are they? Could new tools of computation and visualization enable better models of the behavior of individuals and systems? The future belongs to those brave enough to be humble about how little we know and how much is remaining to be discovered.

Scientific humility is the key that opens a whole new possibility space—a space where being unsure is the norm, where facts and logic are intertwined with imagination, intuition, and play. It's a dangerous and bewildering place, where all sorts of untested and unjustified ideas lurk. What is life? What is consciousness? How can we understand the complex dynamics of cities? Or even my goldfish bowl? One can quickly see why when faced with uncertainty most of us would rather retreat. Don't. This is the space where amazing things happen.

The clearest and most compelling message from the history of science is that old ideas, even very good old ideas, are regularly augmented or even replaced with new ideas. As the case of classical and quantum mechanics shows, examples can be highly counterintuitive. Right now, someone somewhere is beginning to question something we all take for granted, and the result will radically change our future.

INDEX

glassmaking, 239–40
Go (game), 54, 467–68, 482
Gödel, Kurt, 61, 462, 491
Godfrey-Smith, Peter, 374
Goethe, Johann Wolfgang von, 390–91, 392
Goldenfeld, Nigel, 500–501
golden ratio *(phi)*, 491
Goldstein, Rebecca Newberger, 273–76
Goleman, Daniel, 98–99
Golomb, Beatrice, 440–43
Goodall, Jane, 246
Good, I. J., 393
Gopnik, Alison, 325–28
Gould, Stephen Jay, 11–14, 25, 266
Grandin, Temple, 358, 362
gravitation, 15–16
gravitational lensing, 169–71
gravitational radiation, 139–41
Gray, Kurt, 199–201
Greer, Lindsay, 495–96
Grey, CGP, 258
Griffin, Donald, 243, 245, 246, 248
Griffiths, Tom, 377–78
Grove, Andy, 448
Gruber, June, 473–75

habituation, 278–81
hacking, 59–61
Haldane, J. B. S., 328–30
Halley, Edmund, 50, 231
Hall, Lars, 63–64
hallucinations, 129, 353–54
Halvorson, Hans, 159–62
Hardin, Garrett, 201–2
hard problem of consciousness, 165, 211–12
Harlow, Daniel, 293
Harnad, Stevan, 393–94
Harrison, Ross G., 401
Harris, Sam, 83–85
Hart, I. R., 436
Hawking, Stephen, 2–3, 175–76, 183
heart attacks, 439–40, 442
heat engines, 432–33
hedonic adaptation, 280

Heidegger, 478
Heisenberg, Werner, 155, 157, 196–97
Hemingway, Ernest, 116
Henry VIII, 199
herd immunity, 402–5
hermiticity, 166–67
Hershfield, Hal, 48
Hertz, Heinrich, 139
Hidalgo, César, 407–8
Higashida, Naoki, 358–59
Higgs boson, 486
Higgs physics, 269, 271
Highfield, Roger, 416–18
hillclimbing metaphor, 234–37
Hillis, W. Daniel, 118–20
Hippocrates, 121
historiometrics, 370–73
Hitchcock, Diane, 40
Hitler, Adolf, 63, 152, 185
Hochberg, Michael, 379–81
Hoffman, Donald D., 290–93
Hofstadter, Douglas, 109, 461–62
Holloway, Ralph, 153
Holmes, Oliver Wendell, Jr., 362
Holocene epoch, 34–35
holographic principle, 290–93
Holt, Jim, 174–75
homeostasis, 120–22, 128, 130
homology, 264–65
homophily, 312–14
Honig, W. K., 246
Hood, Bruce, 479–81
Hook, Daniel, 166–68
Hooke, Robert, 469
Hoque, Ehsan, 474
Horgan, John, 268–69
Hossenfelder, Sabine, 433–35
Hoven, Friedrich Wilhelm von, 347
Hoyle, Fred, 133, 137
Hrdy, Sarah, 321
Hubble, Edwin, 134, 186
Hubble Space Telescope, 140, 171
Huberman, Bernardo, 446
Hughes, Howard, 329
Hulse, S. H., 246
Human Genome Project, 417
Humboldt, Alexander von, 37

humility, 5–6, 187, 213, 305, 308–9, 501–3
Humphrey, Nicholas, 62–64
hunter-gatherers, 415

iatropic stimulus, 411–13
Ictinus, 289
ideal free distribution, 450–52
ignorance, deliberate, 454–57
Ijjas, Anna, 146
illusion of explanatory depth (IOED), 4–6
illusory conjunction, 352–54
impedance matching, 118–20
impossibility, 431–33
Included Middle theory, 196–98
indexical information, 471–73
industrial revolution, 415–16
inflation of universe, 137–38, 141–46, 172
informational organisms (inforgs), 410–11
information pathology, 408–11
instrumentalism, 273–75
Intel, 448
intellectual honesty, 83–85
International Center for Information Ethics, 410
Internet, 360–61, 444–45
intertemporal choice, 44–47
intuitive beliefs, 318–20
invariance, 174–75
invisibility shields, 228–30
IQ (intelligence quotient), Breeder's Equation, 487–89
isolation mismatch, 207–11
Ito, Joichi, 361–63

Jablonski, Nina, 365–68
Jackson, Matthew O., 312–14
Jacob, François, 24
Jacobs, Jane, 126
Jacquet, Jennifer, 34–36
James, William, 369, 370
Janoff-Bulman, Ronnie, 280
Janus effects, 442–43
Jeffery, Kate, 239–41

Johansson, Peter, 63–64
Jolie, Angelina, 320
Joy of Destruction game, 202–3

Kahneman, Daniel, 131, 188–89
Kane, Gordon, 269–71
Kant, Immanuel, 160
Kauffman, Stuart A., 492–93
Keats, John, 457–58
Keil, Frank, 4, 5
Kelly, Kevin, 234–37
Kelvin, Lord (William Thomson), 194
Kepler, Johannes, 184
King, Martin Luther, Jr., 206
Kinzler, Katherine D., 278–81
Kleinberg, Jon, 421–24
Klein, Gary A., 188–89, 387–90
Knuth, Donald, 482–83
Knutson, Brian, 47–49
Koestler, Arthur, 355–56
Kollmann, Julius, 265–66
Kosko, Bart, 475–78
Kosslyn, Stephen M., 427–29
Krause, Kai, 108–10
Krauss, Lawrence M., 299–302
Kruger, Justin, 5
Krugman, Paul, 232
Krumme, Coco, 56–58
Kruskal, Martin, 483
Kurzban, Robert, 232–34
Kurzweil, Ray, 116
Kuziemko, Ilyana, 204

Lamb, Charles, 354
Laplace, Pierre, 107, 195
Large Hadron Collider (LHC, CERN), 135, 300–301
Larkin, Philip, 250
last-place aversion (Kuziemko), 204
Lavoisier, Antoine, 346–47
law of eponymy (Stigler), 230–31
law of requisite variety (Ashby), 122–24
law of small numbers, 348–50
Lazarsfeld, Paul, 312
Lee, Peter, 390–93
Legare, Cristine H., 322–25

viral information, 15
virial theorem, 14–17
vitalism, 37
voter behavior, 63–64, 300
Vouloumanos, Athena, 337–38
Vrba, Elisabeth, 11–14
Vygotsky, Lev, 337–38

Wallace, Alfred Russel, 259, 262
Watson, James, 455
Waytz, Adam, 4–6
Weinberg, Steven, 173
Weinstein, Eric R., 94–97
welfare programs, 344–45
Wheeler, John Archibald, 150
Whewell, William, 294–96
Whitman, Walt, 156
Wiener, Earl, 389
Wiener, Norbert, 196
Wigner, Eugene, 160
Wilbrecht, Linda, 334–36
Wilczek, Frank, 155–56
Wilkes, Jason, 384–87
Willans, Geoffrey, 390, 392

Williams, George C., 68–69
Wing, Jeannette, 398
witches, 203
Woese, Carl, 501
Wolfram's law, 55
Wolfram, Stephen, 54, 55
Wong, Yan, 8, 10
World War II, 315, 348–50
World Wide Web, 392–93, 394
Wrigley-Field, Elizabeth, 338–42
Wyatt, Victoria, 224–26

xenophobia, 210–11

Yamanaka, Shinya, 69–70
Yanai, Itai, 405–7
Yellin, Dustin, 190–93
Yoshida, Beni, 293

Zaki, Jamil, 474
Zen stone towers, 281–82
Zhang, Xiang, 229–30
zone of proximal development
   (Vygotsky), 337–38

ALSO BY JOHN BROCKMAN

What Should We Be Worried About?

This Explains Everything

This Will Make You Smarter